AF805969

Microfluidic Methods for Molecular Biology

Chang Lu • Scott S. Verbridge
Editors

Microfluidic Methods for Molecular Biology

Springer

Editors
Chang Lu
Department of Chemical Engineering
Virginia Tech
Blacksburg, VA, USA

Scott S. Verbridge
Department of Biomedical Engineering
and Mechanics
Virginia Tech
Blacksburg, VA, USA

ISBN 978-3-319-30017-7 ISBN 978-3-319-30019-1 (eBook)
DOI 10.1007/978-3-319-30019-1

Library of Congress Control Number: 2016938159

© Springer International Publishing Switzerland 2016
This work is subject to copyright. All rights are reserved by the Publisher, whether the whole or part of the material is concerned, specifically the rights of translation, reprinting, reuse of illustrations, recitation, broadcasting, reproduction on microfilms or in any other physical way, and transmission or information storage and retrieval, electronic adaptation, computer software, or by similar or dissimilar methodology now known or hereafter developed.
The use of general descriptive names, registered names, trademarks, service marks, etc. in this publication does not imply, even in the absence of a specific statement, that such names are exempt from the relevant protective laws and regulations and therefore free for general use.
The publisher, the authors and the editors are safe to assume that the advice and information in this book are believed to be true and accurate at the date of publication. Neither the publisher nor the authors or the editors give a warranty, express or implied, with respect to the material contained herein or for any errors or omissions that may have been made.

Printed on acid-free paper

This Springer imprint is published by Springer Nature
The registered company is Springer International Publishing AG Switzerland

Preface

From the optical instruments that first provided a window into the microscopic world, to molecular analysis tools that have helped clarify the genetic underpinnings of life, the most significant advancements in the biological sciences over the past century have largely been driven by the availability of powerful new quantitative tools. A more recent theme in biotechnology has been the miniaturization of analytical tools, enabled in large part by adapting and developing the methods of the microelectronics industry. Microfluidics as a field has experienced a period of rapid development and evolution since the 1980s. Most of the early microfluidics work focused on miniaturization of tools used in analytical chemistry (e.g., chromatography and electrophoresis). Microfluidics provided unparalleled flexibility for miniaturization, integration, and automation. This permitted the creation of devices that were substantially more sophisticated than tools used in conventional analytical chemistry, and at a substantially lower cost of production and operation. Seminal early efforts of the field led to major innovations in both material science (e.g., the wide use of polydimethylsiloxane) and technology (e.g., two-layered pneumatic valves). These important developments greatly expanded the applications of microfluidics and have underpinned a more recent renaissance in microfluidics for biological applications.

With the explosion of genomics in the 1980s (and later of additional "omics" fields), molecular biology has always been an important area of application for miniaturized devices, given these devices' unique access to the size scales relevant to the function of cells. Protein/nucleic acid separation and PCR-based analysis are widely practiced on microfluidic devices. Single-cell analysis has been an intensively explored direction in recent years, due to the unique size advantage associated with microfluidics for single-cell manipulation. With the decreasing cost of next-generation sequencing, recent years have witnessed substantial efforts directed towards genome-wide studies (as opposed to investigations focused on specific loci).

In parallel to new opportunities in basic science, there has been an increasing demand for well-established and robust microfluidic technologies that may have a direct impact on clinical practice and therapies. Personalized Medicine (PM), more recently re-branded as Precision Medicine, provides just such a unique opportunity. PM is based on the premise that every patient is unique at the tissue, cellular, and molecular level. Thus conducting molecular biology tests on patient samples is essential for providing clinicians with genomics, transcriptomics, epigenomics, and proteomics information that will ultimately improve their ability to optimize decision making for individual patients. Microfluidics offers the ideal platforms for handling and analyzing low quantities of cell/molecular samples to enable this exciting personalized approach; however there remains much work to be done before this paradigm is routine practice.

This book volume highlights recent progress on the topic of microfluidics for molecular biology studies. We cover various aspects of current microfluidics research in this growing field, which now spans the disciplines of biology, physics, chemistry, forensics, engineering, earth and atmospheric sciences, and beyond. Chapters are presented on various types of molecular analysis (genetic, epigenetic, proteomic, and next-generation sequencing), use of model organisms and patient materials, analysis at bulk and single-cell levels, techniques on cell culture and isolation, and a variety of different platforms at the technological cutting edge of these respective fields (flow, droplet, and paper-based microfluidics). We are hopeful that new microfluidic tools will continue to enable new insights into basic science as well as technology and biomedicine. To analyze molecular populations at the resolution of single cells or even single molecules will undoubtedly provide new ways of understanding complex biological processes, for example the dynamics of the adaptive immune system, or the role of tumor heterogeneity in cancer. High-resolution tools translated to the clinic could enable entirely new ways to treat a patient's own disease, as opposed to treating a hypothetical average patient as is the current pharmaceutical paradigm. We hope to provide a succinct but comprehensive picture of the state of the art for microfluidic molecular assays, and we look forward to the advancements yet to come in this exciting and rapidly progressing field.

Blacksburg, VA, USA
December 15, 2015

Chang Lu
Scott S. Verbridge

Contents

Contributors

Rafael V. Davalos, Ph.D. Bioelectromechanical Systems Lab, Department of Mechanical Engineering, Virginia

Tech—Wake Forest University, School of Biomedical Engineering and Sciences, Blacksburg, VA, USA

Deepika Devadas, Ph.D. Department of Mechanical and Industrial Engineering, University of Toronto, Toronto, ON, Canada

Temple A. Douglas Bioelectromechanical Systems Lab, Department of Mechanical Engineering, Virginia

Tech—Wake Forest University, School of Biomedical Engineering and Sciences, Blacksburg, VA, USA

Tao Geng, Ph.D. Department of Chemistry, University of California, Berkeley, CA, USA

Environmental Molecular Sciences Laboratory, Pacific Northwest National Laboratory, Richland, WA, USA

Andrew Godwin Pathology and Laboratory Medicine, University of Kansas Medical Center, Kansas City, KS, USA

University of Kansas Cancer Center, Kansas City, KS, USA

Dustin K. Harshman, Ph.D. Biomedical Engineering Graduate Interdisciplinary Program and Department of Agricultural and Biosystems Engineering, The University of Arizona, Tucson, AZ, USA

Mei He, Ph.D. Department of Biological and Agricultural Engineering, Johnson Cancer Research Center, College of Engineering, Kansas State University, Manhattan, KS, USA

Caroline N. Jones, Ph.D. Department of Biological Sciences, Virginia Polytechnic Institute and State University, Blacksburg, VA, USA

Tania Konry, Ph.D. Department of Pharmaceutical Sciences, Northeastern University, Boston, MA, USA

Xinyu Liu, Ph.D. Department of Mechanical Engineering, McGill University, Montreal, QC, Canada

Chang Lu, Ph.D. Department of Chemical Engineering, Virginia Tech, Blacksburg, VA, USA

Hang Lu, Ph.D. Interdisciplinary Program of Bioengineering, Georgia Institute of Technology, Atlanta, GA, USA

School of Chemical and Biomolecular Engineering, Georgia Institute of Technology, Atlanta, GA, USA

Shay Mailloux, Ph.D. Multiplex Biotechnology Laboratory, Department of Chemistry, University at Albany, State University of New York, Albany, NY, USA

Joseph M. Martel-Foley, Ph.D. Department of Biomedical Engineering, Wentworth Institute of Technology, Boston, MA, USA

Richard A. Mathies, Ph.D. Department of Chemistry, University of California, Berkeley, CA, USA

Robert J. Meagher, Ph.D. Biotechnology and Bioengineering Department, Sandia National Laboratories, Livermore, CA, USA

Arnab Mukherjee, Ph.D. Division of Chemistry and Chemical Engineering, California Institute of Technology, Pasadena, CA, USA

Daniel A. Porto Interdisciplinary Program of Bioengineering, Georgia Institute of Technology, Atlanta, GA, USA

Aashish Priye, Ph.D. Artie McFerrin Department of Chemical Engineering, Texas A&M University, College Station, TX, USA

Lisa Ramirez, B.S. Multiplex Biotechnology Laboratory, Department of Chemistry, University at Albany, State University of New York, Albany, NY, USA

Tel M. Rouse Interdisciplinary Program of Bioengineering, Georgia Institute of Technology, Atlanta, GA, USA

School of Chemical and Biomolecular Engineering, Georgia Institute of Technology, Atlanta, GA, USA

Pooja Sabhachandani, M.S. Department of Pharmaceutical Sciences, Northeastern University, Boston, MA, USA

Adriana San-Miguel School of Chemical and Biomolecular Engineering, Georgia Institute of Technology, Atlanta, GA, USA

Department of Chemical and Biomolecular Engineering, North Carolina State University, Raleigh, NC, USA

Saheli Sarkar, Ph.D. Department of Pharmaceutical Sciences, Northeastern University, Boston, MA, USA

Charles M. Schroeder, Ph.D. Department of Chemical and Biomolecular Engineering, Center for Biophysics and Computational Biology, Institute for Genomic Biology/Biosystems Design, University of Illinois at Urbana-Champaign, Urbana, IL, USA

G. Stephanopoulos Department of Chemical Engineering, Massachusetts Institute of Technology, Cambridge, MA, USA

Martin M. Thuo, Ph.D. Department of Materials Science and Engineering, Iowa State University, Ames, IA, USA

Victor M. Ugaz, Ph.D. Artie McFerrin Department of Chemical Engineering, Texas A&M University, College Station, TX, USA

Department of Biomedical Engineering, Texas A&M University, College Station, TX, USA

A.E. Vasdekis Department of Physics, University of Idaho, Moscow, ID, USA

Jun Wang, Ph.D. Cancer Research Center, University at Albany, State University of New York, Rensselaer, NY, USA

Elisa M. Wasson Bioelectromechanical Systems Lab, Department of Mechanical Engineering, Virginia

Tech—Wake Forest University, School of Biomedical Engineering and Sciences, Blacksburg, VA, USA

Meiye Wu, M.A., Ph.D. Biotechnology and Bioengineering Department, Sandia National Laboratories, Livermore, CA, USA

Mengsu Yang, Ph.D. Department of Biomedical Sciences, City University of Hong Kong, Kowloon, Hong Kong, People's Republic of China

Key Laboratory of Biochip Technology, Biotech and Health Centre, Shenzhen Research Institutes of City University of Hong Kong, Shenzhen, People's Republic of China

Jeong-Yeol Yoon, Ph.D. Biomedical Engineering Graduate Interdisciplinary Program and Department of Agricultural and Biosystems Engineering, The University of Arizona, Tucson, AZ, USA

Edmond W.K. Young, Ph.D. Department of Mechanical and Industrial Engineering, University of Toronto, Toronto, ON, Canada

Wanqing Yue, Ph.D. Department of Biomedical Sciences, City University of Hong Kong, Kowloon, Hong Kong, People's Republic of China

School of Sciences, China Pharmaceutical University, Nanjing, People's Republic of China

Yong Zeng, Ph.D. University of Kansas Cancer Center, Kansas City, KS, USA

Department of Chemistry, University of Kansas, Lawrence, KS, USA

Chen Zhao Department of Mechanical Engineering, McGill University, Montreal, QC, Canada

Yan Zhu Department of Chemical Engineering, Virginia Tech, Blacksburg, VA, USA

Chapter 1
Microfluidic Platforms for Quantitative Biology Studies in Model Organisms

Daniel A. Porto*, Tel M. Rouse*, Adriana San-Miguel*, and Hang Lu

Abstract The use of lab-on-chip tools has been adopted in a wide variety of scientific fields. Hundreds of applications that speed up, miniaturize, or enable otherwise unfeasible assays have emerged in the last couple of decades [1, 2]. The microfluidic toolbox offers several advantages which make it a very attractive resource for biological studies: reduced sample volume, control of spatiotemporal chemical compositions, streamlined assays, precise and predictable fluid flow regimes, portability, and integration with sensors, actuators, controllers, and automation systems [1–3]. The main drive of the field has thus far focused on the development of microfluidic tools that replace conventional methods with proof-of-principle applications. However, widespread adoption of these technologies for fundamental research is still in progress. Microfluidics has nonetheless had a significant impact in fundamental biological studies with model organisms [4, 5]. In this article, we provide an overview of the current state of the field, the impacts of microfluidics in model organism research, and the outlook, challenges, and opportunities for the future.

---

*Author contributed equally with all other contributors.

D.A. Porto
Interdisciplinary Program of Bioengineering, Georgia Institute of Technology, Atlanta, GA 30332, USA

T.M. Rouse • H. Lu, Ph.D. (✉)
Interdisciplinary Program of Bioengineering, Georgia Institute of Technology, Atlanta, GA 30332, USA

School of Chemical and Biomolecular Engineering, Georgia Institute of Technology, Atlanta, GA 30332, USA
e-mail: hang.lu@gatech.edu

A. San-Miguel (✉)
School of Chemical and Biomolecular Engineering, Georgia Institute of Technology, Atlanta, GA 30332, USA

Department of Chemical and Biomolecular Engineering, North Carolina State University, Raleigh, NC 27607, USA
e-mail: asanmig@ncsu.edu

© Springer International Publishing Switzerland 2016

C. Lu, S.S. Verbridge (eds.), *Microfluidic Methods for Molecular Biology*,
DOI 10.1007/978-3-319-30019-1_1

Keywords Microfluidics • Model organisms • Quantitative biology • Microscopy • Phenotypes • Genetics • High-throughput • High-content • Screens

1.1 Introduction

The use of lab-on-chip tools has been adopted in a wide variety of scientific fields. Hundreds of applications that speed up, miniaturize, or enable otherwise unfeasible assays have emerged in the last couple of decades [1, 2]. The microfluidic toolbox offers several advantages which make it a very attractive resource for biological studies: reduced sample volume, control of spatiotemporal chemical compositions, streamlined assays, precise and predictable fluid flow regimes, portability, and integration with sensors, actuators, controllers, and automation systems [1–3]. The main drive of the field has thus far focused on the development of microfluidic tools that replace conventional methods with proof-of-principle applications. However, widespread adoption of these technologies for fundamental research is still in progress. Microfluidics has nonetheless had a significant impact in fundamental biological studies with model organisms [4, 5]. In this article, we provide an overview of the current state of the field, the impacts of microfluidics in model organism research, and the outlook, challenges, and opportunities for the future.

Microfluidics takes advantage of the physics that govern fluids at small scales. The most important phenomena that dominate at the micro-scale are laminar flow and surface effects [6, 7] (Fig 1.1). Laminar flow entails a predictable stream behavior and minimal mixing. Manipulation of fluids at the micro-scale thus becomes precise and reproducible. These characteristics are useful to not only control spatiotemporal chemical compositions [8–12] but also to manipulate specimens via the exertion of forces driven by fluid flow [13]. Operating in a laminar flow regime implies that mixing of fluids requires special consideration, as this occurs solely through diffusion and Taylor dispersion [7]. Hence some assays require the incorporation of channel geometries that enable efficient fluid mixing within a desired residence time [14, 15]. Surface effects refer to the high surface-area-to-volume ratio of fluids constrained at the micro-scale [7, 16, 17]. These can be advantageous in the drawing of liquid via capillary forces that act through surface wetting [7, 17], or through field effects in electro-osmotic flows [16]. The high surface-to-volume ratio also facilitates interfacial transport and reactions [17]. At the same time, the small length scales enable fast changes in temperature and chemical composition. These, along with small sample volumes, are some of the main advantages that microfluidics offers. In particular, for model organisms such as bacteria, yeast, cells, nematodes, and some higher-order organism, the length scales of microfluidic channels are comparable to the specimen of interest (~ tens to hundreds of microns), providing the spatial resolution required for individual trapping, handling, culturing, and perturbation.

Early microfluidic devices were made mostly of silicon and glass, as these materials were ubiquitous in the semiconductor industry. Hard and brittle, silicon

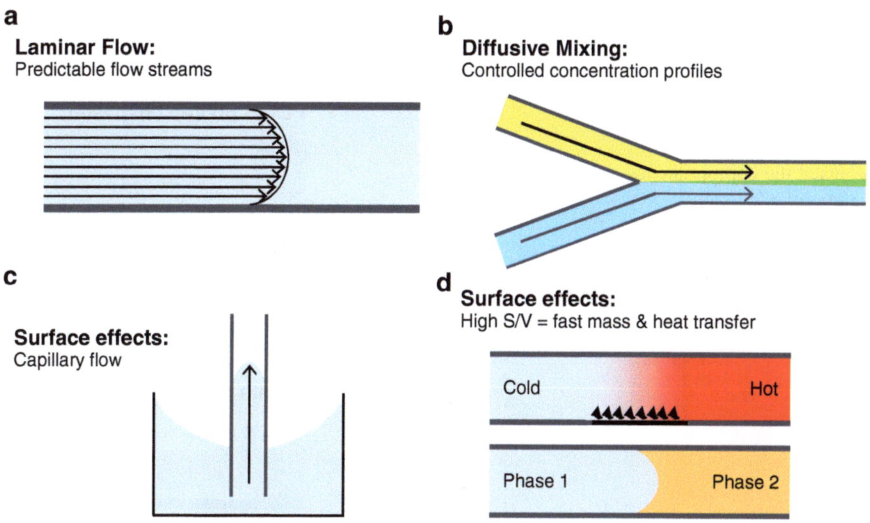

Fig. 1.1 Relevant phenomena in microfluidics. (**a**) Laminar flow. (**b**) Diffusive mixing. (**c**) Surface effects enable draw of liquid by capillary flow through surface wetting. (**d**) The High S/V (surface-to-volume) ratio enables fast heat transfer and easy mass transport across interfaces

and glass microfluidic devices present several drawbacks for biological studies [1]. Microfluidic devices are now mostly fabricated out of elastomers, plastics, or paper. These are often inexpensive, easy to make, and single-use. Polydimethylsiloxane (PDMS) first appeared as an alternative to silicon and glass, which revolutionized the field and greatly popularized soft lithography [18, 19]. The fabrication of PDMS devices is a simple, well-developed process [18]. PDMS enables fast replica molding from silicon or photoresist molds, as it conforms to substrates with micro-scale features. Devices are cast and peeled off from a mold typically made on a silicon wafer by photolithography (Fig. 1.2). After peeling, inlet and outlet ports are created. PDMS devices are then typically bonded to a glass slide or other materials for sealing. Being a flexible, optically transparent, gas permeable, and biocompatible polymer, PDMS presents many advantages that have made it the most popular material for lab-on-chip studies. The flexibility of PDMS enables deflection of membranes, which has proven useful for the development of on-chip valves. For some biological applications, particularly cell-based studies, alternative materials are being explored. Polystyrene [20], polymethyl methacrylate [21], and other thermoplastics [2] have been explored for their ability to reduce transport and adsorption of small molecules through the device. However, these additional properties sometimes come at the cost of losing optical transparency or material flexibility. Paper-based microfluidic chips have also gained great popularity [22, 23]. Typically relying on capillary forces for fluid flow, paper devices have found widest applications in biochemical assays and in inexpensive

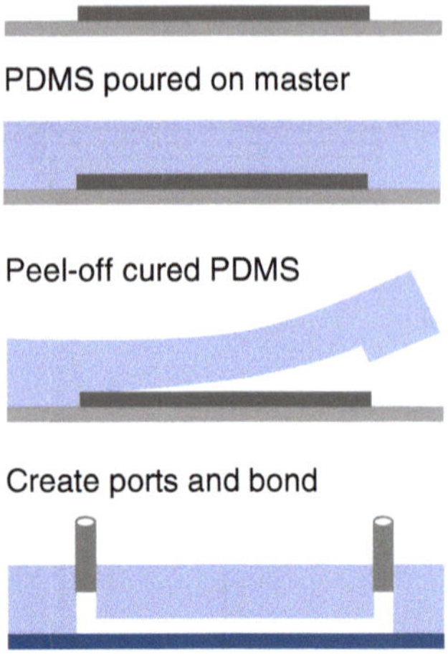

Fig. 1.2 PDMS device fabrication. PDMS pre-polymer is cast on a master mold. After curing, PDMS is peeled off. Ports are created by excising cores at the appropriate inlet/outlet locations. Finally, devices are bonded to a glass slide for sealing and are connected through ports to fluid inlet and outlet lines

and disposable point-of-care (POC) diagnostic devices [24]. Overall, the choice of material will depend on weighing the ease of fabrication and scale-up, component interfacing, and specific assay requirements. Currently, given that PDMS is oxygen permeable and enables the integration of on-chip valves, it is expected to continue being the most widely used material for lab-on-chip experiments with model organisms.

Microfluidic chips have also had a significant impact in many other fields of biological research, in addition to studies with multicellular organisms. These include applications in biochemical assays, single-cell analysis, organ-on-chip platforms, (POC) diagnostics, and studies with cell aggregates [2, 25]. Not only have these platforms enabled previously unfeasible biological studies [5, 26, 27], but they have also facilitated high-throughput, high-content, and large-scale assays [28–32]. Some applications where the field has had more impact hitherto are molecular and biochemical assays, which benefit from reduced sample size requirements, decrease in reaction times, and parallelization and multiplexing capabilities [25]. Some examples include PCR-on-chip [33], DNA sequencing [34], immunoassays and enzymatic analysis [2], transcriptomics [35, 36] and proteomics [37], and POCs [38]. Microfluidic technology has also had a substantial impact in the field of cell biology. Lab-on-chip platforms have been developed for the manipulation, capturing, separation, and tracking of cells [12, 13, 26, 39]. Studies performed with cells in microfluidic devices take advantage of well-controlled chemical environments, mechanical control of samples via fluid flow, and increased surface area. Some examples of fundamental biological studies include chemotaxis and migration [10], cell adhesion [13], biophysics [40], and single-cell culture and tracking [26]. As previously mentioned, microfluidic technologies have recently found applications beyond biochemical assays and cell biology. The application of lab-on-chip platforms for studies with model organisms has experienced a significant rise in popularity, and it is expected to open up a breadth of opportunities to tackle thus far elusive scientific questions at the whole living organism level.

1.2 Model Organisms/Systems Amenable to Microfluidic Manipulations

Model organisms are nonhuman biological specimens that are extensively studied to gain knowledge on biological phenomena. These have been deliberately selected due to their experimental advantages, to serve as a model that enables gaining insight into fundamental biological mechanisms conserved in humans and other higher level organisms. Many model organisms exist, ranging in complexity from simple single-cell bacterium to the rhesus monkey. In this chapter, we will focus on some of the most extensively studied model organisms: the nematode, *Caenorhabditis elegans*, the fruit fly, *Drosophila melanogaster*, and the zebrafish, *Danio rerio* (Fig. 1.3). These share many experimental advantages, such as possessing some of the fastest generation times, being easily and inexpensively cultured, being amenable to easy genetic manipulation, and possessing fully or

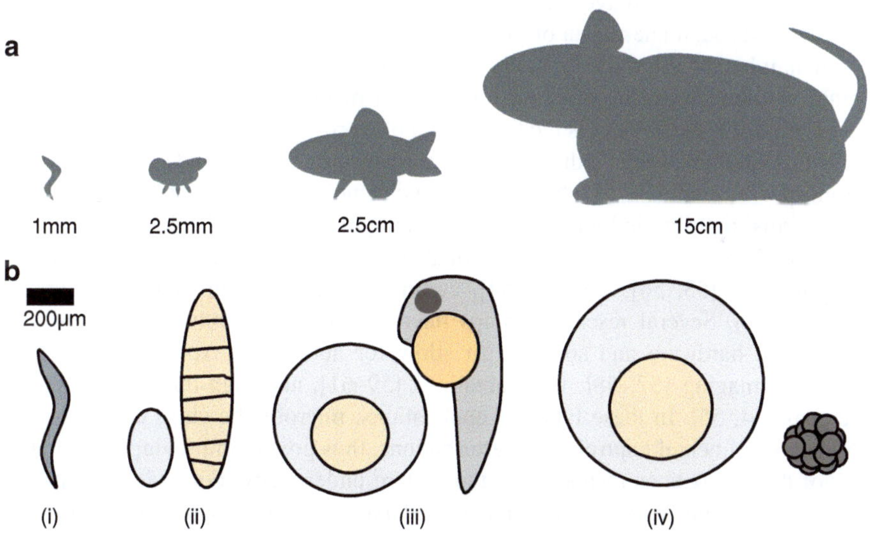

Fig. 1.3 Size scale of model organisms. (**a**) Size of adults of commonly used model organisms: the nematode *Caenorhabditis elegans*, the fruit fly *Drosophila melanogaster*, the zebrafish, *Danio rerio*, and the house mouse *Mus musculus*. Besides *C. elegans*, the other model organisms grow beyond the micro-scale and adults are incompatible with microfluidics. (**b**) Model organism stages that are applicable to microfluidics. *C. elegans* adults (*i*) grow up to 1 mm, and all life stages are compatible with microfluidics. In the case of flies (*ii*) and zebrafish (*iii*), their embryos and larvae have been studied in microfluidics. Mammalian embryos have also been studied in microfluidics, as well as embryoid bodies (*iv*)

partially transparent bodies throughout embryo development or into adulthood [15, 41–45]. Most importantly, many mechanisms discovered in these model organisms show conservation to higher level organisms and even to humans. Notable examples include Nobel Prize-winning discoveries in genetic regulation of organ development and programmed cell death [46–48] and RNA interference [49] in *C. elegans*.

1.2.1 Worms

The nematode, *C. elegans*, is one of the most widely studied model organisms using microfluidics [4, 50]. *C. elegans* has many practical advantages, such as ease of maintenance, short generation times, and optical transparency into adulthood. Additionally, extensive use in genetics studies [46] in recent decades has resulted in well-established genetic manipulation techniques. This has led to the creation of a multitude of transgenic strains, providing some of the latest, most powerful biomolecular toolkits available. However, due to their small size (up to 1 mm in length), nematodes are difficult to handle, which creates a bottleneck in many experimental processes for large population studies. Additionally, their mobility makes it difficult to image them in high magnification, typically requiring immobilization through anesthesia or glues.

A number of microfluidic tools have been developed to overcome these and other challenges, while enabling novel experiments. The primary advantage of using microfluidic devices is the facilitation of worm handling; by taking advantage of differential pressures, individual animals can be directed to and easily recovered from specific chambers [51, 52] or imaging channels [53–56]. In order to overcome the worms' mobility in high magnification imaging, microfluidic devices have also been used to immobilize animals through various methods including mechanical compression [53, 56], cooling [54], carbon dioxide exposure [56], or pluronic gelling [51]. Several research groups have integrated microfluidic devices with peripheral hardware and software to allow for automated experiments, such as calcium imaging [57, 58], laser ablations [59–61], and high-throughput genetic screens [54, 55]. In these types of applications, microfluidic chips enable exceptional experimental control over perturbations, thus greatly improving the reliability of the readouts as responses of the system under study. For example, on-chip calcium imaging experiments enable understanding the quantitative nature of neuronal responses under well-controlled chemical stimulation. The application of microfluidics as controllable experimentation platforms for *C. elegans* thus allow for obtaining deeper insights into the biological phenomena under study.

1.2.2 Flies

Similar to worms, the fruit fly, *D. melanogaster*, is one of the most well-studied model organisms. The fruit fly was one of the first model organisms used for genetic studies; the work by Thomas Hunt Morgan using *Drosophila* established much of our understanding of genetic transmission [62]. Fruit flies are a prominent model organism because of their ease of culture, short life cycle, and simple diets. In contrast to *C. elegans*, fruit fly adults can grow up to 2.5 mm, making them incompatible to work with in microfluidics. However, various research groups have utilized microfluidics to study flies in earlier stages of development (Fig. 1.3). Embryogenesis is widely studied using *D. melanogaster*, which can involve the characterization of morphogen and segmentation patterning through development [62]. In order to study how these patterns are perturbed by external cues, various devices have been designed to enable precise control of the embryo's surrounding environment [63, 64]. Conventionally, embryos are mounted on agar pads and manually handled in order to orient them correctly for imaging. To overcome this laborious and time-consuming procedure, advances have been made using microfluidics to increase the throughput and quality of imaging by using embryo arrays with specific structures [65]. These devices take advantage of the specimen handling via flow control, the similar sizes of microfluidic chambers and fly embryos, and laminar flow to control the spatiotemporal chemical environments that microfluidics enable. In addition, some groups have developed devices to study fly larvae behavior [66, 67].

1.2.3 Fish

The zebrafish, *Danio rerio*, is another widely used model organism in biology [68, 69]. Similar to worms and flies, they produce a large number of progeny, and many transgenic animals have been developed. In contrast to worms and flies, however, zebrafish are vertebrates. This makes them a popular model organism for studying vertebrate development. As with flies, the large size of adult animals prevents the use of microfluidics for studies in full-grown adults. However, zebrafish embryos and larvae are small and optically transparent, making them convenient and easy to work with in microfluidics (Fig. 1.3). Many advances have been made in the application of microfluidics to improve the throughput of conventional embryo culturing techniques [70], as well as the efficiency and quality of imaging [71]. Zebrafish have also been used extensively for drug screens, and various microfluidic devices have been developed to perform these assays in a high-throughput fashion [72].

1.2.4 Mammalian Systems

Although the above-mentioned model organisms are useful for providing insight into conserved genetic mechanisms, there is still a need for studies specific to mammalian systems. As mammals are much larger in comparison and are incompatible with aqueous environments, it is nonsensible to study adult animals in microfluidics (Fig. 1.3). As with flies and fish, some efforts have been made to develop devices for on-chip culturing of mammalian embryos [71, 73]. However, mammalian embryos rapidly grow too large for most microfluidic systems, making it difficult to culture them for a long period of time. In addition to embryos, many research groups have also used embryonic stem cells and embryoid bodies (EBs) to study early developmental biology in mammalian systems (Fig. 1.3). Several research groups have developed devices for improving the generation and handling of EBs, such as micropatterning of cell-adherent areas [74, 75] and physical restriction by microwells [76, 77]. Several devices have been developed which use laminar flow to deliver culture media to embryos or EBs with high spatial and temporal precision [78, 79]. In addition to embryoid bodies, recent works have focused on the development of organs-on-chip. These are systems consisting of cultured living cells in a microfluidic device designed to model particular physiological functions of interest. Microfluidics offers numerous advantages over conventional methods of 3D cell cultures. Notable advantages include the ability to precisely control transport phenomena and mechanical cues on the cells via fluid flow through microfluidic channels, the ability to deliver physiologically relevant cyclical perturbations to systems, and ease of observation and analysis of cell response to perturbations [80].

1.3 Applications

Microfluidics has played a large role in addressing common bottlenecks in model organism studies while also enabling additional experimental capabilities. There are numerous applications where microfluidics has improved the standardization, regulation, throughput, and/or integration of methods and technologies in biological studies on small organisms [4]. We will address some of the most prominent applications of microfluidics in biological studies on model organisms below.

1.3.1 Control of Flow

One of the most useful features of microfluidics is the predictable fluid mechanics at the micro-scale. As stated earlier, laminar flow allows for repeatable and precise stream behaviors, which can be used to deliver spatiotemporally precise chemical

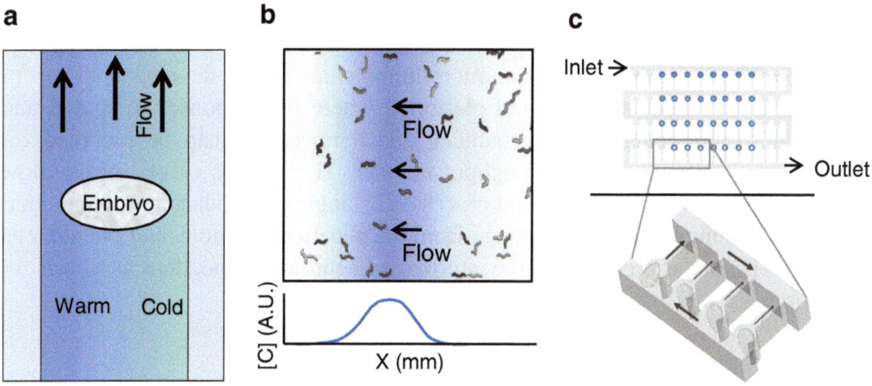

Fig. 1.4 Microfluidic platforms offer numerous advantages in the studies of model organisms. (**a**) Laminar flow regimes in devices offer precise control over spatial gradients, which can be used to study embryo development under differential chemical exposure. (**b**) Peripheral flow control systems allow for temporal control of stimuli, which can be used to study behaviors in model organisms, such as chemotactic responses to stimuli pulses in *C. elegans* in a microfluidic arena. (**c**) Microfluidic devices allow for rapid and delicate control of small model organisms, which enables increased throughput for biological assays and screens. Organisms can be arrayed into dense formations for imaging, long-term monitoring, and perturbation studies

perturbations to model systems, such as spatial gradients of chemicals, odors, or temperature (Fig. 1.4). Microfluidic devices that allow for precise control of stimuli have enabled studies on the environmental effects on development, behavior, and many other biological outputs [57, 81–84]. For example, Albrecht et al. [11] developed a device that employs a mixing tree to create various spatiotemporal chemical patterns. This device has been used to study behavioral responses and neuronal activity in *C. elegans* [85]. The predictable nature of flow patterns allows control not only of chemical composition and shear rates but also of temperature. For instance, the work in Lucchetta et al. uses laminar flow in a Y-junction to deliver cold and warm media to a *Drosophila* embryo to investigate the effect of temperature on morphogen gradients and development [63].

Microfluidic devices have also been useful to study effects of long-term perturbations to model organisms. Although challenging in traditional culture systems, controllable spatiotemporal exposure to stimuli for long periods is possible with microfluidic devices. For instance, chips have been developed for life-long culture and imaging of *C. elegans* under a variety of conditions such as exposure to chemicals, immobilization for high-resolution imaging, and behavior tracking [51, 52, 86–89]. These experiments, however, impose several difficulties. First, as experiments run for longer times, the probability of clogging of micro-channels increases, especially for experiments under constant perfusion of media and when flowing liquids with suspended particles, such as bacteria. The introduction of on-chip filters and bypass channels for periodic flushing and unclogging has been useful for this purpose [51, 52, 89]. Another challenge in using chips for long-term studies is the increased volume of data acquired, as imaging in the timescale of days can require extensive data storage capabilities. In order to accommodate for this, many groups use sparse recording when making observations with low frequency. In spite of these minor difficulties, microfluidic chips can be developed as part of short- and long-term experimental platforms where the responses to spatial and temporal stimuli (chemical, mechanical, and temperature) can be recorded for single individuals in a high-throughput manner. These types of platforms allow us to track responses and biological processes in single individuals throughout their life span, enabling findings that range from the genetics and molecular phenomena that govern cellular processes to the interplay of neuronal connections as a basis for behavior.

1.3.2 Precise Manipulation of Samples

Aside from assisting in imaging and quantitative readouts, microfluidics also offers unique solutions where precision control of samples is required (Fig. 1.4). Examples of precision manipulation include the rapid orientation and packing into dense arrays of fly embryos for imaging [65], the development of an on-chip microinjection system for the genetic manipulation of *C. elegans* [90], and the development of an on-chip laser ablation system for the rapid microsurgery of neurons on *C. elegans* samples for neuroscience studies [30]. Interfacing microfluidics to other technologies has enabled a variety of types of experiments to be performed on chip. For instance, microfluidic devices have been applied to perform optogenetics experiments on chip. Optogenetics involves the use of light-activated ion channels, allowing for optical control of excitable cells like neurons [91]. The use of microfluidic chips for optogenetics experiments enables isolation and tracking of animals, while increasing yield through parallelization. For example, a microfluidic device has been applied to study synaptic transmission in *C. elegans* in a high-throughput fashion [32]. Because most devices are optically transparent, it is easy to deliver light stimulus to samples in chips, while allowing individual tracking and isolation of animals under study. Many of the platforms developed for precision control and manipulation translate very laborious and difficult techniques that few labs possess the skill to perform. These platforms offer the opportunity to deliver simpler, sometimes fully automated, solutions to make these powerful techniques more accessible to the research community.

1.3.3 Observations

One of the key applications of microfluidics is the improvement over conventional methods in both the quantity and quality of observed measurements. Fluorescence imaging enables the capability to contrast biological molecules and structures of interest against background signals in a highly selective manner [92]. In the case of model organism studies where the embryo or specimen of interest is partially or fully transparent, fluorescent dyes and genetically encoded fluorescent proteins enable researchers to noninvasively quantify progressions of events throughout the organisms' tissues in vivo. However, despite partial or full transparency of tissues, it is common that the organism or embryo must be properly oriented to allow optimal imaging of the features. Conventional sample preparation steps for model organisms often involve laborious and time-consuming protocols in which samples must be manually oriented and fixed to a glass slide before imaging the samples individually. Several microfluidic platforms have been developed to address this rate-limiting step. Notable examples include the fly embryo array by Chung et al [65]. A key technique in embryogenesis studies is the imaging of morphogen and segmentation patterns in embryos via fluorescent markers. Conventional methods for imaging embryos are laborious and time-consuming, requiring manual manipulation in order to place them in the desired orientation. In order to improve the throughput and quality of imaging embryos, Chung et al. have developed microfluidic devices for handling and orienting a high number of embryos [65]. The embryo array utilizes hydrodynamic focusing to passively and rapidly position embryos in the vertical orientation necessary for the observation of dorsal/ventral developmental morphogen gradients [65]. There have also been microfluidic platforms designed to rapidly manipulate *C. elegans* into the field of view in a manner that results in preferential orientation of the organism for observation of key neurons or nerve cord puncta [93].

An application of fluorescence imaging that has been extensively used in *C. elegans* is calcium imaging. Calcium imaging involves the use of calcium-binding fluorescent proteins that allow for optical recordings of neuronal activity. Conventionally, calcium imaging recording involves experimentally difficult and laborious techniques, such as immobilizing animals with glue or anesthetics [94–96]. Also, most recordings are done in high magnification, and sometimes require careful manipulation to place animals in the correct orientation. In order to overcome these challenges, various research groups have developed devices for handling and immobilizing samples that enable high-throughput calcium imaging experiments [57]. Additionally, laminar flow can be used to deliver chemical cues [57, 97] or gradients [85, 97] while acquiring neuronal responses to stimuli via calcium imaging (Fig. 1.5).

Model organisms are also commonly used to understand genetic factors affecting behavior. Researchers analyze and quantify behavior patterns in model organisms to better understand the underlying genetic and environmental factors. As behavioral outputs are in many cases stochastic, behavior data tends to be extremely noisy, and small variations in experiments can greatly affect measurements. Many research groups have attempted to minimize this variability by using microfluidics to better control the environment around samples. For instance, in the work developed by Chung et al. [89], an array of traps allows tracking the behavior of multiple worms trapped in individual chambers at the same time, under well-controlled chemical perturbations. In this way, these types of devices enable obtaining reliable behavioral readouts in a high-throughput, quantitative manner.

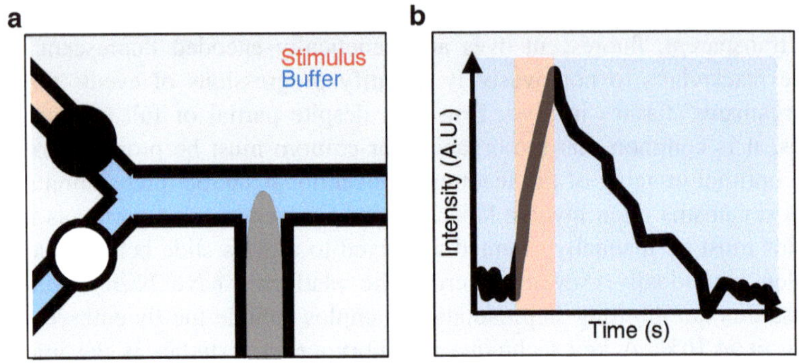

Fig. 1.5 Calcium imaging on-chip. (**a**) Microfluidics can be used to immobilize animals for high magnification imaging of specific neurons. Laminar flow and valves can be used to deliver chemical cues of interest. (**b**) Immobilization of animal allows for extraction of fluorescence intensity information, leading to traces of neuronal activity

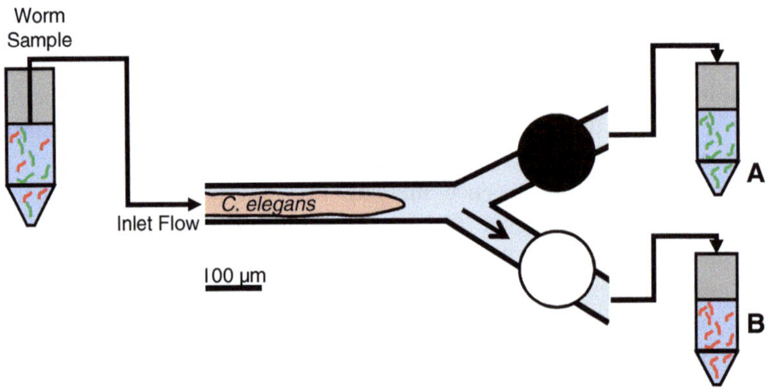

Fig. 1.6 On-chip sorting. Microfluidic platforms offer flexible sorting solutions in which the user can sort worms based on a variety of simple or complex metrics not capable by standard sorting technologies such as flow cytometry. In the example above, a sample of worms in solution is sorted into appropriate destinations through the alternation of on-chip pneumatic valves

1.3.4 Sorting

A special type of assay that has greatly benefited by microfluidic devices is phenotyping and sorting (Fig. 1.6). Through the control provided by on-chip valves and predictable fluid streams, microfluidic devices enable the isolation of specimens of interest by routing them to holding chambers or outlets [31, 50, 54, 93, 98–100]. Sorting is essential in experiments that require the isolation and further processing of organisms with special characteristics. For example, in drug or genetic screens, the objective is to identify animals with a phenotype that differs from a baseline due to drug exposure or a genetic mutation. In these experiments, it is necessary to rapidly assess phenotypes and make fast decisions on whether the animal under analysis should be selected for desired traits. However, a major limitation to throughput in traditional assays involving these model organisms is the difficulty of manipulation, sample preparation for imaging, and phenotype quantification. Traditionally, samples must be exposed to specific solutions, placed on glass slides, and carefully oriented under a microscope to be imaged one at a time. This can be extremely laborious and time-consuming, and can severely limit the feasibility of studies where analysis of many of age-synchronized samples is required [100]. Systems such as flow cytometry have been used for decades to sort specimens (usually cells) based on varying physiological properties or molecular readouts. In some cases, however, these commercial systems prove infeasible for handling, obtaining nonstandard subcellular readouts of interest, or sorting samples based on more complex metrics. An example of microfluidics application for sample sorting was a genetic screen by Crane et al. where *C. elegans* were sorted based on patterns of fluorescently labeled subcellular features, which are difficult to quantify by qualitative inspection [54]. With the emergence of peripheral image processing and behavior analysis systems to supplement microfluidic platforms [11, 32, 101], we foresee future possibilities in animal sorting based on rapidly acquired dynamic behaviors in a high-throughput fashion.

1.3.5 Interfacing

Microfluidics has proven to be a valuable component in the study of biology. It is evident that microfluidic devices are often an integral component of larger, complex platforms that incorporate peripheral technologies. The use of microfluidic devices as true laboratory-on-a-chip platforms require the integration of components necessary for complete analysis, and are hence referred to as Micro-Total-Analysis Systems (μTAS). On- and off-chip sensors, pumps, mixers, membranes, valves, heaters, and detection systems thus require interfacing to microfluidic devices [28, 102]. In particular for biological studies of model organisms, the most important components are on- and off-chip pumps and mixers. The control provided by these systems, in terms of fluid flow and spatiotemporal control of environmental stimuli, enables high-throughput analysis, individual tracking, sorting, dynamic process monitoring, and multiplexing [29, 31, 32, 39, 50, 54, 89, 98, 103]. In order to maximize the benefits of these tasks on chip, automation of assays via external hardware and software is desirable. Experimental control and automation is typically accomplished by several key components: (1) image analysis, to monitor processes and enable decision-making, (2) temperature control platforms, (3) computer-controlled pressure sources for fluids and valves, and (4) custom software typically written in MATLAB® or LabVIEW. These setups have dramatically increased the speed of experimentation while enabling studies otherwise unfeasible with conventional techniques.

1.4 Outlook and Opportunities

The development of microfluidic technologies aimed at facilitating biological studies in model organisms has opened up a large number of opportunities. These include previously elusive scientific questions, novel paradigms for experimental procedures, and unimaginable large-scale approaches for data acquisition. Microfluidic chips have enabled realizing studies with the utmost control of concentrations, reagents, temperatures, and external forces. Specimen handling has become much easier and faster, and the data extracted per experiment has greatly expanded through parallelization. The biggest challenge in the microfluidics field is the widespread adoption of these technologies by classic biology laboratories. The problem lies in nontrivial technology transferability and the common need for experiment-specific tailoring. In order to address these limitations, it is necessary to increase data and tool sharing, open access to chip designs, peripheral equipment and interfacing instructions, and code for control and automation. We expect that microfluidic platforms will become more modularized, allowing an easier development of a tailored platform for the requirements of individual experiments. Moreover, collaborative efforts by engineers and scientists can greatly speed up the development of these tools and their use as a tool for fundamental studies. We believe that the impact of microfluidic devices in the model organism community will only continue to grow, and it will greatly benefit the field by enabling fundamental studies otherwise unapproachable. In the future, it is expected that the most benefit from on-chip biological studies on model organisms will be based on automated experimentation platforms that enable large-scale high-content and high-throughput analysis, venturing in areas of drug screening, large-scale genomics, quantitative biology, large-scale data, and systems and synthetic biology.

References

1. Whitesides GM (2006) The origins and the future of microfluidics. Nature 442:368–373
2. Sackmann EK, Fulton AL, Beebe DJ (2014) The present and future role of microfluidics in biomedical research. Nature 507:181–189
3. Salieb-Beugelaar GB, Simone G, Arora A, Philippi A, Manz A (2010) Latest developments in microfluidic cell biology and analysis systems. Anal Chem 82:4848–4864
4. Hwang H, Lu H (2013) Microfluidic tools for developmental studies of small model organisms—nematodes, fruit flies, and zebrafish. Biotechnol J 8:192–205
5. San-Miguel A, Lu H (2013) Microfluidics as a tool for C. elegans research. WormBook. doi: 10.1895/wormbook.1.162.1
6. Sia S, Whitesides G (2003) Microfluidic devices fabricated in poly(dimethylsiloxane) for biological studies. Electrophoresis 24:3563–3576
7. Squires TM, Quake SR (2005) Microfluidics: fluid physics at the nanoliter scale. Rev Mod Phys 77:977–1026
8. Jeon NL, Dertinger SKW, Chiu DT, Choi IS, Stroock AD, Whitesides GM (2000) Generation of solution and surface gradients using microfluidic systems. Langmuir 16:8311–8316
9. Dertinger SKW, Chiu DT, Jeon NL, Whitesides GM (2001) Generation of gradients having complex shapes using microfluidic networks. Anal Chem 73:1240–1246

10. Irimia D (2010) Microfluidic technologies for temporal perturbations of chemotaxis. Annu Rev Biomed Eng 12:259–284
11. Albrecht DR, Bargmann CI (2011) High-content behavioral analysis of Caenorhabditis elegans in precise spatiotemporal chemical environments. Nat Methods 8:599–605
12. Chingozha L, Zhan M, Zhu C, Lu H (2014) A generalizable, tunable microfluidic platform for delivering fast temporally varying chemical signals to probe single-cell response dynamics. Anal Chem 86:10138–10147
13. Lu H, Koo LY, Wang WM, Lauffenburger DA, Griffith LG, Jensen KF (2004) Microfluidic shear devices for quantitative analysis of cell adhesion. Anal Chem 76:5257–5264
14. Stroock AD, Dertinger SKW, Ajdari A, Mezić I, Stone HA, Whitesides GM (2002) Chaotic mixer for microchannels. Science 295:647–651
15. Williams MS, Longmuir KJ, Yager P (2008) A practical guide to the staggered herringbone mixer. Lab Chip 8:1121–1129
16. Stone HA, Stroock AD, Ajdari A (2004) Engineering flows in small devices. Annu Rev Fluid Mech 36:381–411
17. Atencia J, Beebe DJ (2005) Controlled microfluidic interfaces. Nature 437:648–655
18. Duffy DC, McDonald JC, Schueller OJA, Whitesides GM (1998) Rapid prototyping of microfluidic systems in poly(dimethylsiloxane). Anal Chem 70:4974–4984
19. McDonald JC, Duffy DC, Anderson JR, Chiu DT, Wu HK, Schueller OJA, Whitesides GM (2000) Fabrication of microfluidic systems in poly(dimethylsiloxane). Electrophoresis 21:27–40
20. Wang Y, Balowski J, Phillips C, Phillips R, Sims CE, Allbritton NL (2011) Benchtop micromolding of polystyrene by soft lithography. Lab Chip 11:3089–3097
21. Chen Y, Duan H, Zhang L, Chen G (2008) Fabrication of PMMA CE microchips by infrared-assisted polymerization. Electrophoresis 29:4922–4927
22. Martinez AW, Phillips ST, Butte MJ, Whitesides GM (2007) Patterned paper as a platform for inexpensive, low-volume, portable bioassays. Angew Chem Int Ed 46:1318–1320
23. Cate DM, Adkins JA, Mettakoonpitak J, Henry CS (2015) Recent developments in paper-based microfluidic devices. Anal Chem 87:19–41
24. Martinez AW, Phillips ST, Whitesides GM, Carrilho E (2010) Diagnostics for the developing world: microfluidic paper-based analytical devices. Anal Chem 82:3–10
25. Vyawahare S, Griffiths AD, Merten CA (2010) Miniaturization and parallelization of biological and chemical assays in microfluidic devices. Chem Biol 17:1052–1065
26. Velve-Casquillas G, Le Berre M, Piel M, Tran PT (2010) Microfluidic tools for cell biological research. Nano Today 5:28–47
27. K-i K (2013) Cutting-edge microfabricated biomedical tools for human pluripotent stem cell research. J Lab Automat 18:469–481
28. Thorsen T, Maerkl SJ, Quake SR (2002) Microfluidic large-scale integration. Science 298:580–584
29. Cheong R, Wang CJ, Levchenko A (2009) Using a microfluidic device for high-content analysis of cell signaling. Sci Sig 2:pl2
30. Chung K, Lu H (2009) Automated high-throughput cell microsurgery on-chip. Lab Chip 9:2764–2766
31. Crane MM, Chung K, Lu H (2009) Computer-enhanced high-throughput genetic screens of C. elegans in a microfluidic system. Lab Chip 9:38–40
32. Stirman JN, Brauner M, Gottschalk A, Lu H (2010) High-throughput study of synaptic transmission at the neuromuscular junction enabled by optogenetics and microfluidics. J Neurosci Methods 191:90–93
33. Ottesen EA, Hong JW, Quake SR, Microfluidic LJR, Digital PCR (2006) Enables multigene analysis of individual environmental bacteria. Science 314:1464–1467
34. Blazej RG, Kumaresan P, Mathies RA (2006) Microfabricated bioprocessor for integrated nanoliter-scale Sanger DNA sequencing. Proc Natl Acad Sci 103:7240–7245
35. Streets AM, Zhang X, Cao C, Pang Y, Wu X, Xiong L, Yang L, Fu Y, Zhao L, Tang F, Huang Y (2014) Microfluidic single-cell whole-transcriptome sequencing. Proc Natl Acad Sci 111:7048–7053

36. Macosko EZ, Basu A, Satija R, Nemesh J, Shekhar K, Goldman M, Tirosh I, Bialas AR, Kamitaki N, Martersteck EM, Trombetta JJ, Weitz DA, Sanes JR, Shalek AK, Regev A, McCarroll SA (2015) Highly parallel genome-wide expression profiling of individual cells using nanoliter droplets. Cell 161:1202–1214
37. Freire SLS, Wheeler AR (2006) Proteome-on-a-chip: mirage, or on the horizon? Lab Chip 6:1415–1423
38. Sia SK, Kricka LJ (2008) Microfluidics and point-of-care testing. Lab Chip 8:1982–1983
39. Chung K, Rivet CA, Kemp ML, Lu H (2011) Imaging single-cell signaling dynamics with a deterministic high-density single-cell trap array. Anal Chem 83:7044–7052
40. Zheng Y, Nguyen J, Wei Y, Sun Y (2013) Recent advances in microfluidic techniques for single-cell biophysical characterization. Lab Chip 13:2464–2483
41. Kaletta T, Hengartner MO (2006) Finding function in novel targets: C. elegans as a model organism. Nat Rev Drug Discov 5:387–399
42. Meeker ND, Trede NS (2008) Immunology and zebrafish: spawning new models of human disease. Dev Comp Immunol 32:745–757
43. O'Kane C (2011) Drosophila as a model organism for the study of neuropsychiatric disorders. In: Hagan JJ (ed) Molecular and functional models in neuropsychiatry, 7th edn. Springer, Berlin, pp 37–60
44. Pandey UB, Nichols CD (2011) Human disease models in drosophila melanogaster and the role of the fly in therapeutic drug discovery. Pharmacol Rev 63:411–436
45. Leung MCK, Williams PL, Benedetto A, Au C, Helmcke KJ, Aschner M, Meyer JN (2008) Caenorhabditis elegans: an emerging model in biomedical and environmental toxicology. Toxicol Sci 106:5–28
46. Brenner S (1974) The genetics of Caenorhabditis elegans. Genetics 77:71–94
47. Ellis HM, Horvitz HR (1986) Genetic control of programmed cell death in the nematode C. elegans. Cell 44:817–829
48. Sulston JE (1976) Post-embryonic development in the ventral cord of Caenorhabditis elegans. Phil Trans Roy Soc Lond B Biol Sci 275:287–297
49. Fire A, Xu S, Montgomery MK, Kostas SA, Driver SE, Mello CC (1998) Potent and specific genetic interference by double-stranded RNA in Caenorhabditis elegans. Nature 391:806–811
50. Crane MM, Chung K, Stirman J, Lu H (2010) Microfluidics-enabled phenotyping, imaging, and screening of multicellular organisms. Lab Chip 10:1509–1517
51. Krajniak J, Lu H (2010) Long-term high-resolution imaging and culture of C. elegans in chip-gel hybrid microfluidic device for developmental studies. Lab Chip 10:1862–1868
52. Hulme SE, Shevkoplyas SS, McGuigan AP, Apfeld J, Fontana W, Whitesides GM (2010) Lifespan-on-a-chip: microfluidic chambers for performing lifelong observation of C. elegans. Lab Chip 10:589–597
53. Hulme SE, Shevkoplyas SS, Apfeld J, Fontana W, Whitesides GM (2007) A microfabricated array of clamps for immobilizing and imaging C. elegans. Lab Chip 7:1515–1523
54. Chung K, Crane MM, Lu H (2008) Automated on-chip rapid microscopy, phenotyping and sorting of C. elegans. Nat Methods 5:637–643
55. Rohde CB, Zeng F, Gonzalez-Rubio R, Angel M, Yanik MF (2007) Microfluidic system for on-chip high-throughput whole-animal sorting and screening at subcellular resolution. Proc Natl Acad Sci U S A 104:13891–13895
56. Chokshi TV, Ben-Yakar A, Chronis N (2009) CO2 and compressive immobilization of C. elegans on-chip. Lab Chip 9:151–157
57. Chronis N, Zimmer M, Bargmann CI (2007) Microfluidics for in vivo imaging of neuronal and behavioral activity in Caenorhabditis elegans. Nat Methods 4:727–731
58. Chokshi TV, Bazopoulou D, Chronis N (2010) An automated microfluidic platform for calcium imaging of chemosensory neurons in Caenorhabditis elegans. Lab Chip 10:2758–2763
59. Allen PB, Sgro AE, Chao DL, Doepker BE, Edgar JS, Shen K, Chiu DT (2008) Single-synapse ablation and long-term imaging in live C. elegans. J Neurosci Methods 173:20–26

60. Guo SX, Bourgeois F, Chokshi T, Durr NJ, Hilliard MA, Chronis N, Ben-Yakar A (2008) Femtosecond laser nanoaxotomy lab-on-a-chip for in vivo nerve regeneration studies. Nat Methods 5:531–533
61. Zeng F, Rohde CB, Yanik MF (2008) Sub-cellular precision on-chip small-animal immobilization, multi-photon imaging and femtosecond-laser manipulation. Lab Chip 8:653–656
62. Conlon I, Raff M (1999) Size control in animal development. Cell 96:235–244
63. Lucchetta EM, Lee JH, Fu LA, Patel NH, Ismagilov RF (2005) Dynamics of Drosophila embryonic patterning network perturbed in space and time using microfluidics. Nature 434:1134–1138
64. Lucchetta EM, Munson MS, Ismagilov RF (2006) Characterization of the local temperature in space and time around a developing Drosophila embryo in a microfluidic device. Lab Chip 6:185–190
65. Chung K, Kim Y, Kanodia JS, Gong E, Shvartsman SY, Lu H (2011) A microfluidic array for large-scale ordering and orientation of embryos. Nat Methods 8:171–176
66. Ghannad-Rezaie M, Wang X, Mishra B, Collins C, Chronis N (2012) Microfluidic chips for in vivo imaging of cellular responses to neural injury in Drosophila larvae. PLoS ONE 7: e29869
67. Mishra B, Ghannad-Rezaie M, Li J, Wang X, Hao Y, Ye B, Chronis N, Collins CA. (2014) Using microfluidics chips for live imaging and study of injury responses in Drosophila larvae. J Vis Exp. doi: 10.3791/50998
68. Hill AJ, Teraoka H, Heideman W, Peterson RE (2005) Zebrafish as a model vertebrate for investigating chemical toxicity. Toxicol Sci 86:6–19
69. Lieschke GJ, Currie PD (2007) Animal models of human disease: zebrafish swim into view. Nat Rev Genet 8:353–367
70. Funfak A, Broesing A, Brand M, Koehler JM (2007) Micro fluid segment technique for screening and development studies on Danio rerio embryos. Lab Chip 7:1132–1138
71. Raty S, Walters EM, Davis J, Zeringue H, Beebe DJ, Rodriguez-Zas SL, Wheeler MB (2004) Embryonic development in the mouse is enhanced via microchannel culture. Lab Chip 4:186–190
72. Yang F, Chen Z, Pan J, Li X, Feng J, Yang H (2011) Microfluidic droplet encapsulation of highly motile single zoospores for phenotypic screening of an antioomycete chemical. Biomicrofluidics 5:044103–044111
73. Melin J, Lee A, Foygel K, Leong DE, Quake SR, Yao MWM (2009) In vitro embryo culture in defined, sub-microliter volumes. Dev Dyn 238:950–955
74. Peerani R, Rao BM, Bauwens C, Yin T, Wood GA, Nagy A, Kumacheva E, Zandstra PW (2007) Niche-mediated control of human embryonic stem cell self-renewal and differentiation. EMBO J 26:4744–4755
75. Lee LH, Peerani R, Ungrin M, Joshi C, Kumacheva E, Zandstra PW (2009) Micropatterning of human embryonic stem cells dissects the mesoderm and endoderm lineages. Stem Cell Res 2:155–162
76. Hwang Y-S, Chung BG, Ortmann D, Hattori N, Moeller H-C, Khademhosseini A (2009) Microwell-mediated control of embryoid body size regulates embryonic stem cell fate via differential expression of WNT5a and WNT11. Proc Natl Acad Sci U S A 106:16978–16983
77. Karp JM, Yeh J, Eng G, Fukuda J, Blumling J, Suh K-Y, Cheng J, Mahdavi A, Borenstein J, Langer R, Khademhosseini A (2007) Controlling size, shape and homogeneity of embryoid bodies using poly(ethylene glycol) microwells. Lab Chip 7:786–794
78. Fung W-T, Beyzavi A, Abgrall P, Nguyen N-T, Li H-Y (2009) Microfluidic platform for controlling the differentiation of embryoid bodies. Lab Chip 9:2591–2595
79. Kim C, Lee KS, Bang JH, Kim YE, Kim M-C, Oh KW, Lee SH, Kang JY (2011) 3-Dimensional cell culture for on-chip differentiation of stem cells in embryoid body. Lab Chip 11:874–882
80. Bhatia SN, Ingber DE (2014) Microfluidic organs-on-chips. Nat Biotechnol 32:760–772
81. Chalasani SH, Chronis N, Tsunozaki M, Gray JM, Ramot D, Goodman MB, Bargmann CI (2007) Dissecting a circuit for olfactory behaviour in Caenorhabditis elegans. Nature 450:63–70

82. Gray JM, Karow DS, Lu H, Chang AJ, Chang JS, Ellis RE, Marletta MA, Bargmann CI (2004) Oxygen sensation and social feeding mediated by a C. elegans guanylate cyclase homologue. Nature 430:317–322
83. Zhang Y, Lu H, Bargmann CI (2005) Pathogenic bacteria induce aversive olfactory learning in Caenorhabditis elegans. Nature 438:179–184
84. Zimmer M, Gray JM, Pokala N, Chang AJ, Karow DS, Marletta MA, Hudson ML, Morton DB, Chronis N, Bargmann CI (2009) Neurons detect increases and decreases in oxygen levels using distinct guanylate cyclases. Neuron 61:865–879
85. Larsch J, Ventimiglia D, Bargmann CI, Albrecht DR (2013) High-throughput imaging of neuronal activity in Caenorhabditis elegans. Proc Natl Acad Sci U S A 110:E4266–E4273
86. Cornaglia M, Mouchiroud L, Marette A, Narasimhan S, Lehnert T, Jovaisaite V, Auwerx J, Gijs MAM (2015) An automated microfluidic platform for C. elegans embryo arraying, phenotyping, and long-term live imaging. Sci Rep 5:10192
87. Kopito RB, Levine E (2014) Durable spatiotemporal surveillance of Caenorhabditis elegans response to environmental cues. Lab Chip 14:764–770
88. Krajniak J, Hao Y, Mak HY, Lu H (2013) C.L.I.P.-continuous live imaging platform for direct observation of C. elegans physiological processes. Lab Chip 13:2963–2971
89. Chung K, Zhan M, Srinivasan J, Sternberg PW, Gong E, Schroeder FC, Lu H (2011) Microfluidic chamber arrays for whole-organism behavior-based chemical screening. Lab Chip 11:3689–3697
90. Zhao X, Xu F, Tang L, Du W, Feng X, Liu B-F (2013) Microfluidic chip-based C. elegans microinjection system for investigating cell–cell communication in vivo. Biosens Bioelectron 50:28–34
91. Deisseroth K (2011) Optogenetics. Nat Methods 8:26–29
92. Lichtman JW, Conchello J-A (2005) Fluorescence microscopy. Nat Methods 2:910–919
93. Cáceres IDC, Valmas N, Hilliard MA, Lu H (2012) Laterally orienting C. elegans using geometry at microscale for high-throughput visual screens in neurodegeneration and neuronal development studies. PLoS ONE 7:e35037
94. Kerr R, Lev-Ram V, Baird G, Vincent P, Tsien RY, Schafer WR (2000) Optical imaging of calcium transients in neurons and pharyngeal muscle of C. elegans. Neuron 26:583–594
95. Suzuki H, Kerr R, Bianchi L, Frokjaer-Jensen C, Slone D, Xue J, Gerstbrein B, Driscoll M, Schafer WR (2003) In vivo imaging of C. elegans mechanosensory neurons demonstrates a specific role for the MEC-4 channel in the process of gentle touch sensation. Neuron 39:1005–1017
96. Hilliard MA, Apicella AJ, Kerr R, Suzuki H, Bazzicalupo P, Schafer WR (2005) In vivo imaging of C. elegans ASH neurons: cellular response and adaptation to chemical repellents. EMBO J 24:63–72
97. Luo L, Wen Q, Ren J, Hendricks M, Gershow M, Qin Y, Greenwood J, Soucy ER, Klein M, Smith-Parker HK, Calvo AC, Colon-Ramos DA, Samuel ADT, Zhang Y (2014) Dynamic encoding of perception, memory, and movement in a C. elegans chemotaxis circuit. Neuron 82:1115–1128
98. Lee H, Crane MM, Zhang Y, Lu H (2013) Quantitative screening of genes regulating tryptophan hydroxylase transcription in Caenorhabditis elegans using microfluidics and an adaptive algorithm. Integr Biol 5:372–380. doi:10.1039/C2IB20078C
99. Hu C, Dillon J, Kearn J, Murray C, O'Connor V, Holden-Dye L, Morgan H (2013) NeuroChip: a microfluidic electrophysiological device for genetic and chemical biology screening of Caenorhabditis elegans adult and larvae. PLoS One 8, e64297
100. Chung K, Crane M, Lu H (2009) Automated on-chip imaging and sorting of C. elegans. Protocol exchange
101. Stephens GJ, Johnson-Kerner B, Bialek W, Ryu WS (2008) Dimensionality and dynamics in the behavior of C. elegans. PLoS Comput Biol 4, e1000028
102. Melin J, Quake SR (2007) Microfluidic large-scale integration: the evolution of design rules for biological automation. Annu Rev Biophys Biomol Struct 36:213–231
103. King KR, Wang S, Irimia D, Jayaraman A, Toner M, Yarmush ML (2007) A high-throughput microfluidic real-time gene expression living cell array. Lab Chip 7:77–85

Chapter 2
Microfluidic Methods in Single Cell Biology

Arnab Mukherjee and Charles M. Schroeder

Abstract Stochastic variations within seemingly homogeneous cell populations determine the emergent properties of complex cellular systems such as biofilms, tumors, pluripotent stem cells, and multispecies ecosystems. The advent of microfluidic technologies, coupled with rapid advances in fluorescence-based molecular imaging and genomic, transcriptomic, and proteomic profiling techniques, has spurred a revolution in biological analysis at the level of single cells. Over the past decade, several microfluidic platforms have been developed that enable the isolation, enrichment, and biochemical or genetic analysis of individual cells with high spatiotemporal resolution in a fashion that is not achievable using macroscale methods. In sharp contrast to population-averaged measurements based on bulk-level techniques, microfluidic cell culture platforms permit the acquisition of multiparametric and high-content information while preserving the identity and monitoring the behavior of individual cells over time. In this way, microfluidics has ushered in new frontiers in single cell biology with a direct impact on applied and foundational studies in microbial ecology, systems biology, therapeutics development, and clinical diagnostics. In this chapter, we describe the transformative impact of microfluidics in single cell biology with particular emphasis on the following areas: (1) microfluidic bioreactors for cellular analysis in dynamically changing microenvironments, (2) microfluidic chips for in vitro drug screening, and (3) single cell confinement and isolation microchips for sorting and profiling rare or unculturable cells in complex environmental consortia.

Keywords Biological noise • Stochasticity • Laminar flow • Antibiotic resistance • Single cell genome amplification • Circulating tumor cells • Unculturable microbes • Time-lapse fluorescent microscopy

A. Mukherjee, Ph.D.
Division of Chemistry and Chemical Engineering, California Institute of Technology, 1200 E. California Blvd., 123 Spalding, Pasadena, CA 91125, USA
e-mail: arnabm@caltech.edu

C.M. Schroeder, Ph.D. (✉)
Department of Chemical and Biomolecular Engineering, Center for Biophysics and Computational Biology, Institute for Genomic Biology/Biosystems Design, University of Illinois at Urbana-Champaign, 600 S. Mathews Avenue, Urbana, IL 61801, USA
e-mail: cms@illinois.edu

© Springer International Publishing Switzerland 2016

C. Lu, S.S. Verbridge (eds.), *Microfluidic Methods for Molecular Biology*,
DOI 10.1007/978-3-319-30019-1_2

2.1 Introduction

2.1.1 *Single Cell Biology*

Cells are inherently noisy systems in that they exhibit significant off-mean excursions in protein and RNA levels even in isogenic populations and in homogeneous environments [1, 2]. The stochastic nature of gene expression and the resulting variations in cellular phenotypes are exemplified by several biological phenomena including microbial bet-hedging strategies to survive stressful environments, generation of drug-resistant tumor subpopulations, antibiotic persistence, aging, development of multifunctional phenotypes of cytotoxic tumor-infiltrating lymphocytes (CTLs), stem cell differentiation, and variable metastatic potential within a population of circulating tumor cells (CTCs) derived from the same tumor [1–11]. Stochastic fluctuations in gene expression are strongly affected by the architecture of the underlying genetic network and are particularly relevant at low concentrations of biomolecules. To this end, a quantitative resolution of cell-to-cell variations in gene expression and associated phenotypes for natural and engineered cellular systems constitutes a fundamental challenge for single cell biology [3, 4, 7]. Classically, biological noise has been studied using destructive end-point techniques such as fluorescence-activated cell sorting (FACS) or time-lapse fluorescence microscopy (TLFM) in combination with fluorescently labeled promoters, proteins, or RNA molecules (typically labeled using fluorescence in situ hybridization or FISH) [6, 12, 13]. FACS, however, is limited in its temporal resolution, while conventional TLFM is limited in the throughput and information content accessible from a single experiment. Furthermore, these approaches do not permit facile manipulation of the cellular microenvironment using precise spatiotemporal cues, which is critical for quantitatively interrogating single cell behavior.

2.1.2 *Microfluidic Approaches to Single Cell Biology*

Microfluidics broadly refers to the manipulation of picoliter to nanoliter-scale volumes of fluids in flow channels, reagent wells, and reaction chambers with dimensions ranging from a few to hundreds of micrometers [14–17]. Microfluidics enables miniaturization of biochemical reactions by a factor of 10^3–10^6, which greatly improves the sensitivity (by enhancing signal to background) of conventional cellular and molecular biological assays, while also minimizing sample and reagent consumption and reducing the overall experimental footprint. Microfluidic devices are compatible with widely used assay readout technologies including fluorescence, luminescence, surface plasmon resonance, and even mass spectrometry. Importantly, fluid flow in microfluidic channels tends to be laminar (such that viscous forces dominate over inertial forces), which facilitates the creation of precise gradients in concentrations of reagents, thereby enabling spatially defined manipulations with cellular to subcellular spatial resolution [14–21]. Heat exchange and mass transfer proceed rapidly owing to the large surface-to-volume ratios of microfluidic devices, which is useful for delivering dynamic temporal stimuli to cells. Finally, the small form factor associated with microfluidic devices provides a useful framework to realize multiplexed systems integrating multiple steps in cell biology including cell growth, sorting, lysis, and downstream analysis using quantitative PCR, FISH, and antibody labeling. Integrated microfluidic systems circumvent the need for off-chip procedures, thereby greatly improving speed and precision of biological assays [14, 22, 23]. Furthermore, automation can be readily achieved using inexpensive computer-controlled syringe pumps and air flow controllers, which obviates the need for cost-prohibitive robotic liquid handling systems.

2.1.2.1 Microfluidic Fabrication Techniques

Microfluidic systems can be constructed using a variety of materials including silicon, glass, polycarbonate, parylene, polystyrene, polymethyl methacrylate, water-in-oil droplets, and even paper [16, 24]. The material of choice for the vast majority of microfluidic cell handling platforms, however, continues to be a silicone elastomer—polydimethyl siloxane, commonly referred to as PDMS. PDMS offers unique advantages for biological applications including nontoxicity for cell growth, high gas permeability, optical transparency, and relative inertness to most reagents typically employed in cell culture. In addition, PDMS enables rapid prototyping of microfluidic devices using widely established soft lithographic procedures. In this technique, a microfluidic relief pattern is first patterned on a silicon wafer using a UV light source together with a photomask to spatially control light exposure on a layer of photoresist-coated silicon wafer to produce a master mold. Next, unpolymerized photoresist is dissolved away leaving behind a positive or negative relief pattern on the silicon wafer, known as the master mold. The master mold wafers are treated with a release agent (trichlorosilane) and spin coated with a mixture of degassed PDMS and a curing agent and thermally cured for several hours at 65–80 °C in order to polymerize the PDMS. Cured PDMS can be detached from the silanized wafer and bonded to a glass slide that has been rendered hydrophilic by exposure to oxygen plasma. In this way, a single master mold can be used to rapidly develop up to hundreds of PDMS chips using replica molding [24–27]. Active fluidic valves can be incorporated into PDMS-based microdevices by fabricating multilayer chips, wherein PDMS layers are bonded to each other via plasma treatment [24, 25, 28–30]. Typical multilayer microfluidic devices consist of a fluid layer comprising the microfluidic channel network and arrays of reagent chambers sandwiched between PDMS and glass. Importantly, the low Young's modulus of PDMS can be exploited to enable facile on-chip fluid routing through the use of microfluidic valves, which can be pneumatically or hydraulically actuated. In this approach, a second PDMS layer known as the control layer is aligned and bonded above the fluidic layer. The control layer houses air flow lines that enable the regulation of fluid flow in the fluidic layer by pushing (actuate-to-close vales) or pulling up (actuate-to-open vales) on the underlying PDMS layer at the intersection points of the control and fluid layers [28, 31]. Furthermore, by assembling and actuating valves in specified sequences or programs, peristaltic or rotary pump action can be achieved on-chip. Indeed, soft lithography can be recognized as the single most important factor responsible for the pervasive application of microfluidics by facilitating rapid design-to-prototype transition without the labor and expense associated with silicon microfabrication [16–18, 20, 21, 23–27, 30–32].

In the past decade, the vast majority of conventional macroscale technologies employed in cell and molecular biology studies has been micro-engineered and translated to microfluidic systems. This has led to the development of microfluidic cell sorters, electrophoresis platforms, patch clamps, on-chip thermal regulators, gradient generators, continuous and batch-mode cell culture systems, size-exclusion filters, and cell traps. Recently, microfluidic total analysis systems have been developed as miniaturized, self-contained platforms for cell culture and downstream chemical cytometry [14, 17, 20, 21, 25]. In this chapter, we present a broad overview of major advances in microfluidic applications for single cell biology. In order to highlight the transformative impact of microfluidics, we focus on key applications of novel microfluidic designs in first-of-its-kind studies in synthetic biology, environmental microbiology, and therapeutics development. Specifically, in Sect. 2.2, we describe the development and application of microfluidic bioreactors and single cell traps for investigating temporal patterns in gene expression. In Sect. 2.3, we review the application of microfluidics for developing complex spatial gradients in order to interrogate cellular phenotypes and gene expression in dynamic microenvironments. In Sect. 2.4, we discuss the extension of microfluidics platforms for high throughput and high content in vitro drug screening. Finally, in Sect. 2.5, we shift focus from microfluidic chips for single cell studies to microfluidic platforms that enable single species isolation and analyses. Here, we describe microfluidic devices that have been designed to isolate and profile rare cells (e.g., unculturable microbes) from heterogeneous and multispecies consortia. A unifying theme in the applications described in this chapter is the critical need for single cell isolation and/or characterization, which is uniquely enabled through the application of microfluidic technologies.

2.2 Microfluidic Platforms for Studying Gene Expression

An overarching goal of synthetic and systems biology is to achieve a quantitative understanding of gene expression and cell signaling pathways underlying cellular response in various conditions [14, 20]. At the heart of this approach is the ability to develop rigorous models of transfer functions that map a genetic network's output to a set of input stimuli. Experimentally, this is achieved through the application of defined perturbations to cells followed by a readout of cellular phenotype, typically using fluorescent reporters [3, 7, 12, 13]. Microfluidic lab-on-a-chip platforms provide an ideal test bed for developing and testing such transfer functions because they enable precise and reproducible spatiotemporal control over the cellular microenvironment along with single cell resolution in experimental readout (e.g., using TLFM). Microfluidic approaches have several potential advantages compared to widely used microtiter plate-based assays, flow cytometry, and microarrays, in that microfluidics combines the benefits of fine-scale spatial and temporal control over the cellular microenvironment while alleviating the ensemble-averaging effects of population based assays. For these reasons, microfluidic chips have been extensively used for functional interrogation of engineered or natural genetic circuits in single isolated cells as well as in cell populations subjected to dynamic gradients in space and time [14, 20, 21]. In the following sections, we describe a few seminal studies in systems and synthetic biology that have leveraged the aforementioned benefits of microfluidic technologies.

2.2.1 Microfluidic Dial-a-Wave Chip

A prolific example of the integration of microfluidics and synthetic biology is afforded by the dial-a-wave microfluidic chip designed by Hasty and coworkers [33]. In this design, a population of yeast cells is constrained to grow as a monolayer in a microfluidic chamber, which is connected to feeding channels that supply the cells with nutrients and media (Fig. 2.1). Importantly, the composition of the media can be dynamically tuned using an upstream fluidic switch, which enables the generation of a wide range of stimulus waveforms. The fluidic switch was implemented using the principle of laminar interface guidance in which the laminar interface between two parallel flow streams can be precisely modulated by adjusting the respective flow rates. Hasty et al. used laminar interface guidance to control the composition of the input stream by varying the relative pressure (and thus the flow rates) between two reservoirs feeding the input lines (Fig. 2.2). Downstream of the inlets, complete mixing of the laminar flow streams was achieved using chaotic advection mixers that utilize the Coanda effect. In this way, a sinusoidal waveform of glucose concentrations (ranging from 0 to 0.25 %) against a background of constant galactose concentration (0.2 %) was generated and fed to the cell chambers. Strikingly, the authors demonstrated that the yeast cells respond to dynamic fluctuations in glucose concentrations by acting as low-pass filters, essentially filtering out nutrient fluctuations at frequencies exceeding 0.2 mHz (5.6 rad h^{-1}) (Fig. 2.3).

2.2.2 Microfluidic Chemostat

A persistent issue with microfluidic cell growth chambers is the clogging of microfluidic wells and flow lines by rapid cell growth and biofilm formation.

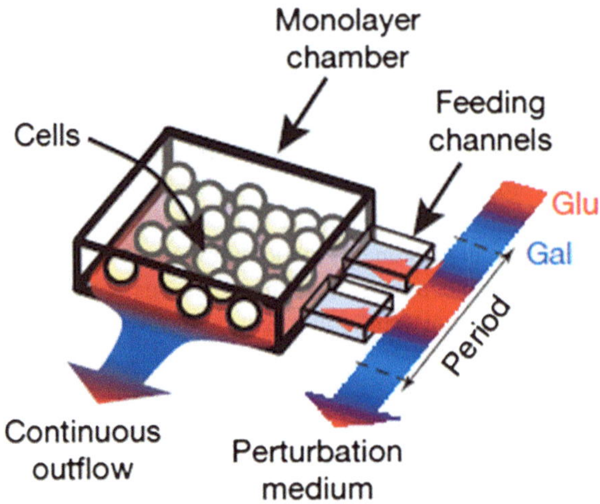

Fig. 2.1 Microfluidic dial-a-wave. The microfluidic dial-a-wave chip can be used to expose a monolayer of yeast cells to precisely defined waveforms in nutrient concentrations via feeding channels that connect the cell growth chamber to a flow channel perfused with media. In this example, Hasty et al. employed the dial-a-wave chip to subject cells to sinusoidal oscillations in glucose concentrations against a constant background of galactose. Reproduced with permission from Ref. [33]

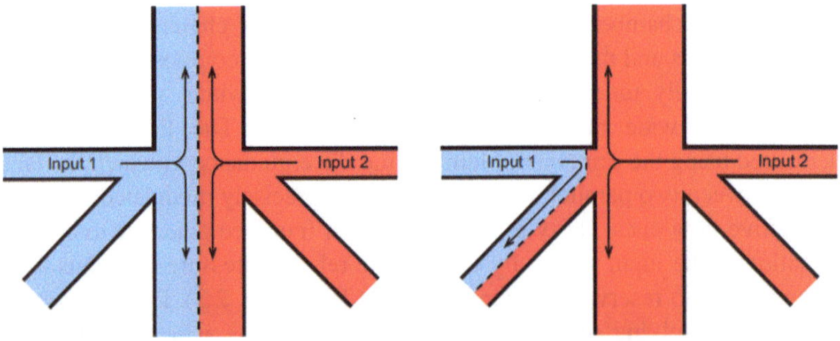

Fig. 2.2 Principle of laminar interface guidance. Laminar interface guidance works by guiding the interface between two laminar input flow streams, across an output channel by adjusting the relative flow rates of the input streams (stream 1 and stream 2 in the above illustration). In the microfluidic dial-a-wave, the flow rates were adjusted by controlling the pressure difference between the source reservoirs feeding input streams 1 and 2. Reproduced with permission from Ref. [33]

With a view toward addressing this issue, Quake and coworkers developed a highly integrated microfluidic chemostat for synthetic biology applications [34]. The microchemostat features 6 nL cell growth chambers where each growth chamber is essentially a microfluidic loop that employs on-chip peristaltic pumps to circulate cells at flow velocities of 250 μm s^{-1}. Each growth chamber was further divided into 16 individually addressable segments. Through the implementation of on-chip

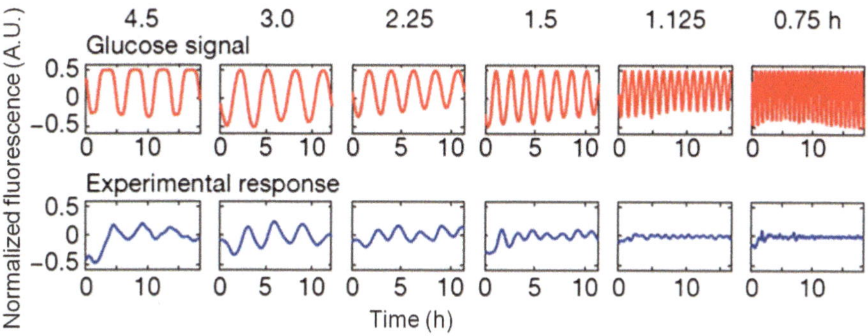

Fig. 2.3 Low-pass filtering action of yeast cells to dynamic fluctuations in glucose concentrations. As the frequency of oscillations in glucose concentrations is increased, the cellular response (quantified as normalized fluorescence on the y-axis) dies out indicating the robust filtering out of high frequency oscillations. Reproduced with permission from Ref. [33]

valves, each segment could be periodically isolated from the rest of the growth loop and washed with lysis buffer and sterile media in order to remove cells including wall-adhering biofilms. In principle, this step served as the microfluidic equivalent of continuous mode operation of a macroscale chemostat. Using the microchemostat, the authors demonstrated robust cell growth at dilution rates ranging from 0.072 to 0.37 h^{-1}. Importantly, the authors leveraged the microchemostat to sustain robust oscillations in a population of *E. coli* cells that had been genetically engineered to express a population control circuit. Specifically, *E. coli* cells were engineered to produce a quorum-sensing molecule (acyl homoserine lactone or AHL), which in turn could activate a toxic ccdB gene in a concentration-dependent manner. Consequently, as cell density increased in the microchemostat, the killer circuit was activated leading to a rapid decline in cell growth. Once the AHL concentrations dropped, the cells could resume growth again. Although several platforms for mimicking chemostat-like conditions on-chip have been described in recent times, the microchemostat by Quake and coworkers remains as one of the most intricate microfluidic cell culture platforms that accurately replicates chemostat-like conditions and enables long-term cell culture.

2.2.3 *Microfluidic "Pin-Ball" Traps for Single Cells*

Microfluidic live cell culture platforms such as the chemostat can be readily adapted for quantifying gene expression with single cell resolution over a statistically significant number of cells. In some applications, however, it is desirable to track gene expression or phenotype in a single cell over an extended duration, for example, in studies on cell aging or antibiotic persistence. Microfluidic

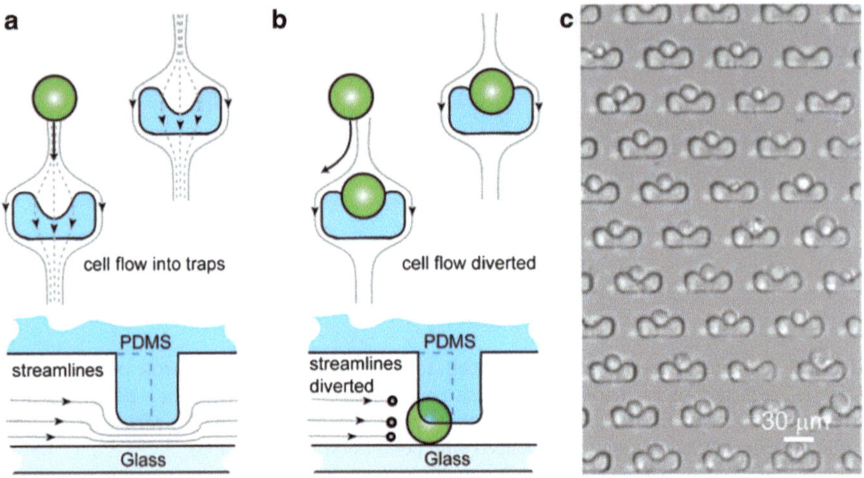

Fig. 2.4 Microfluidic pinball trap for single cells. (**a**) The pinball trap features cup-shaped PDMS obstacles that reduce the channel height to 2 μm at the trapping zones. In the absence of cells, fluid can flow freely through the reduced channel gap. (**b**) Upon encountering a PDMS post, a single cell becomes trapped because the channel height is three to fourfold smaller than the diameter of typical eukaryotic cells. Fluid flow is diverted around a cell trapped at the PDMS post, thereby forming an effective seal at the trapping zone. (**c**) Several PDMS posts are designed on a single chip to develop a single cell trapping array. Reproduced with permission from Ref. [35]

technologies for flow-based cell trapping avoid the need for potentially perturbative cell surface bonding approaches using electrostatic interactions or via chemical cross-linking. To this end, Lee and colleagues developed a microfluidic single cell isolation array that uses cup-shaped PDMS obstacles to function as physical barriers for trapping cells flowed into the device, in a fashion reminiscent of capturing balls in a game of pinball [35]. In this design, cells were flowed in through 40 μm high channels towards the trapping region that featured an array of suspended PDMS obstacles, which reduced the channel height to 2 μm (Fig. 2.4a). As the channel height at the trapping post is significantly smaller than the typical eukaryotic cell diameter (10 μm), cells were hydrodynamically trapped against the suspended PDMS obstacles. Trapping of a cell against a PDMS pillar effectively sealed the 2 μm high channel leading to highly efficient isolation of single cells at each trapping zone (Fig. 2.4b). Furthermore, the depth of the trapping pocket could be varied to tune the number of cells trapped in each obstacle, with greater than 50 % of the traps harboring single cells for a 10 μm deep pocket. Each microchip featured ~3300 PDMS obstacles per mm^2 of the device, which enabled the isolation of several cells on a single chip (Fig. 2.4c). The authors employed the microfluidic trap to isolate single HeLa, 293T, and Jurkat cells. Using this approach, the authors quantified for the first time, intracellular enzymatic activity of carboxylesterases using a fluorogenic substrate [35].

2.2.4 Microfluidic Yeast "Jails"

In order to quantify stochastic variability in the expression of genes implicated in aging in yeast, the ability to track a single yeast mother cell over several cycles of budding is critical. Ryley and Pereira-Smith achieved this by designing PDMS yeast jails to physically trap a yeast cell >4.5 μm in diameter [36]. Optimal trapping of single cells was achieved using square-shaped PDMS posts arranged to enclose a space approximately 4.5 μm wide. Asymmetric cell division of a mother cell trapped in between the square posts resulted in the budding of a daughter cell of diameter < 4.5 μm, which could escape through the jail bars while the parent cell remains trapped (Fig. 2.5). Using the microfluidic yeast jail array, the authors demonstrated up to 92 % cell-to-cell variations in the levels of the RAS2 in yeast cells trapped and observed via TLFM for 18 h. As RAS2 overexpression has been demonstrated to enhance yeast life span by 20–40 %, the cell-to-cell variations in RAS2 levels observed by the authors could explain variations in life spans previously observed in isogenic yeast cell populations [36].

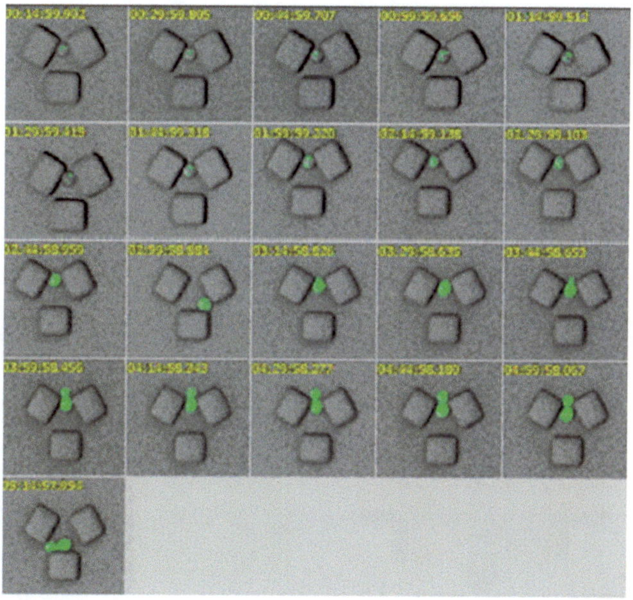

Fig. 2.5 Microfluidic yeast jails. A mother cell (>4.5 μm in diameter) is effectively trapped between the walls of the PDMS blocks, though it is free to move within the space enclosed by the PDMS posts. As the mother cell buds and produces a daughter cell, the smaller daughter cell is able to squeeze in through the gaps between the PDMS pillars leaving the mother cell isolated. Reproduced with permission from Ref. [36]

2.2.5 Single Cell Microbioreactor

A common feature of the vast majority of microfluidic traps involves the mechanical confinement of cells in PDMS microstructures (such as narrow grooves, jails, posts) fabricated on-chip. In contrast, our group reported the development of a contact-free microfluidic confinement approach for single cells and particles [37–39]. Our single cell microbioreactor (SCM) is a two-layered microfluidic device with the lower (fluidic) layer housing two inlet channels that converge at a 60 μm deep cross-slot and diverge through two orthogonal outlet streams, thereby resulting in an extensional flow at the cross slot. As a result, a stagnation point (or point of zero fluid velocity) develops at the cross-slot, and the position of the stagnation point can be precisely regulated along the direction of extensional flow (i.e., the outlet direction) using on-chip valves located above the outlet streams in the control layer (Fig. 2.6). Cells are introduced in the cross-slot through a separate

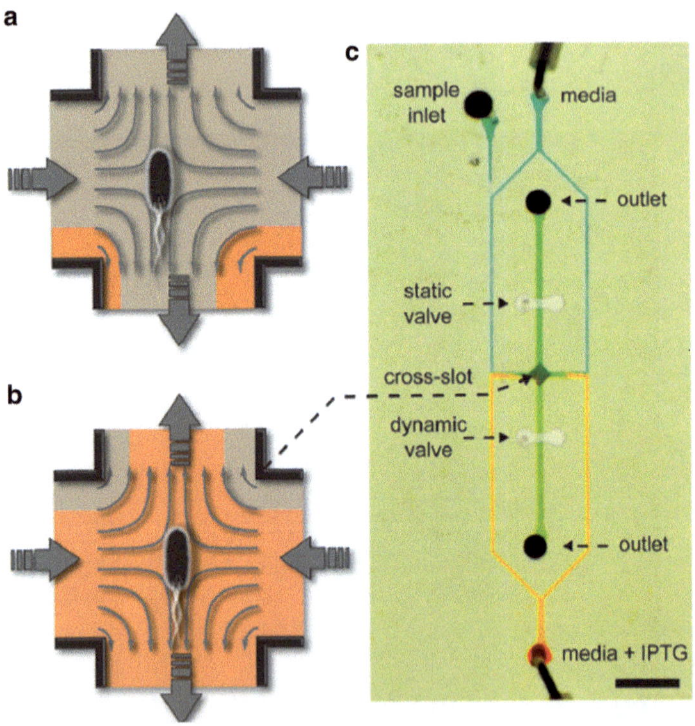

Fig. 2.6 Single cell microbioreactor. (**a**) The single cell microbioreactor traps individual cells in free solution at the saddle point of an extensional flow established in a microfluidic cross-slot. (**b**) Rapid media switching is achieved by changing the composition of the fluid flow in the inlet streams using programmable syringe pumps. (**c**) Micrograph of the single cell microbioreactor showing the fluid flow layers (filled with food coloring) and the control valves (in *white*) that are used to control the position of the stagnation point to actively maintain a cell in the trapping zone. Reproduced with permission from Ref. [39]

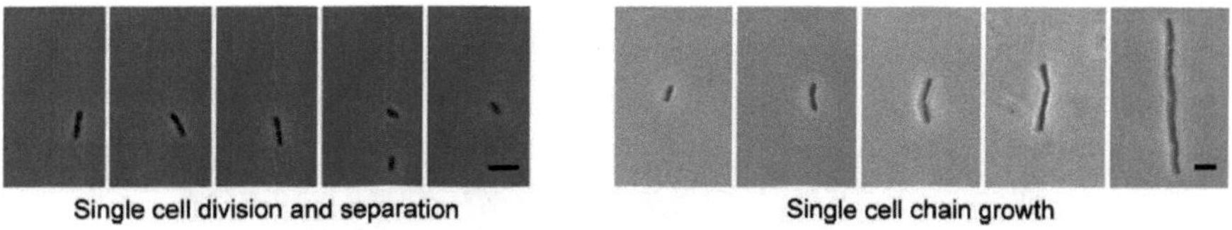

Fig. 2.7 Cell division and growth in the single cell microbioreactor. Single cells that were hydrodynamically trapped at the cross-slot are observed to grow and divide normally in case of cells harvested from exponentially growing cultures. In some cases, particularly for cells harvested from late stationary phase cultures, linear microcolonies resulting from chain-like growth were seen to develop at the cross-slot. Reproduced with permission from Ref. [39]

inlet stream connected to one of the input lines. Robust and long-term trapping (>3 h) of single cells at the cross-slot is achieved using a proportional feedback controller that pneumatically regulates pressurization and depressurization of the on-chip valve to precisely manipulate the position of the stagnation point within the cross-slot. The microfluidic SCM was used to quantify cell elongation and division rates in *E. coli* cells. Interestingly, we observed that single cells of *E. coli* adopt a filament-like growth morphology (where cells string up without splitting) when the mother cell is derived from a stationary phase culture whereas mother cells from exponential cultures symmetrically divide into two daughter cells (Fig. 2.7). A key advantage of the SCM is the ability to rapidly deliver nutrients and reagents on demand using automated programmable syringe pumps at the two inlets, which can be programmed to deliver precise stimulus waveforms. Media switching is accomplished significantly faster (≈1 s) compared to macroscopic methods for buffer exchange and other existing microfluidic cell culture chips, many of which rely on diffusive rather than convective mixing to accomplish fluid exchange. Using our approach, we investigated reporter gene expression in *E. coli* cells treated with a step increase in inducer (isopropyl thiogalactopyranoside or IPTG) concentration as well as a periodic square-wave form of IPTG concentration. We demonstrated that cellular response to a periodic forcing function differs from its response to a step function with respect to a delayed onset of gene expression as well as a slower rate of expression in the former case. Finally, we demonstrated that the SCM could be used to observe protein–DNA interactions in single cells and in real time. Specifically, *E. coli* cells were engineered to express the TetR repressor fused to a fluorescent protein and localized to a tandem TetR operator sequence array integrated in the chromosome. Focal spots of fluorescence localized at the chromosome could be clearly identified in single cells trapped at the cross-slot. Upon rapid (≈9 s) stimulation with anhydrotetracycline, the fluorescently tagged TetR proteins could be observed to unbind from the DNA and diffuse throughout the cytoplasm (Fig. 2.8) [39]. Overall, the SCM enables highly precise estimation of the time points of intracellular protein unbinding and diffusion, and we anticipate that the SCM can be a valuable tool for probing intracellular dynamics with exquisite temporal resolution.

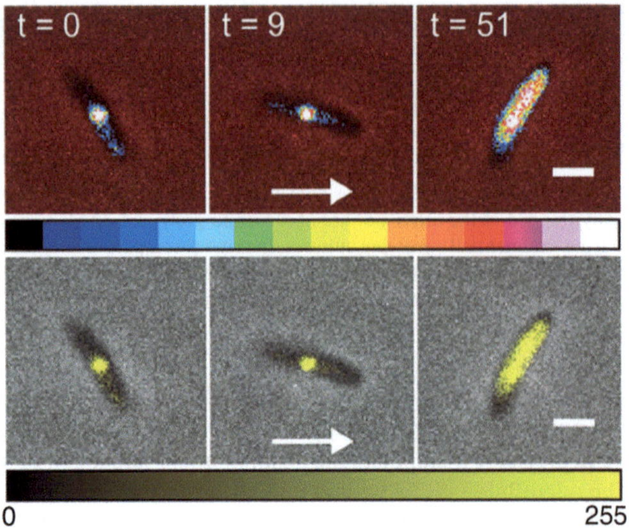

Fig. 2.8 Fluorescently tagged TetR proteins are observed to localize at the tetO binding array in the chromosome in a single trapped *E. coli* cell ($t = 0$ s). In approximately 9 s, the cell was treated with anhydrotetracycline (aTc) by rapid media switching. aTc binds to the TetR protein and causes it to unbind from its binding site at the DNA. At $t = 51$ s, the unbound TetR is seen to freely diffuse throughout the cell cytoplasm. Reproduced with permission from Ref. [39]

2.2.6 *Microfluidic Worm Trap*

Although the focus of the chapter is on single cells, we note that microfluidic traps have also been used to trap and study nematodes, zebrafish, and fly embryos [40–43]. For example, Chronis and colleagues described a microfluidic device that featured tapering channels, which gradually decreased in width from 70 to 40 μm and enabled trapping young adult worms by slightly constricting the thickest part of the worm body while leaving the head and tail free (Fig. 2.9) [43]. Microfluidic worm traps such as the one described by Chronis et al. are less perturbative than existing trapping techniques that employ potentially toxic glues to seal the worm head to agar supports. The gently compressed worms in the microfluidic trap retained the ability to propagate anterior and posterior traveling sinusoidal body waves. In addition, the authors demonstrated that anterior traveling body waves (which correspond to backward locomotion during crawling on agar plates) coincided with peaks in calcium concentrations in the AVA interneurons. The same work also described an olfactory chip that was used to locally stimulate the chemosensory neurons in the olfactory region of a trapped worm to varying stimuli (Fig. 2.10). The microfluidic device comprised four flow channels to subject the worm head to stimulus or buffer solutions (Fig. 2.10a, b). Fluidic switching between stimulus and buffer was achieved using two control channels on the side that delivered dye filled fluids to push the buffer or stimulus flows toward or away

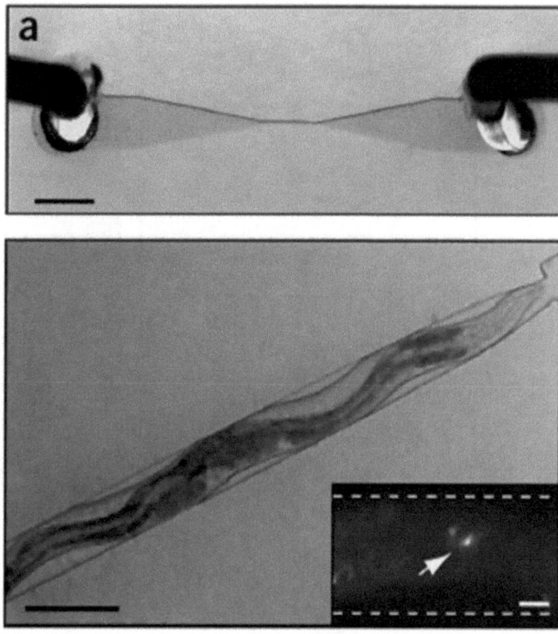

Fig. 2.9 Microfluidic worm behavior chip. A young adult worm is trapped at the constriction zone of a microfluidic channel using gentle constriction applied at the thickest part of the worm body. The worm head and tail are free to move. (*inset*) Fluorescence expression from a calcium indicator (GCaMP) could be readily detected in the worm AVA interneurons and correlated with the development of an anterior traveling wave along the worm body. Scale bars: 1 mm, (*inset*) 20 μm. Reproduced with permission from Ref. [43]

from the worm head toward an outlet (Fig. 2.10c). Using the olfactory chip, the authors demonstrated calcium-dependent increase in fluorescence of a calcium sensor expressed in the ASH neurons of the worm upon the delivery of 15 and 30 s square pulses of a hyperosmotic stimulus.

2.3 Microfluidic Platforms for Generating Spatially Controlled Microenvironments

Although the microfluidic platforms described above are well suited for delivering temporal variations in stimuli, a key appeal of microfluidic platforms for single cell studies relies on the ability to generate highly precise and complex spatial gradients of soluble and surface-adsorbed molecules [19, 21]. Both microbial and eukaryotic cells frequently encounter complex spatial gradients in their native niches, which can be accurately recreated using microfluidics. From this perspective, the ability to generate controlled spatial microenvironments is critical for advancing fundamental understanding of several physiological processes such as chemotaxis in immune cells and bacteria, wound healing, inflammation, cancer metastasis, axonal growth, and stem cell migration. In the next few sections, we describe commonly used microfluidic gradient generators for cell biology studies.

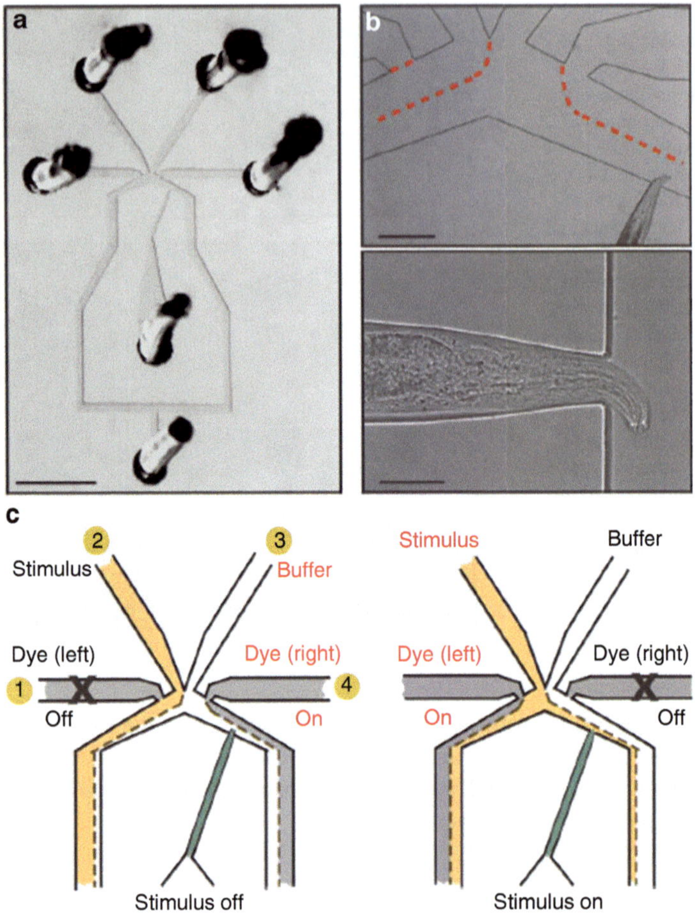

Fig. 2.10 Microfluidic olfactory chip. The microfluidic olfactory chip has four flow lines that (**a**) flow in media to (**b**) subject the head of a trapped worm to a defined odorant. (**c**) Rapid switching between stimuli is achieved via two flow lines on the side that are used to redirect stimulus toward or away from the trapped worm. Reproduced with permission from Ref. [43]

2.3.1 Microfluidic PARTCELL

Concentration gradients in space are typically generated by exploiting the laminar nature of microfluidic flows, which permits fluid streams containing different concentrations of reagents to flow in parallel over several hundreds of micrometers with minimal diffusive mixing. Parallel laminar flow streams can therefore result in stable gradients perpendicular to the flow direction. Flow rates required to maintain stable gradient shapes in microfluidic channels can be determined by the Peclet number, which quantifies the ratio of convective to diffusive flows and is determined by the flow velocity, channel geometry, and the diffusion coefficient of the molecules in question. Furthermore, laminar flow also allows the width of each of the parallel

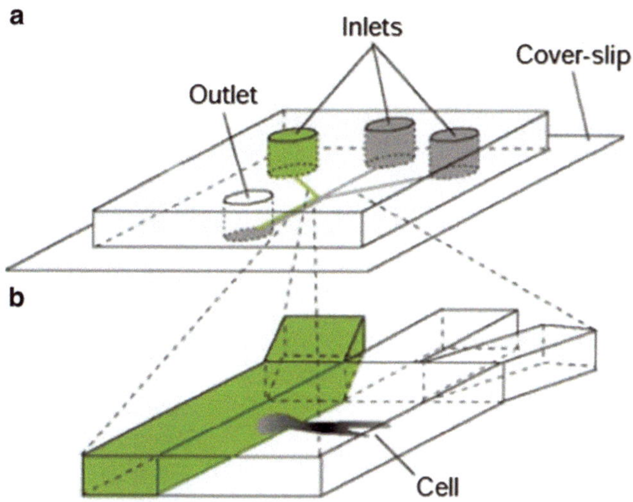

Fig. 2.11 Microfluidic PARTCELL. (**a**) A Y-shaped channel is used to establish parallel laminar flows of different reagents through multiple inlets. (**b**) The parallel streams, which mix minimally owing to the laminar flow regime, result in a concentration gradient perpendicular to the direction of flow. A cell adhered to a cover slip is subjected to the concentration gradient resulting in spatially selective exposure of the cell to various reagents. Reproduced with permission from Ref. [44]

streams to be precisely controlled by adjusting the relative flow rates, thereby enabling an unprecedented degree of spatial control over length scales typical of most eukaryotic cells (10–50 μm) [14, 16, 29]. Whitesides and colleagues first demonstrated this principle by developing the microfluidic PARTCELL (partial treatment of cells using laminar flows) to selectively treat a part of a surface attached to a bovine capillary endothelial (BCE) cell with various dyes and chemicals [44, 45]. In brief, Takayama et al. used a Y-shaped microfluidic channel to flow in two reagents in parallel streams over a surface-anchored BCE cell located $\approx$ 50–300 μm downstream from the fluid inlets (Fig. 2.11). At these length scales, the width of the interface ranged between 1 and 10 μm. PARTCELL was used to selectively label mitochondria on each side of the cell with different colored dyes. Progressive intracellular mixing of the dyes throughout the cell was observed to proceed at rates determined by the intracellular diffusion coefficient of the dye ($\approx$1 $\mu m^2\ s^{-1}$). Interestingly, the same work also demonstrated spatially localized disruption of the cellular cytoskeleton by exposing the right side of the cell to latrunculin B, which is a membrane permeable actin-disrupting molecule (Fig. 2.12).

2.3.2 Y-Step Temperature Gradient Generator

Ismagilov and colleagues leveraged the Y-shaped microfluidic gradient generator design described above to develop temperature gradients across a *Drosophila* embryo attached to the floor of a microfluidic chip [40]. In particular, the authors

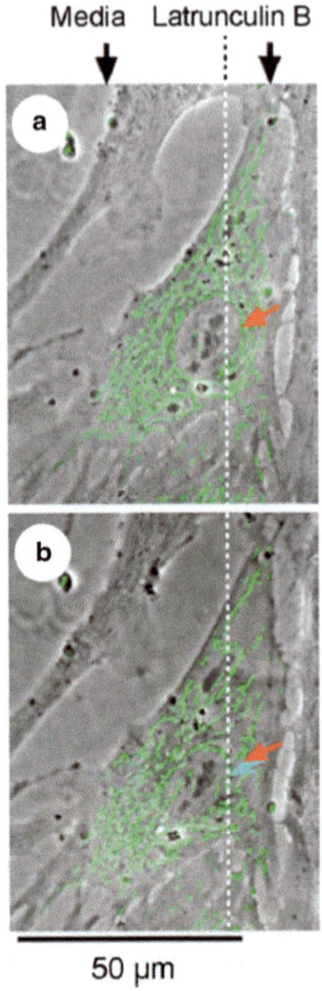

Fig. 2.12 Spatially selective cell treatment using PARTCELL. (**a**) A bovine capillary endothelial cell is adhered to the surface and treated with media flowing in parallel laminar streams. Mitochondria are labeled green and the nucleus is labeled with an arrow. In (**a**), the composition of both media streams is the same. In (**b**), the media on the right side of the cell is supplemented with latrunculin B, which disrupts the cellular actin cytoskeleton on the right side resulting in relocation of the mitochondria and the nuclei toward the left of the cell. The relocated position of the nucleus is indicated by a *blue arrow*. Reproduced with permission from Ref. [44]

employed the Y-shaped microfluidic gradient generator to flow two converging streams of liquid maintained at different temperatures, thereby subjecting each half of the *Drosophila* embryo to a different temperature (Fig. 2.13). Thermal diffusion across the 500 μm wide embryo was calculated to be limited to ~50 μm. In addition, most of the heat was lost through the 1.5 μm thick eggshell, resulting in the development of a tight step gradient in temperature across the width of the embryo. By introducing two parallel streams maintained at 20 °C and 27 °C, the authors demonstrated that the warmer half of the embryo developed more rapidly as evinced by the greater nuclear density in the warmer region of the embryo. Next, the authors followed dynamic expression of a key *Drosophila* developmental gene (known as even-skipped) in the embryo subjected to the 20–27 °C step gradient in temperature. In normally developing embryos, the even-skipped gene expression

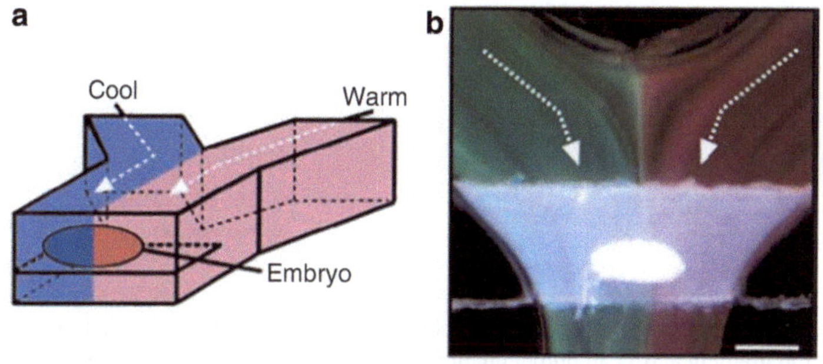

Fig. 2.13 Microfluidic temperature gradient generator. (**a**) By flowing cool and warm media through a Y-shaped microfluidic channel, Ismagilov and colleagues were able to subject a surface-attached *Drosophila* embryo to a step temperature gradient across the anterior–posterior axis of the embryo. (**b**) The temperature gradients were visualized using thermochromic crystals. The *green stream* is at 21 °C and the *red stream* is at 24 °C. Reproduced with permission from Ref. [40]

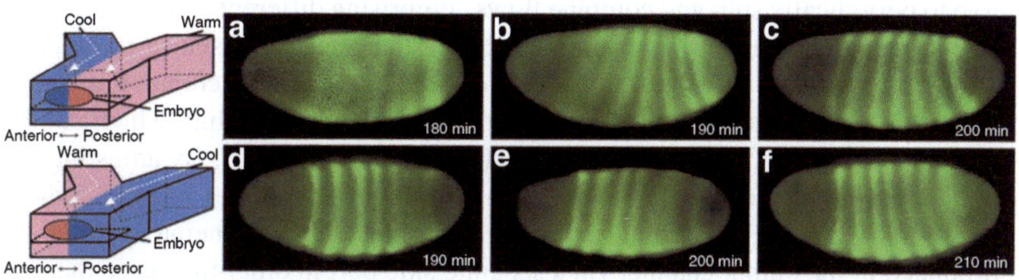

Fig. 2.14 Expression of a *Drosophila* developmental gene (known as even-skipped) in embryos subjected to a temperature gradient. Even-skipped gene expression (revealed by the striated fluorescent patterns) was found to develop preferentially in the warmer half of the embryo first. Strikingly, despite the deviation in the timing of even-skipped gene expression compared to a normal embryo, the final expression pattern was consistent with embryos maintained at a uniform temperature, pointing toward a regulated compensatory mechanism in cells. Reproduced with permission from Ref. [40]

results in a striped pattern across the length of the embryo with the stripes appearing in a precise temporal order. Strikingly, Luchetta et al. demonstrated that stripes resulting from the even-skipped gene expression consistently developed in the warmer half of the embryo first, although eventually the stripes developed across the embryo in a normal fashion pointing toward the presence of a compensating regulatory mechanism in the embryos (Fig. 2.14).

2.3.3 *Microfluidic "Christmas Tree" Gradient Generator*

A widely used microfluidic gradient generator design, first described by Jeon et al., involves a network of serpentine channels interspersed between channel branch points arranged in a pyramidal or Christmas tree pattern [46]. The branch points

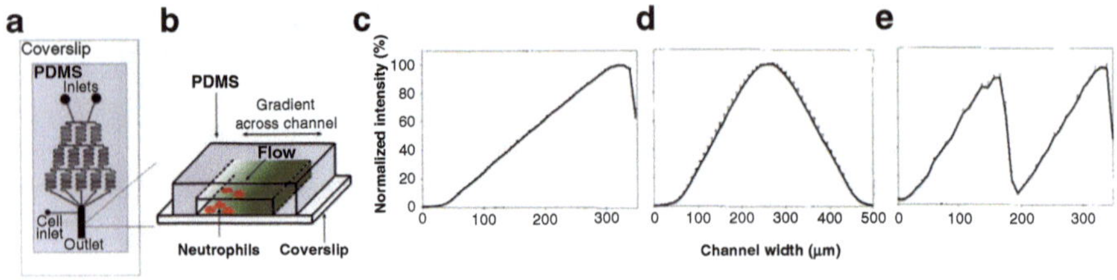

Fig. 2.15 Microfluidic Christmas tree gradient generator. (**a**) A network of serpentine channels interspersed between fluid branch points periodically split, mix, and recombine flows to develop a concentration gradient across an output channel. (**b**) Neutrophils seeded on a glass coverslip at the outlet channel are subjected to a concentration gradient across the channel. (**c**) Linear concentration gradients are readily developed and stably maintained across a 500 μm wide output channel. (**d**) Hill-shaped and (**e**) cliff-shaped gradients can be generated by placing two or more gradient generators in parallel in a head-to-head or a head-to-tail configuration, respectively. Reproduced with permission from Ref. [46]

serve to periodically split and combine flows comprising different concentrations of reagents while the serpentine channels increase the effective distance in order to ensure homogeneous mixing of flows between the branching junctions (Fig. 2.15). In this way, several different concentrations are generated on-chip, which can be directed to an output channel to produce stable spatial gradients. By adjusting the inlet concentrations, channel geometries, and flow rates, complex spatial gradient waveforms including linear and polynomial profiles can be generated. Cells are introduced in the output channel via a separate side port in order to expose them to the spatial gradient. Whitesides and colleagues used the Christmas tree gradient generator to characterize neutrophil chemotaxis in varying gradient waveforms of a soluble chemokine, IL-8. First, the authors quantified chemotactic flux in linear gradients ranging from 0 to 5, 50, and 500 ng/mL IL-8 across an output channel width of 500 μm (Fig. 2.15b). Next, the authors developed more complex hill-like and cliff-like gradient patterns by juxtaposing two gradient generators in parallel in head-to-tail (cliff) or head-to-head (hill) configurations (Fig. 2.15d, e). It was demonstrated that neutrophils encountering a gradual drop in gradient steepness (in case of a hill waveform, Fig. 2.15d) often overshot their preferred localization zone at the region of highest concentration before returning. In contrast, a precipitous change in gradient (cliff configuration, Fig. 2.15e) revealed a tighter chemotactic control mechanism that resulted in the cells localizing at the highest concentration zone without any overshoot. It is noteworthy that bulk chemotactic assays involving concentration gradients developed in agar or agarose might have failed to detect the overshoot response of a small number of neutrophils in gradually varying spatial gradients that was observed in this study. The aforementioned gradient generator represents a cornerstone of several highly significant microfluidics-based studies on directed cell migration in tumors, stem cells, motile microbes, and leukocytes.

2.3.4 *Microfluidic "Honeycomb" Chip for Antibiotic Resistance Studies*

A riveting example of the application of microfluidics to create complex spatial environments is provided by the Goldilocks chip, which was developed by Robert Austin and colleagues to charter the evolution of antibiotic resistance in bacteria in a complex fitness landscape [8]. Austin's device consisted of a honeycomb-shaped microfluidic chip with 1200 hexagonal wells etched in a silicon wafer. The wells were connected to their nearest neighbors via six microchannels, which allowed motile bacteria to travel between the wells. Nutrient and media were flowed through nanoslits etched in the side walls of the wells at the edge of the microfluidic array. A stable gradient of a genotoxic antibiotic (ciprofloxacin) was established across the array by pumping counter-flowing solutions of antibiotic-free bacterial growth media and ciprofloxacin-containing medium at the opposite peripheries of the array. GFP-labeled bacteria were inoculated at the center of the microfluidic array. It was observed that within hours of inoculation, nutrient depletion along with bacterial chemotaxis tended to concentrate an initial population of cells near the peripheries of the array where the nutrient concentration was maximum. At these so-called Goldilocks points, the stress gradients were the highest owing to the convergence of the antibiotic-free and antibiotic-containing streams. Strikingly, it was observed that antibiotic-resistant mutants fixed rapidly (within 10 h) at these Goldilocks points and eventually resulted in bacterial invasion of the entire chip. Removing the antibiotic gradient abolished emergence of resistance. Overall, the unique microfluidic approach described by Austin et al. demonstrates the alarming possibility of rapid development and fixation of antibiotic-resistant mutants in complex fitness landscapes as are likely encountered by invading pathogens in thc human body [8].

2.4 Microfluidic Platforms for Drug Screening

A critical step in drug discovery and development involves the *in cellulo* toxicological and pharmacological evaluation of lead candidates [32]. In cellulo evaluation of potential therapeutics is conventionally achieved using high throughput cell culture platforms together with expensive robotic manipulations for changing media, adding reagents, and measuring cell viability or gene expression. Microfluidics holds significant promise for revolutionizing traditional drug screening by allowing assays to be conducted at physiologically relevant microscale dimensions and using ultra-low quantities of cells and test chemicals. Minimizing sample and reagent consumption is especially crucial as lead drug candidate molecules are typically available in miniscule quantities [47]. Furthermore, reduced material requirement is highly desirable for screening rare or less-readily available cell types such as patient-derived tissue constructs and hematopoietic stem cells.

Finally, as discussed in Sect. 2.1, the low form factor of microfluidic devices enables parallelization and multiplexed evaluation of several potential drugs and combinations thereof in a convenient on-chip format that has significantly less space and is cost-intensive compared to 384 or 1536 well based screens.

2.4.1 Microfluidic Living Cell Array

The advantages of applying microfluidic systems to drug screening are well illustrated by the microfluidic living cell array described by King et al., which enables simultaneous functional interrogation of eight cell lines individually subjected to eight different chemical stimuli (thus, a total of 64 distinct experimental conditions) (Fig. 2.16a) [48]. To enable multiplexed measurements, the living cell array utilizes a two-layered microfluidic design with the underlying fluid layer housing the cell growth chambers, which are fed by a set of eight orthogonally arranged reagent delivery channels. Each cell chamber consists of a 2×2 subarray allowing for quadruplicate measurements. Each chamber is isolated from its neighbors using two sets of valves fabricated in the control layer that rests above the fluid layer (Fig. 2.16a). In contrast to widely used actuate-to-close valves (see Sect. 2.1), King et al. employed actuate-to-open valves that can be opened by applying negative pressure in dead-end control lines (Fig. 2.16b, c). The living cell array is primed by first actuating the seeding valves and flowing in cells through eight individual cell loading lines using a syringe pump. Subsequently, the negative pressure (vacuum) applied to the seeding valves is released, which isolates the individual cell chambers. Each group of eight cell growth chambers can then be subjected to a distinct chemical stimulus by actuating a second set of stimulation control valves and flowing in the chemical reagent in a fashion analogous to cell loading (Fig. 2.16d, e). Using this approach, the authors quantified dynamic responses of rat hepatoma cell lines stably expressing GFP under the regulatory control of common eukaryotic transcription factors (NF-κB, STAT3, GRE) and upon treatment with pro- and anti-inflammatory cytokines and dexamethasone, a steroid-based drug. Importantly, the use of actuate-to-close valves enables facile operation of the microfluidic living cell array without the need for bulky syringe pumps, which significantly improves device portability and ease-of-use.

2.4.2 Microfluidic Antibiotic Susceptibility Testing

We recently developed a microfluidic antibiotic susceptibility testing chip that incorporates actuate-to-open valves are screening for bacterial pathogens (Fig. 2.17) [49]. Our device consists of a 4-by-6 array of microwells in the fluid layer that can be loaded by actuating normally closed valves housed in the control layer (labeled as 1 in Fig. 2.17c, d) [49, 50]. Each well is further subdivided into two

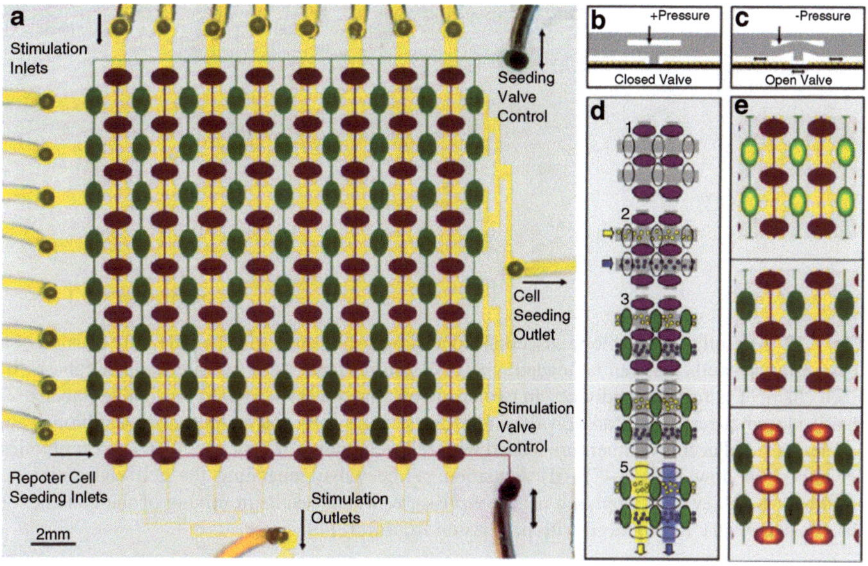

Fig. 2.16 Microfluidic living cell array. (**a**) The microfluidic array consists of a 16 × 16 array of cell growth chambers (shown in yellow) 50 μm in height and 420 μm in diameter that are supplied by eight distinct flow lines for delivering stimuli. Cells can be drawn into the chambers through eight inlets by applying negative pressure at the common outlet. Reagents or small molecule drugs can be loaded in an analogous fashion. Each 2 × 2 subarray of cell growth chambers can be isolated from its neighbors using two sets of control valves (shown in *purple* and *green*) fabricated in the control layer. (**b**) The valves are designed in an actuate-to-open configuration. In the rest state or under positive pressure, the PDMS membrane serving as the valve is in contact with the glass coverslip sealing off adjacent chambers. (**c**) Under the application of negative pressure, the membrane deflects upward opening the valve. (**d**) In *step 1*, the device is first seeded with different cell lines by releasing the cell loading valves (*green*) and flowing in up to eight different cell lines (*step 2*). The cell loading valves are then allowed to close isolating each cell growth chamber from its neighbors (*step 3*). After the cells are attached to the surface, the stimulus loading valves (*purple*) are released (*step 4*) and up to eight different reagents are flowed in to the cell chambers (*step 5*). (**e**) Optical micrographs of the microfluidic living cell array in the cell seeding (*green cell loading valves open*), neutral (*all valves closed*), and stimulation (*purple valves open*) modes of operation. Flow lines are filled with synthetic coloring to enable visual observation. Reproduced with permission from Ref. [48]

compartments isolated by a normally closed mixing valve (labeled as 2 in Fig. 2.17c, d). Cells (e.g., derived from a clinical sample such as blood or sputum) can be loaded in each of the half-wells by applying negative pressure (vacuum) to release the cell loading valves. Antibiotics can be individually loaded in the remaining four half-wells by actuating the antibiotic loading valves (similar in principle to the stimulation valves described by King et al.). Finally, actuation of the mixing valve brings the two half-wells in contact allowing the sample to be individually tested against six antibiotics in quadruplicates (Fig. 2.17d). We showed that on-chip mixing could be enhanced simply by repeated actuation of the mixing valve (to introduce eddies) thereby obviating the need for complex microfluidic

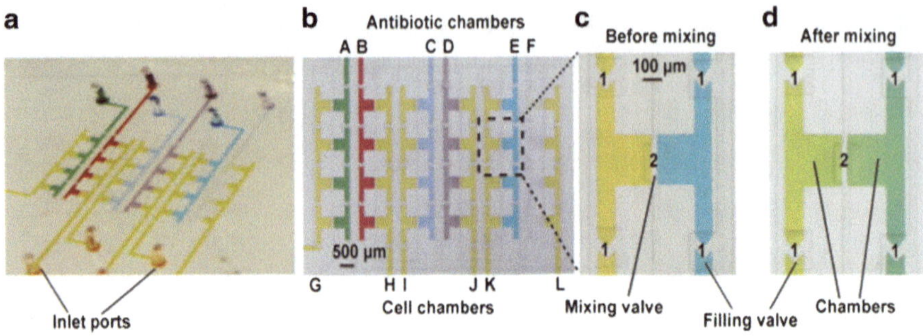

Fig. 2.17 Microfluidic antibiotic susceptibility testing (AST) chip. (**a**) The AST chip consists of a 4 × 6 array of wells that can be loaded with cells and reagents using actuate-to-open valves. (**b**) Each chamber is further subdivided in two chambers that are (**c**) separated by a valve (labeled *2*) that is closed at rest. The loading valves used for seeding cells and adding media/antibiotics are labeled as *1*. Adjacent chambers are loaded with cells (chambers in flow lines *G–L*) and antibiotics (chambers in flow lines *A–F*). (**d**) Actuation of the valve separating the cell and antibiotic chambers is achieved by applying negative pressure, which results in mixing of the contents of the adjacent wells. Reproduced with permission from Ref. [49]

mixers. Using this simple device design, we were able to infer microbial antibiotic resistance profiles (also known as antibiograms) by quantifying the expression of GFP in *E. coli* cells treated with various antibiotics and antibiotic combinations [49]. Strikingly, microfluidic implementation of antibiotic susceptibility testing enabled a tenfold improvement in turnaround time and a greater than 10^4 fold improvement in sensitivity compared to current macroscale antibiotic susceptibility testing methods. In addition, we demonstrated that antibiotic action is not strictly synergistic as a combination of multiple antibiotics does not necessarily translate to improved bactericidal activity [49]. From a clinical standpoint, the finding is significant as it implies that therapeutic interventions involving the use of multiple antibiotics in case of drug resistant infections may not always prove effective.

2.4.3 Organ-on-a-Chip Platforms for Drug Screening

Recently, microfluidic organ-on-a-chip devices have been garnering increasing attention as potential test beds for pharmacokinetic and pharmacodynamic (PK/PD) modeling of drug candidates. In their simplest form, organ-on-a-chip microfluidics seeks to simulate the physiological distribution of a drug molecule within various tissue compartments along with the pharmacological effects of the drug on the tissue of interest. To achieve this, microfluidic chambers are constructed to mimic tissues of interest (such as the liver, lungs, bone marrow, tumor) with respect to conservation of liquid residence times and volumetric ratios of the various organs. The compartments are then seeded with a physiologically relevant density of a specific cell type (e.g., hepatic cells for liver-on-a-chip)

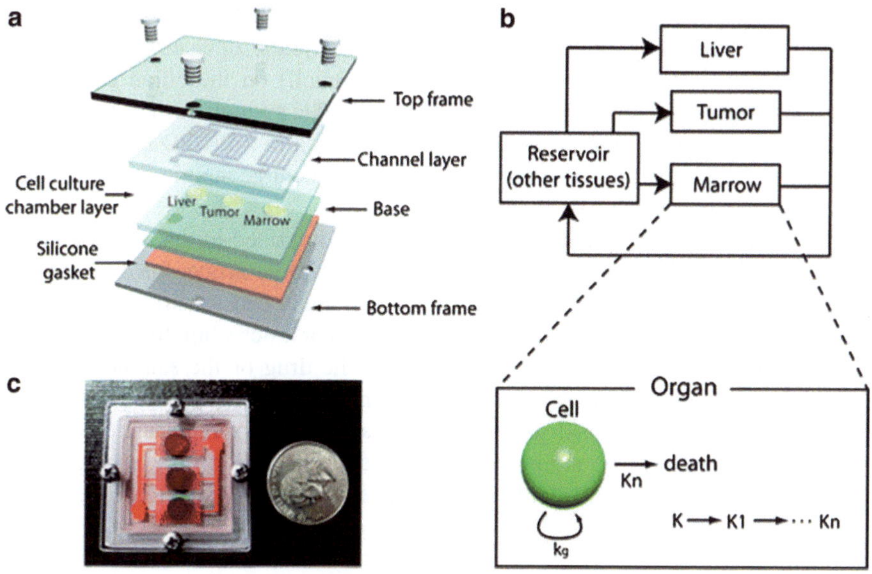

Fig. 2.18 Microfluidic cell culture analog (μCCA). (**a**) The multilayer μCCA is assembled by sealing a PDMS layer housing flow lines over a cell culture chamber. (**b**) Chambers and flow lines are designed to mimic the distribution of fluid flow (and drug) between the various tissue types as well as the pharmacodynamic effect of the drug on a specific tissue type. (**c**) The final assembled device is small, portable, and can be operated without the need for syringe pumps or vacuum manifolds. Reproduced with permission from Ref. [52]

[32, 51]. The tissue-mimetic compartments are interconnected using microfluidic channels designed to simulate physiological rates of fluid transfer between various organs. Laminar flow rates ensure that the shear stress experienced by cells was well within physiological levels (<2 dyn cm^{-2}). In this way, the action of a candidate drug molecule on target cells in the chip is affected by its absorption, distribution, and metabolism within the various organ mimetic compartments, thereby providing a reasonably accurate representation of the drug's PK/PD parameters. For example, the micro-cell culture analog (μCCA) designed by Shuler and colleagues simulates three organs on a chip: liver, a colon tumor, and bone marrow (Fig. 2.18) [52]. Each tissue type is represented by a micro-compartment that is molded in silicone and seeded with cells encapsulated in a hydrogel matrix. The cell-loaded silicone layer is sealed against a PDMS layer that houses flow lines for delivering media and drugs to the individual cell growth chambers. The flow lines are constructed to mimic the relative partitioning of blood between the tumor, liver, and bone marrow (24, 58, and 18 %, respectively). Reservoirs containing pure media are located on the chip and fluid flow within and between compartments is achieved using gravity-induced flow developed by periodically titling the microfluidic chip on a rocking platform. By avoiding the use of external pumps and pneumatic valves, the authors improve usability and portability of the platform

as well as eliminate the notorious problem posed by air bubble formation in flow lines. Shuler and coworkers employed their microfluidic device to quantify the PK/PD profiles of a chemotherapeutic (5-fluorouracil) on the colon cancer cell lines. Moving ahead, one can envision that microfluidic systems such as these could serve as critical platforms for early stage drug testing in a PK/PD-based manner, thereby avoiding the high attrition rates and consequent economic burden associated with the drug discovery pipeline.

In summary, the simple microfluidic platforms described in this section underscore several key advantages of microfluidic cell culture arrays for drug testing, including rapid assaying of different drugs with single cell sensitivity and under physiologically relevant conditions in a convenient microchip format (~1 cm^2 working area) using only a few microliters of the drug or the sample. Looking ahead, microfluidic screening chips would be immensely beneficial for preclinical drug screening applications, particularly in cases where sample and/or reagent consumption needs to be minimized (as in the case for lead drug candidates or clinical specimens) as well as for field deployment of microfluidic screening devices.

2.5 Microfluidic Technologies for Single Cell Isolation and Analysis

Selective isolation and analysis of single cells has remained a long-standing challenge for sorting and studying rare cell types such as circulating tumor cells (CTCs), fetal cells in maternal blood, and drug-resistant microbial persister cells [9, 53, 54]. The ability to isolate and profile single cells also underpins escalating efforts aimed at identifying and genotyping unculturable microbes from microbiome collections. In contrast to macroscale sorting technologies, microfluidics affords three key advantages for analyzing complex consortia to isolate infrequently occurring cell types: (1) microfluidic platforms are ideal for handling small sample volumes that are typically obtained from clinical biopsies and environmental sample collections, (2) nanoliter-scale volumes of microfluidic chambers combined with the ability to integrate several thousands of wells on a single chip enables stochastic partitioning of individual cells from the parent population, and (3) the isolation of single cells in nanoliter-sized wells greatly improves the speed and sensitivity of analytical biochemical reactions such as PCR and FISH.

2.5.1 Microfluidic Platforms for Isolating "Rare" Eukaryotic Cells

Rare eukaryotic cells such as fetal cells in maternal blood, drug-resistant hematopoietic stem cells in chronic myeloid leukemia patients, and circulating tumor cells (CTCs) in cancer patients comprise between 10^{-4} and 10^{-7} % of the total cell population in blood. Nevertheless, these rare cells are of tremendous diagnostic and prognostic significance, respectively for prenatal diagnosis and for predicting metastatic burden in cancer patients. Microfluidics has recently emerged as a valuable platform for isolating and profiling clinically and therapeutically relevant rare eukaryotic cells.

2.5.1.1 Microfluidic CTC Isolation Platforms

CTCs are responsible for establishing metastatic tumors at sites secondary to the primary tumor and are recognized as the major cause of mortality in cancer patients. For this reason, improved methods for the isolation and characterization of CTCs from patient whole blood are critically required. The low concentration of CTCs, however, makes their detection extremely challenging. Furthermore, even CTCs originating from the same primary tumor vary considerably in their metastatic potential and proteomic profiles, which highlights the need for single cell profiling. To this end, several microfluidic platforms have been described for achieving isolation of pure CTCs [10, 11, 55]. Here, we focus on the CTC chip described by Toner and colleagues and a more recent integrated CTC-secretome profiling microfluidic device described by Shi and coworkers.

The CTC chip, described by Toner et al., features a dense array of 78,000 microposts (100 μm in height and width) engraved using deep reactive ion etching (DRIE) on a silicon wafer with an effective surface area of 970 mm^2 (Fig. 2.19) [10]. The microposts are chemically functionalized with anti-epithelial cell adhesion molecule (EpCAM) antibodies to facilitate affinity capture of CTCs, which are known to overexpress the EpCAM cell surface antigen (Fig. 2.19d). The chip design was informed by model-guided simulations and optimized to maximize contact time between the CTCs and the capture antibodies as well as minimize shear stress at the microposts, which is detrimental to CTC adhesion. Accordingly, an equilateral triangle arrangement of microposts was employed and shown to reduce the shear stress to 0.4 dyn cm^{-2} for a volumetric flow rate of 1 mL h^{-1}, thereby maximizing CTC capture with minimal shear-induced cell damage. Using the optimized design, the authors successfully identified CTCs in 115 patient blood samples in concentrations ranging from 5 to 1281 CTCs per mL of whole blood and with purities of 52–67 % as determined by CTC-specific antibody staining. Interestingly, the high sensitivity of the CTC chip enabled identification of CTCs in patients with clinically localized prostate cancer vis-à-vis metastatic prostate cancer. Importantly, the isolated CTCs could be used for off-chip biochemical analysis via immunostaining for cancer biomarkers (such as the prostate-specific antigen) and RT PCR. The CTC chip outscores immunomagnetic bead-based isolation methods commonly used in clinical applications in terms of the former's high sensitivity (1 in 10^9 cells), specificity (an improvement of two orders of magnitude), and overall CTC yield (99 %). It is conceivable that in coming years, CTC chip would become the platform of choice for detecting metastatic cancers and

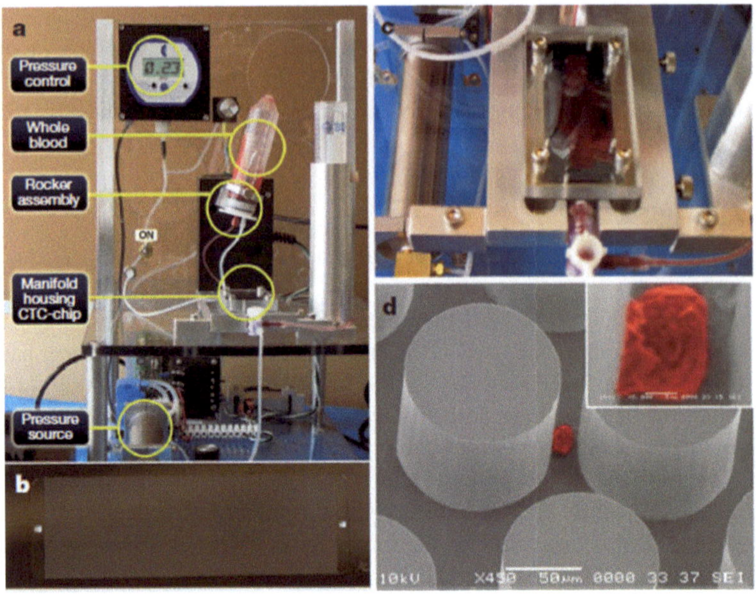

Fig. 2.19 Microfluidic CTC chip. (**a**) Circulating tumor cells are isolated using a simple workstation in which whole blood (mixed continuously on a rocking platform) is pumped through the chip (shown in **b**) using pressure-driven flow. (**c**) Whole blood-loaded chip. (**d**) Scanning electron micrograph of the CTC chip showing a lung cancer cell captured at an anti-EpCAM functionalized post. Reproduced with permission from Ref. [10]

monitoring treatment progression, thereby obviating the need for cumbersome tissue biopsies.

Shi and colleagues recently described a highly integrated microfluidic device for the isolation and secretomic profiling of CTCs (Fig. 2.20) [11]. Fabricated as a multicomponent glass-PDMS multilayer chip, the device combines four previously described microfluidic designs to achieve integrated CTC isolation and analysis, including selective isolation of CTCs, size-based removal of the smaller red blood cells, immunomagnetic removal of leukocytes and granulocytes, and single cell bar coding of the isolated CTCs to detect secreted cytokines (secretome profiling). For the isolation of CTCs from blood, CTCs were selectively labeled using anti-EpCAM antibodies (analogous to the capture antibody described above) that were chemically conjugated to single-stranded DNA (ssDNA) using a photocleavable linker. ssDNA-encoded antibody labeled cells were then affinity-captured by flowing the samples through a microfluidic channel functionalized with complementary ssDNA. DNA hybridization was enhanced by incorporating a herringbone chip design that maximized mixing by generating microvortices (Fig. 2.20b–d). Erythrocytes were removed via size-based exclusion by flowing the samples through spiral microfluidic channels. White blood cells were removed using on-chip immunomagnetic pull-down, and CTCs were released from the ssDNA capture probes using brief UV illumination to cleave the photolabile linker. Finally,

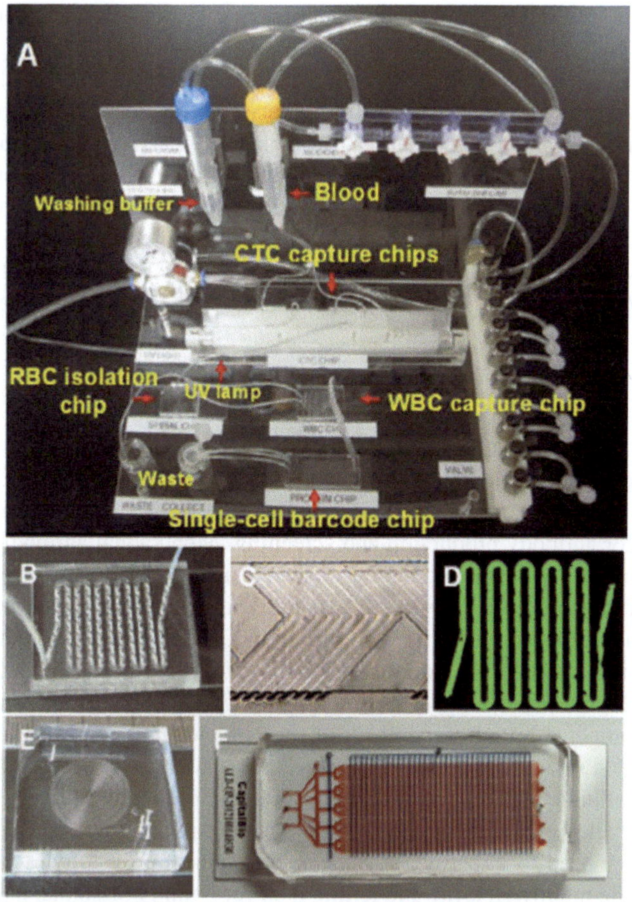

Fig. 2.20 Microfluidic CTC isolation and secretome analysis workstation. (**a**) The integrated CTC chip isolates CTC(s) from whole blood, while removing RBC(s) and WBC(s) and routes the CTC (s) to a single cell barcoding chip for analyzing secreted proteins. (**b–d**) Herringbone patterns on the chip generate microvortices, which improve efficiency of ssDNA hybridization-mediated capture of CTC(s). Hybridized DNA (fluorescently tagged) in shown in (**d**) and indicates a high density of DNA patterning on-chip. (**e**) A size-exclusion spiral chip is used to separate out the RBC (s). (**f**) The single cell barcoding chip is basically an array of single cell isolation compartments that are patterned with a panel of antibodies to identify the proteins secreted by the trapped cell. Reproduced with permission from Ref. [11]

the released CTCs were routed to a single cell bar coding (SCBC) microfluidic chamber for secretomic profiling (Fig. 2.20f). The SCBC, first described by Heath and colleagues [56], is a microfluidic device consisting of an array of microchambers (1 nL volume per chamber) for isolating single cells. Each microchamber also contains a micropatterned antibody barcode attached to the surface via DNA hybridization. CTCs are loaded and partitioned between the antibody barcoded chambers following Poisson statistics (<2 cells per chamber).

Poly-L-lysine coating of the glass layer improves CTC retention in a nanochamber. Following cell culture, the secreted proteins are immobilized at the specific antibody sites in each reaction chamber. Detection is achieved by loading the nanochambers with biotinylated secondary antibodies and probing with streptavidin-conjugated fluorophores. In this way, each CTC is quantitatively barcoded in terms of its secretome. The small volume of the reaction chambers coupled with the high efficiency of antibody capture enables the detection of secreted cytokines with exquisite sensitivity and specificity. Using this approach, the authors demonstrated a high degree of heterogeneity between secreted levels of IL-8 and VEGF among CTCs isolated from the blood of a cancer patient.

2.5.1.2 Microfluidics for Monoclonal Antibody Production

As a final example of the burgeoning importance of microfluidics for the isolation and analysis of single eukaryotic cells, we describe the application of microfluidic platforms for antibody production and screening. Monoclonal antibodies are high-value bioproducts as therapeutics in a broad range of diseases. Current antibody production technologies, however, generate polyclonal mixtures of antibody-secreting hybridomas, which need to be laboriously screened to isolate the clone producing the antibody of interest. Furthermore, during the lengthy screening and clonal expansion process, clones of interest that produce the desired antibody are often outcompeted by other members of the polyclonal consortia that have faster growth rates. Love and colleagues addressed this challenge by developing a PDMS-based microfluidic platform comprising 25,000 micro-engraved wells, each 100 μm in diameter and height and spaced 100 μm apart from neighboring wells [57]. The wells were loaded by pipetting a dilute suspension of mouse hybridomas on the surface of the microwell array and letting it sit for 3 min. Poissonian partitioning of the cells ensured that the majority of the wells received between 0 and 3 cells. Excess liquid was removed using filter papers, and the microfluidic array was inverted and clamped against a glass slide and incubated at 37 °C for a few hours. The glass slide was pretreated and covalently functionalized with antigens specific to the antibody of interest. Under these conditions, antibodies secreted by cells in each microwell will adhere to the glass substrate only if it is specific to the antigen coating of the glass substrate. Detection is achieved using fluorescently labeled secondary antibodies and the glass slide is scanned using a microarray scanner to identify the location of the fluorescent signal, which is spatially registered with the corresponding antibody secreting monoclonal cell bearing microwell (Fig. 2.21). Following detection, the hybridoma cells of interest can be clonally expanded for scaling up antibody production. In this way, a relatively simple microfluidic platform (single layer and without valves, mixers, and pumps, fluid loading by pipetting) greatly simplified a long-standing challenge for therapeutic antibody production [57].

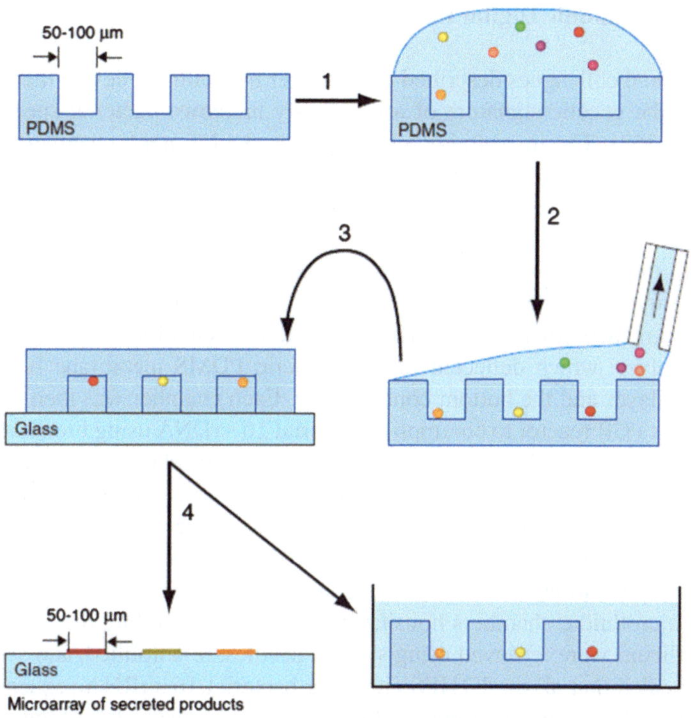

Fig. 2.21 Microfluidic hybridoma screening chip. A microfluidic array of wells enabled the isolation of single hybridoma cells based on Poisson partitioning. The seeded microfluidic array was dewetted using a filter paper and inverted over a glass slide functionalized with antibodies to detect a secreted antigen of interest. The glass slides were scanned using a microarray scanner to identify the spatial position corresponding to the hybridoma secreting an antibody of interest. Following detection, the hybridoma chip can be reused for detecting additional antigen producing cells as well as to scale up production. Reproduced with permission from Ref. [57]

2.5.2 Isolation and Genetic Analysis of Uncultivated Microbes

It is estimated that the vast majority of microorganisms are not amenable to laboratory-scale cultivation techniques that frequently fail to faithfully replicate the microbe's native niche (e.g., a nontrivial dependence on host-derived factors) [54]. In addition, cultivation of heterogeneous microbial consortia usually selects for fast-growing strains, which rapidly outcompete other species. Unculturable microbes, sometimes referred to as biological dark matter, represent an untapped and poorly studied resource that is of tremendous clinical as well as industrial significance. While metagenomic sequencing enables genetic analysis of heterogeneous environmental samples, the identification of species-specific genes and genome sequences from complex metagenomic data is an intense and computationally challenging endeavor. Recently, several microfluidics-based cell isolation platforms have significantly accelerated the discovery of novel microbes from complex multispecies communities, as is described below.

2.5.2.1 Microfluidic Digital PCR Chip

Leadbetter and colleagues described a multilayer microfluidic device that was used to discern the genetic signature of a previously unknown bacterial species in the termite gut [58]. The microfluidic chip comprised 1176 nano-chambers (6.25 nL volume per chamber) separated by actuate-to-close micromechanical valves connected by a single feed line. Termite hindgut luminal contents were diluted in PCR buffer and loaded in the microfluidic chip to achieve a Poissonian distribution of cells in the chambers, thereby ensuring that the majority of chambers housed between 0 and 1 bacterial cell. Subsequent to loading, the individual chambers were isolated from each other by pneumatic or hydraulic actuation of micromechanical push up valves, which deflected the elastomeric PDMS membrane between the upper flow layer and the bottom control layer. Each chamber was then used as an independent PCR reactor to co-amplify bacterial 16 s rDNA using broad-specificity bacterial primers as well as the gene for formyltetrahydrofolate synthetase (FTHFS), which is a key metabolic enzyme involved in bacterial homoacetogenesis and hitherto unassociated with a specific species or clade in the termite hindgut microbiome. Colocalization of spectrally separated fluorescence signals derived from the amplification of bacterial rDNA as well as the FTHFS gene were used to identify microfluidic chambers housing bacterial cells encoding the FTHFS gene. PCR amplicons were retrieved using syringe needles, re-amplified, and sequenced, which revealed that all the FTHFS-encoding bacterial 16 s rRNA sequences clustered in four closely related groups within the *Spirochaete* cluster. Members of this newly identified clade were not observed in microfluidic PCR chambers that failed to amplify the FTHFS gene, suggesting that the *Spirochaete* cluster is the predominant species responsible for homoacetogenic metabolism in the termite hindgut. In this way, a relatively simple microfluidic design capable of achieving single cell confinement and digital PCR enabled the clade-specific association of a key bacterial enzyme affecting energy requirements of the microbiome's termite host.

2.5.2.2 Single Cell Genome Amplification Chip

Quake and coworkers developed a more integrated version of the microfluidic digital PCR chip detailed in the prior section in order to isolate and sequence the genome of a novel bacterial species from a human subgingival swab sample (Fig. 2.22a) [59]. The microfluidic platform integrated morphology-based cell sorting, lysis, neutralization, and whole genome amplification in a single chip. In this design, cell sorting, lysis, and neutralization chambers (each 3.5 nL in volume) along with a larger (50 nL volume) whole genome amplification (WGA) chamber are arranged sequentially and isolated from each other through the use of micromechanical valves similar to the ones described above (Fig. 2.22b). Single bacterial cells with a characteristic rod-like shape typical of the uncultivated TM7 phylum are manually identified using microscopy and mechanically trapped in the

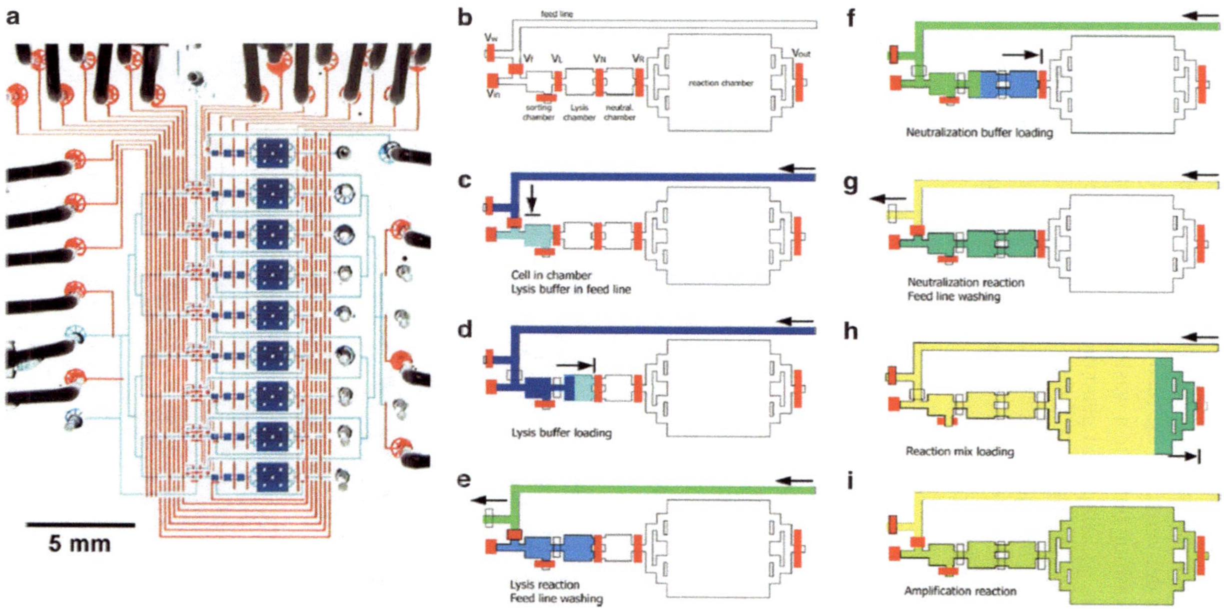

Fig. 2.22 Single cell genome amplification chip. (**a**) Digital image of the single cell whole genome amplification chip capable of processing eight samples in parallel. The flow and control lines are respectively filled with blue and red food coloring. (**b**) Each processing unit comprises separate chambers for cell sorting, lysis, neutralization, and whole genome amplification. Reagents are loaded through the feed line to the various chambers or routed to the waste line when the valve V_w is open. A single cell is introduced into the sorting chamber by opening valve V_{in}. The sorting, lysis, and neutralization chambers are separated by valves V_L, V_N, and V_R. Valve V_{out} can be opened to collect the amplified DNA directly in a microcentrifuge tube. (**c**) A single cell (identified based on morphology) is loaded into the sorting chamber and the feed line is filled with lysis buffer (shown in dark blue). (**d**) The lysis buffer is used to guide the cell into the lysis chamber by opening valve V_L. (**e**) As the cell is being lysed, the feed line is rinsed with buffer (by opening valve V_w). (**f**) Neutralization buffer (shown in *green*) is loaded in the feed line and used to flow the lysate into the neutralization chamber by actuating valve V_N. (**g**) As the lysate is neutralized, the feed line is rinsed and (**h**) loaded with reagents for whole genome amplification, which flows the neutralized lysate into the reaction chamber. (**i**) Genome from a single cell is isolated in the reaction chamber and the products collected. Reproduced with permission from Ref. [59]

sorting chamber (Fig. 2.22c). A feed line is used to deliver lysis reagents to the cell chamber and the lysis reaction mix is routed to the downstream lysis chamber via actuation of an intervening valve (Fig. 2.22d). Following lysis, the reaction mixture is neutralized (Fig. 2.22f) and routed to the reaction chamber for whole genome amplification of a single bacterial cell (Fig. 2.22h). Reagents for neutralization and genome amplification are delivered via the feed line, which is flushed in between each step by routing the washing buffer to a waste stream through the actuation of a separate valve (Fig. 2.22e, g). Amplified genomic DNA is retrieved and used for 16S rRNA sequencing to identify the clade corresponding to the isolated rod-shaped bacterium. Using microfluidic sorting, genome amplification, retrieval, and sequencing, the authors were able to assemble the genome sequence of the first representative species of the TM7 phylum.

2.5.2.3 Microfluidic "SlipChip"

Recently, Ismagilov and coworkers adapted the SlipChip microfluidic platform originally developed for protein crystallization and digital PCR to facilitate the cultivation of a representative gut microbiome species (Fig. 2.23) [60]. The SlipChip platform consists of two glass plates housing nanowells along with microfluidic channels for loading the nanowells. The two plates can be moved

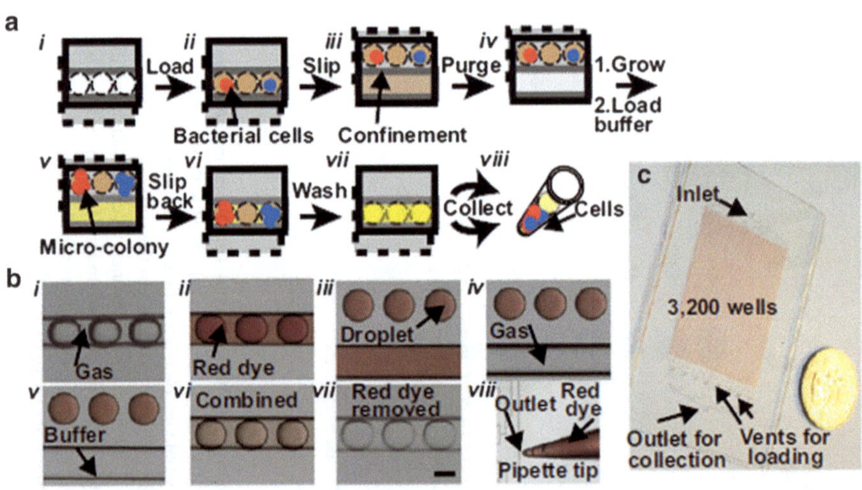

Fig. 2.23 Microfluidic SlipChip. (**a**) Wells engraved in one of the plates of the SlipChip can be loaded with cells by flowing a cell-loaded solution through channels engraved in the other plate. Slipping the plates relative to each other confines the cells in a microwell to form microcolonies. In order to isolate the microcolonies, the plates are slipped back again to align the channels with the wells. Colonies are collected by flowing a collection buffer through the channels. (**b**) Operation of the schematically illustrated steps in (**a**) are visualized using a red dye. (**c**) The SlipChip is slightly larger than a US quarter and features 3200 wells for isolating and growing single cells. Reproduced with permission from Ref. [60]

relative to one another enabling the user to (1) align the nanowells in one plate with microchannels in the other plate to accomplish well loading and (2) to bring the contents in nanowells housed in the upper plate in contact with nanowells in the lower plate to enable reagent mixing by diffusion (Fig. 2.23a, b). A key feature of the SlipChip platform is that it allows fluidic routing without the need for pumps and valves. SlipChips developed for microbial cultivation comprise approximately 3200 nanowells (6 nL volume) engraved in photoresist in a chromium-coated glass substrate (Fig. 2.23c). In a typical operation of the SlipChip, the microfluidic channels and nanowells are loaded with a bacterial sample from a clinical biopsy (e.g., from the mucosal lining of human cecum). Slipping the bottom plate relative to the upper plate results in confinement of the stochastically partitioned bacterial cells in each nanowell, following which the cells are allowed to grow and establish microcolonies. To collect the cells, the lower plate is slipped in order to realign the wells with the channel and a buffer solution is flowed through the channel to collect pooled bacterial colonies, which can be analyzed off-chip for 16S rRNA ribotyping. A key feature of the SlipChip platform is that the top and bottom plates can be separated from each other following the growth of microcolonies. In this way, bacterial populations are partitioned between nanowells in the two plates, one of which can be used for destructive genetic profiling off-chip while the other plate can be retained to enable macroscale cell cultivation based on the genotyping results, thereby providing a platform for targeted microbial cultivation. An added advantage of SlipChip is that it expedites the investigation of diverse culture conditions. This is readily achieved by slipping the bottom plate relative to the upper plate to align the nanowells in the lower plate with the channels in the upper plate, which can then be loaded with a specific conditioning reagent to facilitate cell culture. As a powerful demonstration of the potential utility of the SlipChip cell culture platform, Ma et al. employed SlipChip to cultivate the first species of the *Oscillobacter* genus (classified among the "most wanted taxa" of the human microbiome project) by conditioning the culture medium with autoclaved fluid collected from the sampling site in a human cecum [60].

2.6 Conclusions and Future Outlook

In the preceding sections, we have described several significant advances in various fields of biology that have been enabled and enhanced by microfluidic technologies. Through these examples, we have attempted to highlight key advantages of microfluidics that remain difficult or impossible to achieve using macroscale technologies—for example, single cell resolution, precise spatiotemporal control over reagent delivery, and on-chip integration of multiple steps. Nonetheless, the application of microfluidic devices for studying single cell biology is not without limitations. First, the laminar flow regime that predominates in microfluidic channels may pose a significant challenge in situations where rapid mixing is required. Efforts to circumvent diffusion-limited mixing include the incorporation of passive on-chip mixers such as herringbone grooves or serpentine channels that maximize contact time for mixing. Second, the large surface-to-volume ratios of microfluidic devices result in enhanced absorption and adsorption of hydrophobic molecules to the PDMS surface, which can perturb the concentrations of drugs and reagents. A variety of surface passivation strategies have been described for minimizing biomolecular adhesion in microfluidic devices. Finally, despite the relative ease of replica molding techniques, soft lithography is not yet established as a routine technology in most biology laboratories. This often accounts for a certain degree of inertia in adopting microfluidic technology as an everyday tool to answer the biologist's questions [15, 24]. Several efforts to address this have been undertaken including the organization of microfluidic workshops aimed at biologists, centralized microfluidic fabrication facilities, as well as a rapidly escalating number of companies that design, develop, and deliver microfluidic chips and flow control apparati for academic research applications. However, for microfluidics-based platforms to be adapted as everyday tools in cell and molecular biology labs, it is imperative that newer as well as established microfluidic devices are used beyond proof-of-principle experiments to tackle key challenges in systems and synthetic biology, clinical diagnostics, and applied microbiology [15, 17].

In closing, the examples described in this chapter highlight several applications and new discoveries that were uniquely enabled by microfluidics, the vast majority of which could not have been readily realized using existing macro or microscale alternatives. Taken together with escalating advances in high-throughput sequencing, transcriptome profiling, and genome editing and the development of improved genetically encoded probes for functional imaging, it is clear that microfluidics is ideally poised to address exciting new frontiers in the field of single cell biology.

References

1. Samoilov MS, Price G, Arkin AP (2006) From fluctuations to phenotypes: the physiology of noise. Sci Signal 2006(366):re17–re17
2. Raj A, van Oudenaarden A (2008) Nature, nurture, or chance: stochastic gene expression and its consequences. Cell 135(2):216–226
3. Thattai M, van Oudenaarden A (2001) Intrinsic noise in gene regulatory networks. Proc Natl Acad Sci U S A 98(15):8614–8619
4. Paulsson J (2004) Summing up the noise in gene networks. Nature 427(6973):415–418
5. Ozbudak EM, Thattai M, Kurtser I, Grossman AD, van Oudenaarden A (2002) Regulation of noise in the expression of a single gene. Nat Genet 31(1):69–73
6. Maamar H, Raj A, Dubnau D (2007) Noise in gene expression determines cell fate in Bacillus subtilis. Science 317(5837):526–529
7. Elowitz MB, Levine AJ, Siggia ED, Swain PS (2002) Stochastic gene expression in a single cell. Science 297(5584):1183–1186
8. Zhang Q et al (2011) Acceleration of emergence of bacterial antibiotic resistance in connected microenvironments. Science 333(6050):1764–1767
9. Balaban NQ, Merrin J, Chait R, Kowalik L, Leibler S (2004) Bacterial persistence as a phenotypic switch. Science 305(5690):1622–1625
10. Nagrath S et al (2007) Isolation of rare circulating tumour cells in cancer patients by microchip technology. Nature 450(7173):1235–1239

11. Deng Y et al (2014) An integrated microfluidic chip system for single-cell secretion profiling of rare circulating tumor cells. Sci Rep 4:7499
12. Shaner NC, Steinbach PA, Tsien RY (2005) A guide to choosing fluorescent proteins. Nat Methods 2(12):905–909
13. Mukherjee A, Schroeder CM (2015) Flavin-based fluorescent proteins: emerging paradigms in biological imaging. Curr Opin Biotechnol 31:16–23
14. Zare RN, Kim S (2010) Microfluidic platforms for single-cell analysis. Annu Rev Biomed Eng 12:187–201
15. Whitesides GM (2006) The origins and the future of microfluidics. Nature 442(7101):368–373
16. Gravesen P, Branebjerg J, Jensen OS (1993) Microfluidics-a review. J Micromech Microeng 3 (4):168
17. Beebe DJ, Mensing GA, Walker GM (2002) Physics and applications of microfluidics in biology. Annu Rev Biomed Eng 4:261–286
18. Sia SK, Kricka LJ (2008) Microfluidics and point-of-care testing. Lab Chip 8(12):1982–1983
19. Kim S, Kim HJ, Jeon NL (2010) Biological applications of microfluidic gradient devices. Integr Biol 2(11-12):584–603
20. Breslauer DN, Lee PJ, Lee LP (2006) Microfluidics-based systems biology. Mol Biosyst 2 (2):97–112
21. Bennett MR, Hasty J (2009) Microfluidic devices for measuring gene network dynamics in single cells. Nat Rev Genet 10(9):628–638
22. Toriello NM et al (2008) Integrated microfluidic bioprocessor for single-cell gene expression analysis. Proc Natl Acad Sci U S A 105(51):20173–20178
23. Thorsen T, Maerkl SJ, Quake SR (2002) Microfluidic large-scale integration. Science 298 (5593):580–584
24. Whitesides GM, Ostuni E, Takayama S, Jiang X, Ingber DE (2001) Soft lithography in biology and biochemistry. Annu Rev Biomed Eng 3(1):335–373
25. Sia SK, Whitesides GM (2003) Microfluidic devices fabricated in poly(dimethylsiloxane) for biological studies. Electrophoresis 24(21):3563–3576
26. McDonald JC, Whitesides GM (2002) Poly(dimethylsiloxane) as a material for fabricating microfluidic devices. Acc Chem Res 35(7):491–499
27. Berthier E, Young EW, Beebe D (2012) Engineers are from PDMS-land, biologists are from polystyrenia. Lab Chip 12(7):1224–1237
28. Unger MA, Chou H-P, Thorsen T, Scherer A, Quake SR (2000) Monolithic microfabricated valves and pumps by multilayer soft lithography. Science 288(5463):113–116
29. Suh YK, Kang S (2010) A review on mixing in microfluidics. Micromachines 1(3):82–111
30. de Jong J, Lammertink RG, Wessling M (2006) Membranes and microfluidics: a review. Lab Chip 6(9):1125–1139
31. Grover WH, Ivester RH, Jensen EC, Mathies RA (2006) Development and multiplexed control of latching pneumatic valves using microfluidic logical structures. Lab Chip 6(5):623–631
32. Dittrich PS, Manz A (2006) Lab-on-a-chip: microfluidics in drug discovery. Nat Rev Drug Discov 5(3):210–218
33. Bennett MR et al (2008) Metabolic gene regulation in a dynamically changing environment. Nature 454(7208):1119–1122
34. Balagaddé FK, You L, Hansen CL, Arnold FH, Quake SR (2005) Long-term monitoring of bacteria undergoing programmed population control in a microchemostat. Science 309 (5731):137–140
35. Di Carlo D, Aghdam N, Lee LP (2006) Single-cell enzyme concentrations, kinetics, and inhibition analysis using high-density hydrodynamic cell isolation arrays. Anal Chem 78 (14):4925–4930
36. Ryley J, Pereira-Smith OM (2006) Microfluidics device for single cell gene expression analysis in Saccharomyces cerevisiae. Yeast 23(14-15):1065–1073
37. Tanyeri M, Ranka M, Sittipolkul N, Schroeder CM (2011) A microfluidic-based hydrodynamic trap: design and implementation. Lab Chip 11(10):1786–1794

38. Tanyeri M, Johnson-Chavarria EM, Schroeder CM (2010) Hydrodynamic trap for single particles and cells. Appl Phys Lett 96(22):224101
39. Johnson-Chavarria EM, Agrawal U, Tanyeri M, Kuhlman TE, Schroeder CM (2014) Automated single cell microbioreactor for monitoring intracellular dynamics and cell growth in free solution. Lab Chip 14(15):2688–2697
40. Lucchetta EM, Lee JH, Fu LA, Patel NH, Ismagilov RF (2005) Dynamics of Drosophila embryonic patterning network perturbed in space and time using microfluidics. Nature 434 (7037):1134–1138
41. Levario TJ, Zhan M, Lim B, Shvartsman SY, Lu H (2013) Microfluidic trap array for massively parallel imaging of Drosophila embryos. Nat Protoc 8(4):721–736
42. Chung K et al (2011) A microfluidic array for large-scale ordering and orientation of embryos. Nat Methods 8(2):171–176
43. Chronis N, Zimmer M, Bargmann CI (2007) Microfluidics for in vivo imaging of neuronal and behavioral activity in Caenorhabditis elegans. Nat Methods 4(9):727–731
44. Takayama S et al (2003) Selective chemical treatment of cellular microdomains using multiple laminar streams. Chem Biol 10(2):123–130
45. Takayama S et al (2001) Laminar flows: subcellular positioning of small molecules. Nature 411(6841):1016–1016
46. Li Jeon N et al (2002) Neutrophil chemotaxis in linear and complex gradients of interleukin-8 formed in a microfabricated device. Nat Biotechnol 20(8):826–830
47. Thorson MR et al (2011) A microfluidic platform for pharmaceutical salt screening. Lab Chip 11(22):3829–3837
48. King KR et al (2007) A high-throughput microfluidic real-time gene expression living cell array. Lab Chip 7(1):77–85
49. Mohan R et al (2013) A multiplexed microfluidic platform for rapid antibiotic susceptibility testing. Biosens Bioelectron 49:118–125
50. Schudel BR, Tanyeri M, Mukherjee A, Schroeder CM, Kenis PJ (2011) Multiplexed detection of nucleic acids in a combinatorial screening chip. Lab Chip 11(11):1916–1923
51. Sin A et al (2004) The design and fabrication of three-chamber microscale cell culture analog devices with integrated dissolved oxygen sensors. Biotechnol Prog 20(1):338–345
52. Sung JH, Kam C, Shuler ML (2010) A microfluidic device for a pharmacokinetic-pharmacodynamic (PK-PD) model on a chip. Lab Chip 10(4):446–455
53. Hou HW et al (2011) Microfluidic devices for blood fractionation. Micromachines 2(4):319–343
54. Vincent ME, Liu W, Haney EB, Ismagilov RF (2010) Microfluidic stochastic confinement enhances analysis of rare cells by isolating cells and creating high density environments for control of diffusible signals. Chem Soc Rev 39(3):974–984
55. Gleghorn JP et al (2010) Capture of circulating tumor cells from whole blood of prostate cancer patients using geometrically enhanced differential immunocapture (GEDI) and a prostate-specific antibody. Lab Chip 10(1):27–29
56. Ma C et al (2011) A clinical microchip for evaluation of single immune cells reveals high functional heterogeneity in phenotypically similar T cells. Nat Med 17(6):738–743
57. Love JC, Ronan JL, Grotenbreg GM, van der Veen AG, Ploegh HL (2006) A microengraving method for rapid selection of single cells producing antigen-specific antibodies. Nat Biotechnol 24(6):703–707
58. Ottesen EA, Hong JW, Quake SR, Leadbetter JR (2006) Microfluidic digital PCR enables multigene analysis of individual environmental bacteria. Science 314(5804):1464–1467
59. Marcy Y et al (2007) Dissecting biological "dark matter" with single-cell genetic analysis of rare and uncultivated TM7 microbes from the human mouth. Proc Natl Acad Sci U S A 104 (29):11889–11894
60. Ma L et al (2014) Gene-targeted microfluidic cultivation validated by isolation of a gut bacterium listed in Human Microbiome Project's Most Wanted taxa. Proc Natl Acad Sci 111(27):9768–9773

Chapter 3
Convective PCR Thermocycling with Smartphone-Based Detection: A Versatile Platform for Rapid, Inexpensive, and Robust Mobile Diagnostics

Aashish Priye and Victor M. Ugaz

Abstract Development of portable PCR-based diagnostic instruments has potential to greatly expand availability of advanced health care technologies. But issues associated with cost, complexity, electrical power requirements, and product detection continue to make PCR challenging to deploy outside of conventional laboratory environments. Here we review exciting recent progress toward development of convective thermocycling approaches that promise to overcome these limitations, laying a foundation for a new generation of inexpensive and greatly simplified diagnostic tools that can be readily deployed in a host of field-based settings.

Key words Point of care • PCR • Portable diagnostics • Smartphone

3.1 Introduction

The polymerase chain reaction (PCR) is an established routine tool in molecular biology, and provides the basis for a multitude of "gold standard" medical diagnostic assays. The PCR process exploits polymerase-assisted replication of DNA achieved by repeatedly heating and cooling reagents through denaturing (95 °C), annealing (55–65 °C), and extension (72 °C) temperatures, resulting in an exponential increase in the concentration of a target sequence. Although reaction

A. Priye, Ph.D.
Artie McFerrin Department of Chemical Engineering, Texas A&M University, College Station, TX 77843-3122, USA
e-mail: priye.tamu@gmail.com

V.M. Ugaz, Ph.D. (✉)
Artie McFerrin Department of Chemical Engineering, Texas A&M University, College Station, TX 77843-3122, USA

Department of Biomedical Engineering, Texas A&M University, College Station, TX 77843-3122, USA
e-mail: ugaz@tamu.edu

© Springer International Publishing Switzerland 2016

C. Lu, S.S. Verbridge (eds.), *Microfluidic Methods for Molecular Biology*,
DOI 10.1007/978-3-319-30019-1_3

kinetics at each stage of the process are relatively fast [1], traditional thermocycling instruments are limited in their ability to continually change the reaction temperature, resulting in slow operation (timescales of ~ 1 h or more) and considerable electrical power consumption. These considerations often render PCR unfeasible to implement in a portable format, despite its potential to provide important diagnostic information.

Efforts to overcome these limitations in thermocycling technology have largely focused on miniaturization, with the goal of enhancing the heat transfer characteristics to enable rapid thermocycling and reduced power operation. To this end, a number of microfluidics-based approaches have been reported to enable stationary [2, 3], continuous flow [4, 5], and thermal convection-driven [6–8] PCR. Continuous flow systems offer advantages by enabling isothermal operation whereby the instrument need only maintain spatially distinct surfaces at constant temperature, as compared with traditional stationary designs where static reagent volumes are cyclically heated and cooled. Isothermal operation significantly reduces power consumption, while the miniaturized design simultaneously accelerates the reaction due to more efficient heat transfer. However, the external components required to mechanically pump reagents through the temperature zones relevant for PCR are challenging to miniaturize and introduce additional complexity.

Thermal convection-driven approaches overcome these limitations by applying a temperature gradient to drive a microliter-scale convective flow that transports reagents through the denaturing, annealing, and extension temperature zones without the need for external mechanical pumping, offering the advantages of isothermal operation in a greatly simplified format (Fig. 3.1a). This approach promotes rapid thermocycling because fluid elements within the reagent volume quickly attain thermal equilibrium with their surroundings as they circulate through the thermal gradient. Greatly improved performance is therefore achievable because the time and energy required to heat and cool additional inactive instrument components is eliminated. Implementations of this simple yet effective concept have been reported by establishing vertical temperature gradients (i.e., Rayleigh-Bénard convection) [6, 9–11] (Fig. 3.1b), horizontal temperature gradients [7, 12, 13] (Fig. 3.1c), and in closed loop convection systems (i.e., thermo-siphons) [8, 14–17] (Fig. 3.1d, e). More recently, capillary convective PCR platforms have been introduced that impose vertical thermal gradients via heating the side walls of thin capillary tubes containing the PCR reagent mixture [18–20] (Fig. 3.1f).

In all these convective thermocycling systems, a crucial roadblock to widespread adoption has been the complex interplay that exists between the imposed temperature gradient and the reactor geometry, both of which govern the internal convective flow field. In other words, this interplay implies that the system must be redesigned to perform optimally each time a new primer set is used owing to the different thermal driving force imposed at the new annealing temperature. In this chapter, we describe recent efforts in our research group aimed at overcoming these limitations to enable robust PCR performance in a greatly simplified instrument format. When coupled with fluorescence detection using a standard smartphone camera, convective PCR lays a foundation to deliver a versatile DNA analysis platform that is ideally suited for portable for point-of-care applications.

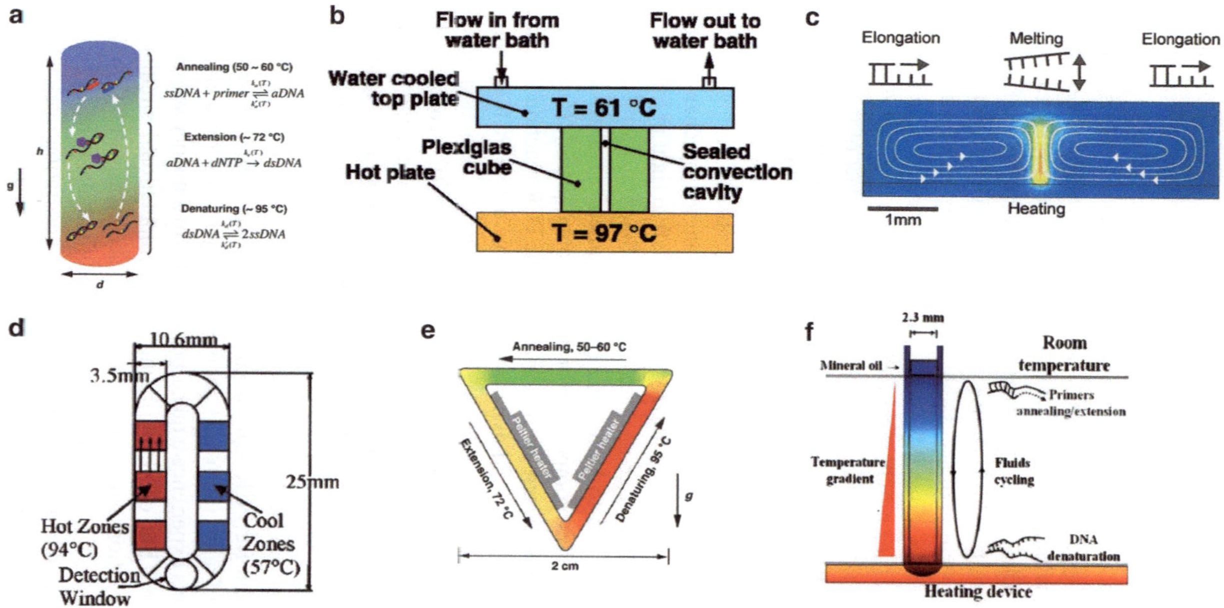

Fig. 3.1 Overview of convective PCR thermocycling systems. (**a**, **b**) Rayleigh-Bénard systems execute thermocycling in a cylindrical reactor geometry by imposing a vertical temperature gradient between the top and bottom surfaces [6, 9]. (**c**) Focused infrared heating generates a horizontal thermal gradient circulating reagents radially outward from the hot center region to the cold periphery [7]. Closed-loop convective systems have also been demonstrated that transport reagents between PCR temperature zones in (**d**) oval [8] and (**e**) triangular [14] flow loops. (**f**) Thermal convection in capillary tubes heated from below has also been harnessed to perform PCR thermocycling [18]. Panel (**a**) reprinted with permission from Ref. [9]. Copyright 2013 American Chemical Society. Panel (**b**) reprinted from Ref. [6]. Reprinted with permission from AAAS. Panel (**c**) reprinted with permission from Ref. [7]. Copyright 2003 by the American Physical Society. Panel (**d**) reprinted with permission from Ref. [8]. Copyright 2013 American Chemical Society. Panel (**e**) reprinted with permission from Ref. [14]. Copyright 2007 John Wiley & Sons. Panel (**f**) reprinted with permission from Ref. [18]. Copyright 2009 BioTechniques

3.2 Design Considerations

To understand the interdependence between the internal flow field and the reactor geometry, we consider reagents confined within a cylindrical cavity of diameter d and height h that is heated from below (i.e., a Rayleigh-Bénard system). The interplay between the destabilizing buoyancy forces and restoring action of thermal and viscous diffusion within the cavity is expressed in terms of a dimensionless Rayleigh number as $Ra = [g\beta(T_2 - T_1)h^3]/\nu\alpha$, where β is the fluid's thermal expansion coefficient, g is the gravitational acceleration, T_1 and T_2 are the temperatures of the top (cold) and bottom (hot) surfaces, respectively, α is the thermal diffusivity, and ν is the kinematic viscosity. This quantity, together with cavity aspect ratio (h/d), uniquely specifies the spectrum of accessible flow states (Fig. 3.2a). For a given h/d, the buoyancy driving force at low Ra is not strong enough to drive the flow. As Ra increases at constant h/d (i.e., moving vertically from bottom to top in Fig. 3.2a), a critical Rayleigh number (Ra_c) is reached beyond which the onset of flow occurs accompanied by axis-symmetric periodic flow trajectories. Upon further increasing Ra, the flow becomes non-axis-symmetric and quasi-periodic, followed by a transition to chaos where fluid elements no longer follow closed periodic paths. Multiple flow trajectories emerge beyond this point, eventually giving way to unsteady convective turbulence. Flow states can also be altered by changing the aspect ratio at constant Ra (i.e., moving horizontally from left to right in Fig. 3.2a). Convective flow at small h/d (wider geometries) generates fluid trajectories that are more disordered when compared to higher h/d (taller geometries), attributable in part to a stronger transverse flow component (in the horizontal plane) (Fig. 3.2b).

To elucidate the interplay between the flow field and the PCR reaction kinetics, we developed a 3D computational fluid dynamics-based convective PCR model to resolve the time evolution of DNA replication. The model yields predictions of the DNA replication rate in terms of a characteristic doubling time (the concept of a "cycle time" in convective PCR differs from conventional thermocycling because all reagents do not follow identical flow paths through the temperature gradient). The efficiency of convective PCR in different reactor geometries can then be quantified in terms of a generation rate (i.e., number of doubling events per hour) [9]. This analysis leads to counterintuitive design rules for selection of reactor geometries optimally suited for PCR. One would naively expect that closed periodic flows characteristic of taller (high h/d) geometries may be best for convective PCR because fluid elements experience a quasi-periodic thermal history similar to a conventional thermocycler. But our experiments and computational simulations surprisingly reveal that reactions are executed more efficiently in wider (low h/d) geometries where chaotic flow trajectories appear to lead fluid elements to experience more disordered thermal histories. More importantly, generation rates achievable in the chaotic flow regime remain essentially constant over nearly two orders of magnitude in Ra (Fig. 3.2c), encompassing virtually any realistic combination of temperature and reactor volume likely to be encountered during PCR. These new results imply a level of versatility previously thought to be unachievable in convective PCR, enabling a single reactor geometry to function optimally in much the same manner as standard PCR tubes and multiwell plates.

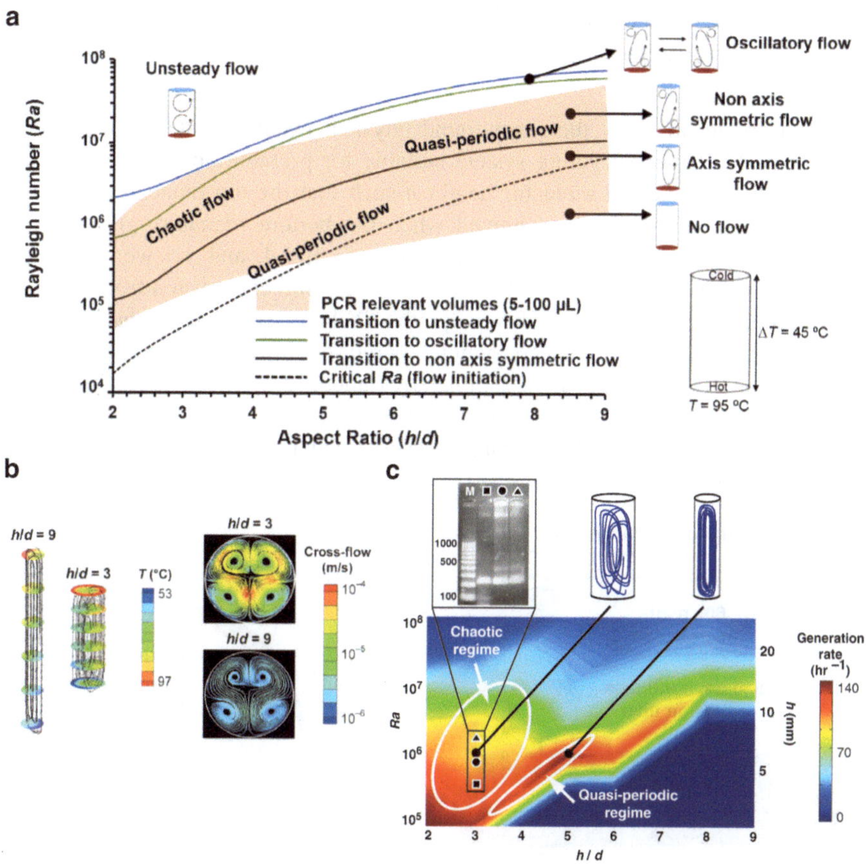

Fig. 3.2 Microliter-scale Rayleigh-Bénard convection embeds a surprisingly complex interplay between reaction and flow. (**a**) 3-D computational fluid dynamics simulations enable a *Ra* versus *h*/*d* parametric flow map to be constructed, revealing a spectrum of accessible convective flow states. The *shaded* region in the map represents cylindrical reactors with volumes relevant for PCR. Within this region, different combinations of size ($Ra \propto h^3$) and shape (h/d) of the cylinder yield flows displaying periodic, quasi-periodic, and chaotic trajectories. (**b**) A strong transverse "cross-flow" component in small aspect ratio cylinders gives rise to more disordered flow trajectories as compared with high *h*/*d* [9]. (**c**) Microscale chaotic advection enables rapid DNA replication to be robustly achieved. Simulated replication timescales are expressed in terms of a generation rate (i.e., number of doubling events per hour). A zone in the chaotic flow regime enables accelerated DNA replication to be stably achieved over a span of two orders of magnitude in *Ra*, suggesting that reactors designed within this zone are capable of delivering consistent replication performance under virtually any combination of temperature and reactor volume associated with the PCR [9]. Panels (**b**) and (**c**) reprinted with permission from Ref. [9]. Copyright 2013 American Chemical Society

3.3 An Inexpensive, Portable Convective PCR Thermocycler

We next sought to apply these newfound physical insights to design a low-cost convective PCR thermocycling system suitable for portable operation (Fig. 3.3). Specifically, we selected a reactor height h such that the top (annealing) surface temperature is automatically achieved when the bottom (denaturing) surface is heated to 95 °C (results reported here in Figs. 3.3, 3.4, and 3.5 were obtained using cylindrical reactors of $h = 10$ mm and $d = 2.5$ mm constructed from 1 in dia. polycarbonate rod stock). In this way, the PCR is actuated isothermally by maintaining a single heater at a constant temperature, drastically reducing electrical consumption to a level that can be supplied by standard 5 V USB sources that power ordinary consumer mobile devices and solar battery packs. Thermal management is achieved by using off the shelf ceramic resistors which convert current to heat via

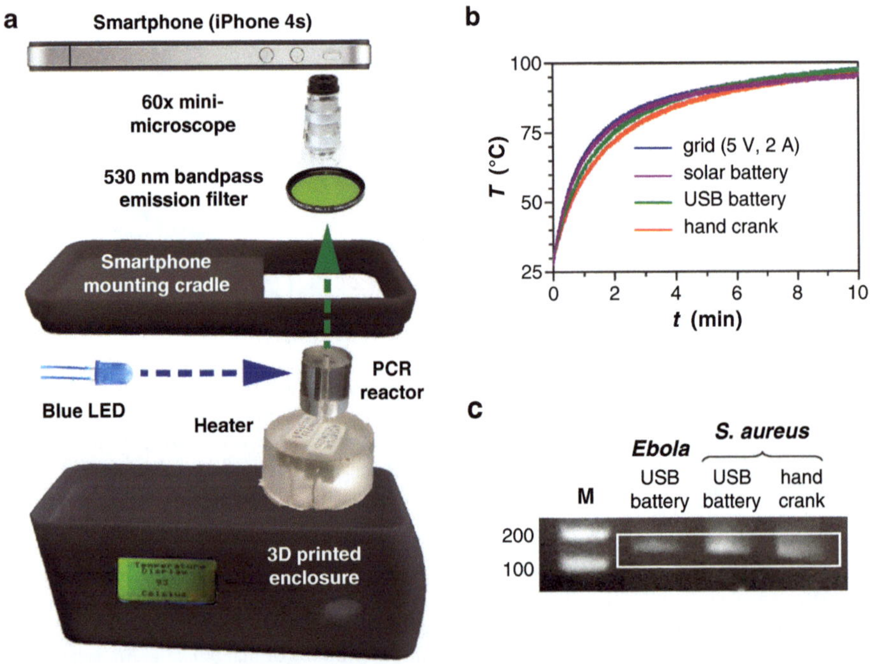

Fig. 3.3 An inexpensive low-power convective PCR thermocycler with smartphone-based fluorescence detection. (**a**) A simplified design enables convective thermocycling to be performed using only a single heater held at constant temperature. (**b**) The resulting low electrical power consumption enables the instrument to be operated using a variety of on- and off-grid power sources while achieving comparable heater performance. The device can operate using any USB-compatible 5 V power source used to charge mobile phones. (**c**) This simplified platform enables successful replication of a 150 bp *Ebola* target and a 147 bp *S. aureus* target after 20 min reaction time (lane M, FlashGel DNA marker) using off-grid USB battery and hand crank power sources

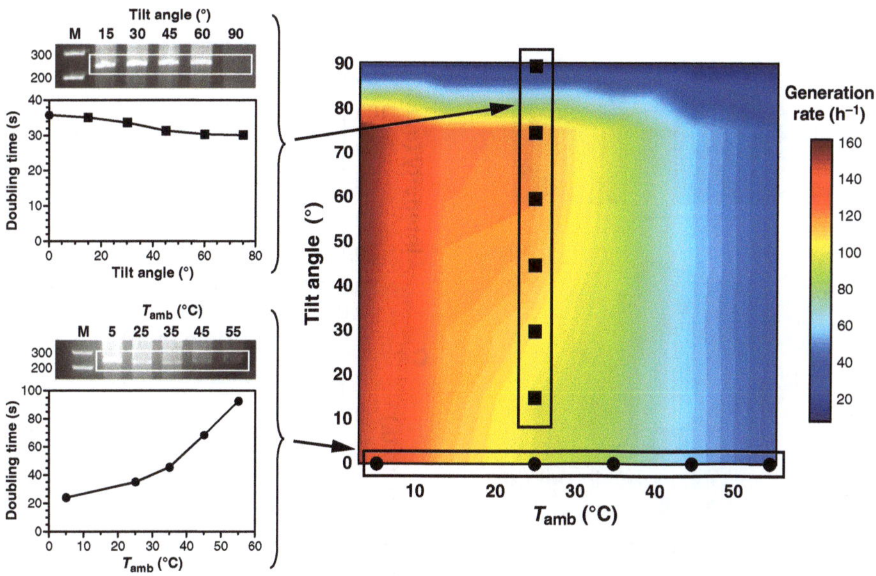

Fig. 3.4 A rugged and robust platform. Experiments and simulations show that product yields are insensitive to the reactor's orientation with respect to the vertical direction (*tilt angle*), as well as the ambient temperature, over a wide range. Gel images show 25 min replication experiments of a 237 bp target sequence from a λ-phage DNA template (*lane M*, FlashGel DNA marker). The doubling time data (*left*) and parametric plot (*right*) show performance predicted via 3-D computational fluid dynamics simulations of a coupled reaction-flow model

joule heating to build a simple mini-hot plate. A sensor monitors the temperature of the resistor and regulates the flow of current through a MOSFET-based feedback loop programmed to maintain the temperature at 95 °C by means of an Arduino-based microcontroller. A polydimethylsiloxane (PDMS) casing around the resistor ensures insulation and efficient heat transfer to the reactor bottom. The circuit also powers a blue LED light source for detection of PCR products in real time. The conventional USB interface makes it possible to employ a variety of power supplies suitable for use in remote settings, including battery pack, hand crank, and solar chargers. Consistent reaction yield and heater performance are achievable regardless of electrical power source.

Another counterintuitive aspect of the convective thermocycler design is its inherent robustness. The interplay between the thermal and flow conditions would appear to imply sensitivity to factors such as ambient environment temperature and orientation of the reactor relative to the vertical direction associated with the gravitational driving force. But remarkably, we find that the opposite is true (Fig. 3.4). Ambient temperature conditions impact thermocycling performance by virtue of their influence on the thermal gradient established within the reactor under the single heating source employed in our design. Experiments and computational simulations both reveal that our system operates consistently across ambient temperatures ranging from 5 to 35 °C. Reactions proceed most efficiently at 5 °C, with

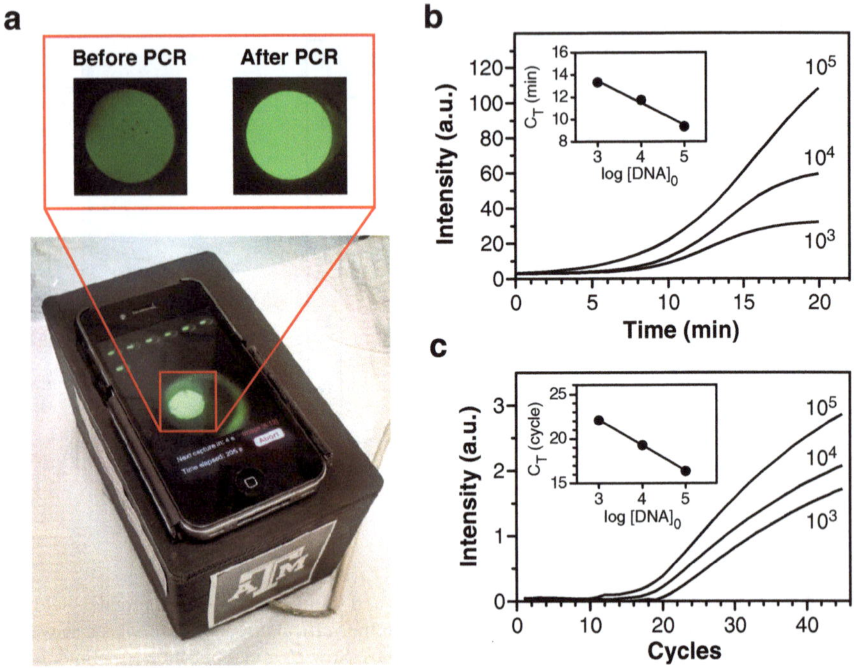

Fig. 3.5 Smartphone-based fluorescence detection offers portability and connectivity. (**a**) A smartphone and accompanying *PCR to Go* analysis app (https://itunes.apple.com/us/app/pcr-to-go/id909227041?mt=8) enable control of image acquisition, processing, and data analysis. (**b**) Smartphone-based quantification is demonstrated by replication of a 237 bp target sequence from a λ-phage DNA template. Sigmoidal fits are applied to the smartphone acquired real-time data, and a standard curve is constructed using reaction times when fluorescence exceeds a threshold value of 20 U (*inset*, $C_T = 9.4$, 11.8, and 13.3 min for starting DNA concentrations of $[DNA]_0 = 10^5$, 10^4, and 10^3 copies/μL, respectively). (**c**) A benchtop real-time PCR instrument (LightCycler 96, Roche) generates comparable results with a nearly identical standard curve

performance dropping above 35 °C where it becomes difficult to maintain the spatially distinct temperature zones required to actuate each stage of the reaction. The reactor's orientation with respect to the vertical direction would also intuitively be viewed as an important consideration given the gravitational origin of the driving force for convective transport. However, coordinated experiments and simulations both confirm that this is not the case, revealing no appreciable change in reaction performance as the reactor's tilt angle is increased, even to nearly horizontal levels. This invariance with spatial orientation is a reflection of the chaotic advection regime's inherent robustness, where consistent flow phenomena are predicted over orders of magnitude in the thermal driving force [9]. The combined influence of these factors can be quantified by performing simulations to determine generation rates across a spectrum-relevant tilt angles and ambient temperatures, yielding a parametric plot of reaction rate across the entire operational envelope.

3.4 Smartphone-Based Detection and Analysis

Complete diagnostic utility requires the ability to perform quantitatively accurate product analysis, but fluorescence-based approaches are often challenging to implement in a portable format due to issues associated with size, cost, and ruggedness. Smartphones have become an ubiquitous devices that are not only an essential part of daily communication but also provide state-of-the-art imaging, computation, and data analysis capabilities. They are also inherently familiar to most users, reducing the need for specialized training. These advantages have enabled smartphones to emerge as promising analysis platforms for point-of-care diagnostic systems. We incorporated a smartphone to image the top of the convective reactor to perform SYBR Green-based fluorescence detection of PCR products (Fig. 3.5a). The reactor is illuminated with a blue LED light source (480 nm, 5 mW) coupled with an excitation band pass filter (480 nm), and the resulting fluorescence signal is filtered through emission band pass filter (520 nm). A companion smartphone-based image analysis application *PCR to Go* was created to enable fluorescence analysis either in an end point mode (yielding a yes-no result) or in real time (providing product quantification) (Fig. 3.5b, c). The app accesses the phone's optical hardware to set the camera exposure time and focal length ensuring consistent readings, a feature often disabled for built-in point and shoot photography applications. The camera's CMOS sensor array acquires images looking down at the top of the convective PCR reactor, which is enclosed in an opaque 3D printed case to provide a dark room environment for consistent lighting. Imaging from the top of the reactor also ensures strongest fluorescent signal due to lower (annealing) temperatures at this location. An interchangeable smartphone cradle ensures proper alignment of the camera and optical components independent of the specific mobile device employed.

3.5 Convective PCR Overcomes Cost and Performance Barriers

Table 3.1 lists several commercially available thermocycling systems aimed toward decentralizing molecular diagnostics from dedicated medical facilities to on-site scenarios via compact and cost-effective platforms that can be operated without specialized training. In environments with limited availability of electrical power, battery-operated instruments limit reliance on unpredictable power grids resulting in less downtime. The ability to deploy the instrument for field-based analysis is also desirable because it lowers chances for contamination and degradation associated with transportation of collected materials and reduces sample to answer time scales leading to faster decision-making.

Instruments such as the GeneExpert [21, 22] (Cepheid), Liat Analyzer [23] (Roche), ML Platform [24] (Enigma), and FilmArray [25] (BioFire Defense)

Table 3.1 Portability assessment of selected PCR instruments. (Instruments listed in order of decreasing portability score)

No. (Fig. 3.6)	Instrument	Portability score[a]	Thermocycling method	PCR time (min)	Product detection	Form factor	Power source	Sample preparation	Cost ($ US)
1	DNA to Go [34]	9	Convective	20	Real time (fluorescence)	0.8 lb 6 × 12 × 8 cm	Battery (USB, V, 1 A)	Off-device	50
2	Palm PCR [30] (Ahram Biosystems)[b] www.ahrambio.com	8	Convective	30	End point (off-device)	1 lb 7 × 12 × 5 cm	Battery (15 V, 1 A)	Off-device	3,450
3	Liat Analyzer [23] (Roche) molecular.roche.com/instruments/Pages/cobasLIATsystem.aspx	8	Mechanical transport between temperature zones	30	Real time (fluorescence)	16 × 12 × 30 cm[c]	External (grid) (110/240 V)	Integrated	>1,000[c]
4	Razor EX [31] (BioFire Defense)[b] biofiredefense.com/razorex/	7	Stationary	30	Real time (fluorescence)	11 lbs 25 × 11 × 19 cm	Battery	Off-device	38,500
5	GeneXpert [21, 22] (Cepheid)[d] www.cepheid.com/us/	7	Stationary	35	Real time (fluorescence)	10 lbs 10 × 30 × 30 cm	External (grid) (110/240 V)	Integrated	17,000
6	T-COR 8 (Tetracore)[d] www.tetracore.com	6	Stationary	45	Real time (fluorescence)	10 lbs	Battery (4 PCR runs)	Off-device	>1,000[c]
7	SolarDX [26] (AAS, Inc.) www.aasinc.co	6	Continuous flow between temperature zones	45	End point (off-device)	15 × 15 × 30 cm[c]	Solar and battery	Off-device	<1,000[c]

8	RAPID [32] (BioFire Defense)[b] biofiredefense.com/rapid/	6	Stationary	25	Real time (fluorescence)	50 lbs 26 × 36 × 49 cm	External (grid) (110/220 V)	Off-device	>1,000[c]
9	OpenPCR (open source)[b] openpcr.org	5	Stationary	35	End point (off-device)	7.5 lbs 25 × 13 × 20 cm	External (grid) (110/240 V)	Off-device	600
10	Open qPCR (Chai Biotechnologies)[b] www.chaibio.com/openqpcr	5	Stationary	35	Real time (fluorescence)	10 lbs 28 × 24 × 19 cm	External (grid) (110/240 V)	Off-device	2,500
11	ML Platform [24] (Enigma)[d] www.enigmadiagnostics.com	5	Stationary	45	Real time (fluorescence)	22 lbs 35 × 31 × 27 cm	External (grid) (110/240 V)	Integrated	>1,000[c]
12	Film Array [25] (BioFire Defense)[d] filmarray.com	5	Stationary	60	Real time (fluorescence)	20 lbs 40 × 25 × 16 cm	External (grid) (220 V)	Integrated	>1,000[c]
13	POCKIT [19, 33] (GeneReach)[b] www.genereach-us.com/product.php	4	Convective	60	End point (off-device)	4.6 lbs	External (grid) (110/240 V)	Off-device	8,000

[a]Methodology for assignment of portability score is described in Table 3.2
[b]Instrument information obtained from the product website under "Specification" section
[c]Cost and form factors estimated based upon data from comparable instruments
[d]Instrument information obtained from product brochure or press release data

Table 3.2 Methodology for assignment of portability scores in Table 3.1

Portability points	PCR time (min)	Detection	Form factor (lb)	Power source	Sample preparation
2	≤30	Integrated	≤1	Battery	Integrated
1	30–45	Off-device	1–10	External (grid)	Off-device
0	≥45		>10		

perform conventional stationary PCR thermocycling in a format incorporating integrated sample preparation and real-time fluorescence detection. Although these devices provide a complete diagnostic solution, they are relatively expensive, complex, and require specialized training, making them less attractive for deployment in resource-limited settings. A particularly novel approach recently reported leverages a solar focusing lens to establish spatially distinct temperature zones needed to perform continuous flow PCR [26]. Although heating occurs passively, electrical power is required to mechanically circulate reagents through the microfluidic network using a syringe pump, post-PCR detection is achieved off-chip using a benchtop microscope, and effective operation is contingent upon availability of sufficient ambient sunlight. The suitcase-sized POCKIT system (GeneReach) is a commercial capillary convective thermocycler, but is of limited utility for portable diagnostics due to lack of an integrated battery and limited endpoint fluorescence detection. The Palm PCR instrument (Ahram Biosystems), a competing capillary convection-based device, offers battery power but lacks integrated fluorescence detection.

All the above-mentioned instruments offer potential to deliver advanced diagnostic capabilities, but to be applicable in low resource field-based settings, they also need to be affordable, rugged, and easy to use. We evaluated their ability to satisfy these constraints by quantifying portability using a weighing function that considers form factor (size, weight), electrical power requirements (battery or grid-based operation), speed, integrated detection, and integrated sample preparation (Table 3.2). A cumulative portability score was assigned to each instrument and cross-plotted against cost (Fig. 3.6). Despite their broader capabilities, systems incorporating integrated sample preparation generally fall in the top-left quadrant of the map due to a combination of high cost and reduced portability. Other adaptations of convective PCR achieve moderate portability at the expense of cost and integrated analysis functions (scattered across the middle of the map). The $50 ($US) hardware cost of our instrument breaks existing price barriers by several orders of magnitude. When combined with a simplified rugged design and smartphone-based detection, our convective PCR approach delivers an ideal combination of low cost and high portability (lower-right quadrant of the map). The simplified imaging and device agnostic design makes it possible to leverage a growing used smartphone marketplace to deliver affordable fluorescence detection. Smartphone-based analysis also introduces extraordinary connectivity, making it possible to tap into existing communication networks for real-time delivery of analysis results.

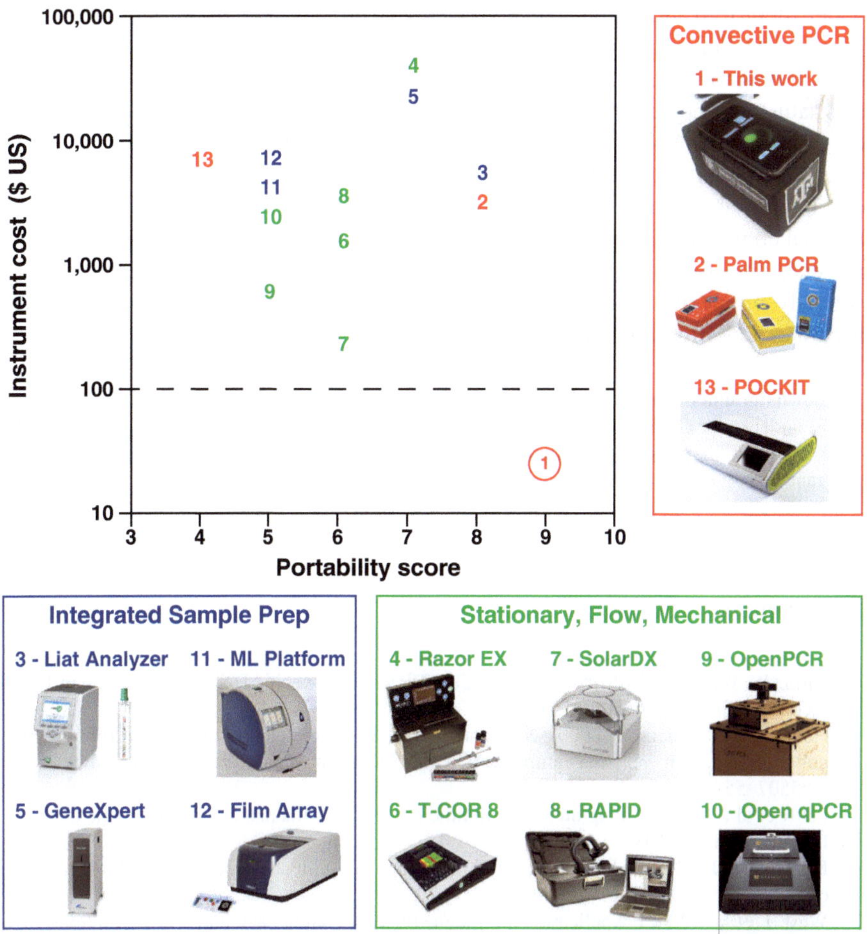

Fig. 3.6 Convective PCR thermocycling offers an attractive combination of portability and cost. A selection of commercial systems were assigned a portability score based on form factor, power requirements, run time, sample preparation, and detection capabilities (see Tables 3.1 and 3.2). Mapping the cumulative portability scores against instrument cost reveals an opportunity for convective technologies to surpass existing price, performance, and portability barriers. Instruments are categorized according to those employing thermal convection (*red*), those incorporating integrated sample preparation (*blue*), and systems involving stationary, flow, or mechanical thermocycling (*green*). The numbers assigned to each instrument correspond to the entries in Table 3.1

3.6 Conclusion and Outlook

Heightened concerns regarding recent infectious disease outbreaks [27] have led to renewed calls for development of portable, rapid, and versatile diagnostic platforms. Successful implementation of PCR-based nucleic acid diagnostics to meet these needs in low resource settings requires that they be inexpensive enough to be widely deployed and simple to use with a high degree of ruggedness and robustness. The United States clinical laboratory improvement amendments (CLIA) classify clinical diagnostic tests as either *high*, *moderate*, or *waived* complexity based upon the nature of the test performed [28, 29]. PCR-based diagnostics are currently classified as high complexity due to prerequisite operational training and sophisticated instrumentation, thereby making them expensive and impractical for mass distribution in portable applications. New instrumentation leveraging low-cost thermal convection-based thermocycling with integrated smartphone-enabled fluorescence detection offers potential to enable PCR to be classified in the moderate or waived complexity categories, opening the door for a new generation of fast, accurate, and affordable diagnostic tools.

Acknowledgments This work was supported in part by the US National Science Foundation (NSF) under grant CBET-1034002. We thank Magda Lagoudas for supporting an undergraduate team to assist in prototype instrument development via the Texas A&M AggiE-Challenge program.

References

1. Farrar JS, Wittwer CT (2015) Extreme PCR: efficient and specific DNA amplification in 15–60 seconds. Clin Chem 61:145–153
2. Hühmer A, Landers J (2000) Noncontact infrared-mediated thermocycling for effective polymerase chain reaction amplification of DNA in nanoliter volumes. Anal Chem 72:5507–5512
3. Pal R et al (2005) An integrated microfluidic device for influenza and other genetic analyses. Lab Chip 5:1024–1032
4. Kopp MU, De Mello AJ, Manz A (1998) Chemical amplification: continuous-flow PCR on a chip. Science 280:1046–1048
5. West J et al (2002) Application of magnetohydrodynamic actuation to continuous flow chemistry. Lab Chip 2:224–230
6. Krishnan M, Ugaz VM, Burns MA (2002) PCR in a Rayleigh-Bénard convection cell. Science 298:793–793
7. Braun D, Goddard NL, Libchaber A (2003) Exponential DNA replication by laminar convection. Phys Rev Lett 91:158103
8. Wheeler E et al (2004) Convectively driven polymerase chain reaction thermal cycler. Anal Chem 76:4011–4016
9. Priye A, Hassan YA, Ugaz VM (2013) Microscale chaotic advection enables robust convective DNA replication. Anal Chem 85:10536–10541
10. Muddu R, Hassan YA, Ugaz VM (2011) Chaotically accelerated polymerase chain reaction by microscale Rayleigh–Bénard convection. Angew Chem Int Ed 50:3048–3052
11. Yao D-J, Chen J-R, Ju W-T (2007) Micro–Rayleigh-Bénard convection polymerase chain reaction system. J Micro Nanolith Mem 6:043007
12. Hennig M, Braun D (2005) Convective polymerase chain reaction around micro immersion heater. Appl Phys Lett 87:183901
13. Braun D (2004) PCR by thermal convection. Mod Phys Lett B 18:775–784
14. Agrawal N, Hassan YA, Ugaz VM (2007) A pocket-sized convective PCR thermocycler. Angew Chem Int Ed 46:4316–4319

15. Zhang C, Xing D (2009) Parallel DNA amplification by convective polymerase chain reaction with various annealing temperatures on a thermal gradient device. Anal Biochem 387:102–112
16. Chen Z, Qian S, Abrams WR, Malamud D, Bau HH (2004) Thermosiphon-based PCR reactor: experiment and modeling. Anal Chem 76:3707–3715
17. Chung KH, Park SH, Choi YH (2010) A palmtop PCR system with a disposable polymer chip operated by the thermosiphon effect. Lab Chip 10:202–210
18. Chou WP et al (2011) Rapid DNA amplification in a capillary tube by natural convection with a single isothermal heater. BioTechniques 50:52–57
19. Chang HFG et al (2012) A thermally baffled device for highly stabilized convective PCR. Biotechnol J 7:662–666
20. Hsieh Y-F et al (2013) A real-time convective PCR machine in a capillary tube instrumented with a CCD-based fluorometer. Sensors Actuators B Chem 183:434–440
21. Ulrich M et al (2006) Evaluation of the Cepheid GeneXpert® system for detecting Bacillus anthracis. J Appl Microbiol 100:1011–1016
22. Raja S et al (2005) Technology for automated, rapid, and quantitative PCR or reverse transcription-PCR clinical testing. Clin Chem 51:882–890
23. Tanriverdi S, Chen L, Chen S (2010) A rapid and automated sample-to-result HIV load test for near-patient application. J Infect Dis 201:S52–S58
24. Goldenberg SD, Edgeworth JD (2015) The Enigma ML FluAB-RSV assay: a fully automated molecular test for the rapid detection of influenza A, B and respiratory syncytial viruses in respiratory specimens. Exp Rev Mol Diagn 15:23–32
25. Poritz MA et al (2011) FilmArray, an automated nested multiplex PCR system for multi-pathogen detection: development and application to respiratory tract infection. PLoS ONE 6, e26047
26. Jiang L et al (2014) Solar thermal polymerase chain reaction for smartphone-assisted molecular diagnostics. Sci Rep 4:4137
27. Cecilia D et al (2015) Development of a multiplex real-time RT-PCR assay for simultaneous detection of dengue and chikungunya viruses. Arch Virol 160:323–327
28. Ehrmeyer SS, Laessig RH (2007) Point-of-care testing, medical error, and patient safety: a 2007 assessment. Clin Chem Lab Med 45:766–773
29. Niemz A, Ferguson TM, Boyle DS (2011) Point-of-care nucleic acid testing for infectious diseases. Trends Biotechnol 29:240–250
30. Lim S, Nan H, Lee M-J, Kang SH (2014) Fast on site diagnosis of influenza A virus by Palm PCR and portable capillary electrophoresis. J Chromatogr B 963:134–139
31. Mölsä M et al (2012) Detection of influenza A viruses with a portable real-time PCR instrument. J Virol Methods 181:188–191
32. Christensen DR et al (2006) Detection of biological threat agents by real-time PCR: comparison of assay performance on the RAPID, the LightCycler, and the Smart Cycler platforms. Clin Chem 52:141–145
33. Tsai Y-L et al (2012) Development of TaqMan probe-based insulated isothermal PCR (iiPCR) for sensitive and specific on-site pathogen detection. PLoS One 7:e45278
34. Priye A et al (2016) Lab-on-a-drone: toward pinpoint deployment of smartphone-enabled nucleic acid-based diagnostics for mobile health care. Anal Chem 88. doi: 10.1021/acs.analchem.5b04153

Chapter 4
Forensic Typing of Single Cells Using Droplet Microfluidics

Tao Geng and Richard A. Mathies

Abstract Short tandem repeat (STR) typing based on polymerase chain reaction (PCR) is a powerful tool for forensic identification. However, it is challenging to analyze low-quantity and low-quality evidence samples comprising mixture of cells and/or DNA from multiple contributors. Herein we present a droplet microfluidics-based technology for single-cell STR typing with high sensitivity, throughput, and fidelity. In this approach, single cells are initially compartmentalized within nanoliter agarose droplets using a microfluidic droplet generator. The transformation of the agarose into microgels enables the diffusion of cell lytic reagents into the droplets for single-cell genomic DNA extraction, while PCR inhibitors are washed away. Following the introduction of PCR components, massively parallel multiplex droplet PCR is performed in oil emulsions to transfer the STR information from an individual cell onto the co-encapsulated primer-functionalized bead. The amplicon-carrying beads are then statistically diluted and reamplified in a secondary PCR. The resulting transfer of STR products to free solution facilitates the detection by conventional capillary electrophoresis (CE) fragment size analysis. This technology is effective in preserving single-genome integrity when analyzing diverse cellular materials yet insensitive to background DNA contamination. Target cells are successfully identified from heterogeneous cell populations with a high (up to 10:1) background, mixtures of cells and cell-free DNA, and samples contaminated with high concentrations of PCR inhibitors.

Keywords Droplet microfluidics • Single-cell analysis • Digital droplet PCR • Forensics • Short tandem repeat (STR) typing • PCR inhibition

---

T. Geng, Ph.D.
Department of Chemistry, University of California, Berkeley, CA 94720, USA

Environmental Molecular Sciences Laboratory, Pacific Northwest National Laboratory, 902 Battelle Blvd., Richland, WA 99354, USA
e-mail: taogeng3@gmail.com

R.A. Mathies, Ph.D. (✉)
Department of Chemistry, University of California, Berkeley, CA 94720, USA
e-mail: ramathies@berkeley.edu

© Springer International Publishing Switzerland 2016

C. Lu, S.S. Verbridge (eds.), *Microfluidic Methods for Molecular Biology*,
DOI 10.1007/978-3-319-30019-1_4

4.1 Introduction

Driven by the remarkable advances in modern molecular biology technologies, forensic DNA analysis has been emerging as an essential and routine tool to assist in identifying perpetrators, victims, and suspects in crimes [1–5]. Over the past decades, diverse DNA typing approaches have been validated for robust analysis of the genetic polymorphisms of materials collected in criminal investigations, among which short tandem repeat (STR) typing has become the gold standard since its first implementation in 1991 [1]. STRs, also termed microsatellites, are noncoding DNA regions consisting of contiguously repeated short sequences that are typically 2–6 base pairs in length, and account for ~3 % of human genomic DNA [3]. The number of repeat units in a STR locus is highly variable among different individuals, and hence STR typing is especially suitable for human identity testing [3]. In the USA, the FBI has established a Combined DNA Index System (CODIS) containing 13 core STR loci as standard markers for forensic identification. Simultaneous typing of all these STR loci plus a sex marker Amelogenin dramatically reduces the random match probability in the whole human population, thereby improving the power of discrimination [3].

Current forensic STR typing relies on multiplex polymerase chain reaction (PCR)-based DNA amplification followed by capillary electrophoretic separation and fluorescent detection to determine the size of STR amplicons and the repeat numbers [3, 4]. Conventional PCR techniques yield reliable results from concentrated and purified DNA samples. However, it is challenging to apply standard protocols to biological evidence samples that are found in complex environments at a crime scene. Such samples are usually compromised in both quantity and quality, and contaminated or degraded by environmental factors and certain potent PCR inhibitors [6–8]. These challenging samples lead to partial STR profiles or even false-negative results due to different degrees of PCR failure such as peak imbalance, and allelic dropout and drop-in [8]. In addition, mixed specimens are frequently involved in common forensic samples, such as those collected from sexual assault scenes, and touch evidences that are often contaminated with cellular materials from other contributors [6, 7]. Although several statistical methods have been employed to interpret the DNA mixture profiles, it is technically difficult to type trace samples due to stochastic allele height variation and samples involving a large number of contributors [9, 10].

To address these limitations, a variety of strategies have been developed for forensic DNA profiling. Novel DNA purification techniques have been developed to remove impurities prior to STR amplification [11–22]. PCR processes have also been modified to mitigate PCR inhibition effects through dilution of DNA templates [23, 24], utilization of DNA polymerases resistant to PCR inhibitors and optimized buffer systems [25–29], and addition of amplification facilitators [23, 24, 30, 31]. Moreover, target cells have been sorted from mixtures prior to DNA processing using diverse cell separation techniques to eliminate the complicated procedure of mixture interpretation [32–36]. However, these analytical techniques require a large amount of cellular material, and still suffer from limitations such as low sensitivity, low throughput, less universality, high sample contamination possibility, and high process complexity.

Single-cell analysis is a powerful approach for probing individual cells within heterogeneous cell samples [37–39]. Forensic typing of single cells represents a conceptually straightforward and versatile method for forensic identification of size-limited and mixed samples. Micromanipulation [40–42] and laser capture microdissection [43–45] have been used to retrieve single cells from forensic samples to perform microliter-scale PCR reactions for DNA profiling at the single-cell level. However, the two techniques are of low throughput (a few cells per hour), and require manual operation and complex instrumentation. The large reaction volume of PCR also compromises its efficiency when starting with a single cell or a low-copy sample. Furthermore, these single-cell PCR reactions are sensitive to minute amounts of DNA contaminants, causing amplification problems.

Recent innovation in microfluidic technologies has opened up a wealth of new opportunities in single-cell isolation, manipulation, and analysis because of comparable dimension (micrometers) and volume scales (picoliters to nanoliters) to a single cell [37–39]. Microfluidics offers inherent advantages for single-cell analysis including high throughput, improved sensitivity and reaction efficiency, and process integration. In this process, single cells are compartmentalized by either single-phase microfluidic systems incorporated with integrated microvalves [46] or two-phase microdroplets [39]. Although the microsystems featuring nanoliter chambers created by microvalves have great analytical power and now have been commercialized by Fluidigm, the complicated fabrication procedures and high cost hinder its wider applications. In contrast, droplet-based microfluidic techniques enable the handling of individual cells with higher throughput, enhanced accuracy, and reduced reagent and fabrication cost [39, 47, 48]. The basic principle of this method is that highly uniform water-in-oil droplets can be produced from microfluidic channels, and subsequently each droplet serves as an isolated miniaturized reactor and a large number of independent reactions can be performed in parallel. The well-defined droplet dimension enables precisely tuned and quantitative analysis of droplet content, while the compartmentalization by oil reduces the risk of sample contamination. These unique characteristics make droplet microfluidics a valuable tool for diverse single-cell applications, including enzymatic reaction assays, drug screening, and genetic analysis [49–59].

In this chapter, we describe a microfluidic droplet-based method for compartmentalizing single cells and performing forensic STR typing at single-cell resolution. This method is built upon our previous work on single-cell and single-molecule PCR in microfluidic nanoliter droplets that combines the benefits of emulsion PCR and microbead-based solid-phase PCR [60–63]. This technology has shown effectiveness and fidelity in analyzing cell mixtures, cell/DNA mixtures, and PCR inhibitor-contaminated cell samples, thereby providing a novel tool for analyzing low-abundance, low-quality, and mixed forensic samples [64, 65].

4.2 Droplet-Based Single-Cell STR Typing

The droplet-based single-cell STR typing method comprises multistep molecular biological reactions involved in standard PCR-based STR typing (i.e., cell separation, cell lysis and DNA extraction, multiplex PCR, and post-PCR cleanup) in an array of suspended nanoliter-volume droplets. The core of this technology is the utilization of thermosensitive agarose droplets that are produced by microfluidics and then serve as reaction compartments. The analytical process, outlined in Fig. 4.1, can be generally divided into three stages: (1) single-cell compartmentalization and genetic sample preparation; (2) massively parallel emulsion droplet PCR for multiplex STR target amplification; and (3) detection of single-cell amplicons. The total procedure could be completed in about 22 h including 3.5 h of working time and 18.5 h of waiting time for cell lysis and DNA extraction (~10 h), droplet PCR (3.5 h), secondary PCR (3 h), and capillary electrophoresis (CE) fragment analysis (2 h). For more detailed description of experimental procedure and results see our previous papers [64, 65].

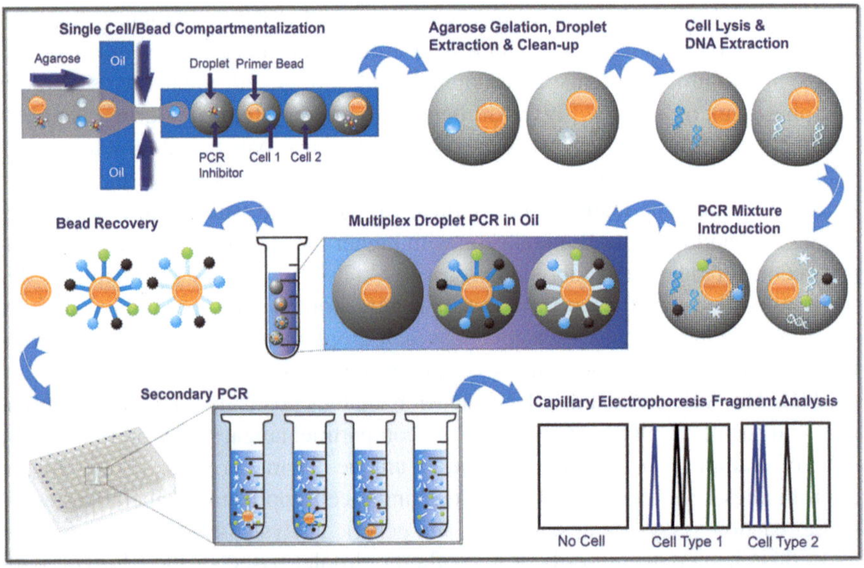

Fig. 4.1 Workflow for single-cell forensic STR typing. Single cells along with primer-functionalized microbeads are first compartmentalized in agarose droplets during droplet generation on a microfluidic chip. After agarose gelation, the droplets are extracted from oil, and PCR inhibitors are cleaned up. Single-cell genomic DNA is then released upon cell lysis and entrapped in the gel matrix. Following the introduction of PCR components, multiplex droplet PCR is performed in oil in a highly parallel manner. The beads are recovered by melting the agarose, and reamplified under statistical dilute conditions for a low-gain secondary PCR that is insensitive to intrinsic or background DNA contamination and inhibitors. Fragment analysis using a conventional CE system reveals that each cell type yields distinct single-cell STR profiles, while empty beads and reactions generate null results

In the first stage, monodisperse aqueous agarose droplets are continuously produced in fluorinated oil using a polydimethylsiloxane (PDMS)-glass microfluidic droplet generator constructed with standard soft lithography. The architecture of the microfluidic device features a flow-focusing channel geometry consisting of a cross-junction created by three input channels and one output channel. Pre-warmed agarose solution admixed with cells and beads is injected into the central input channel, while the immiscible fluorinated oil mixture flows into the two orthogonally oriented side channels. The two phases are then forced to flow through a nozzle channel located at the downstream of the three channels where the inner agarose stream is hydrodynamically focused into a thin thread and then broken owing to the pressure and viscous stresses exerted by the outer oil streams, and the droplets flow out of the chip through the output channel (Fig. 4.2a). Droplet size and generation frequency are dependent on the channel geometry, the volumetric flow rates of the two streams, and the physical properties of the liquids such as viscosity. Stable fluidic physics in the device ensures high uniformity of droplets. Nozzle channels with a dimension of 125 μm (width) × 130 μm (height) can generate 1.5 nL droplets containing 1.5 % (w/v) of agarose at a high rate of 444 droplets per second at the flow rates of 40 μL/min for aqueous phase and 100 μL/min for oil phase. Meanwhile, single cells (or DNA fragments) along with

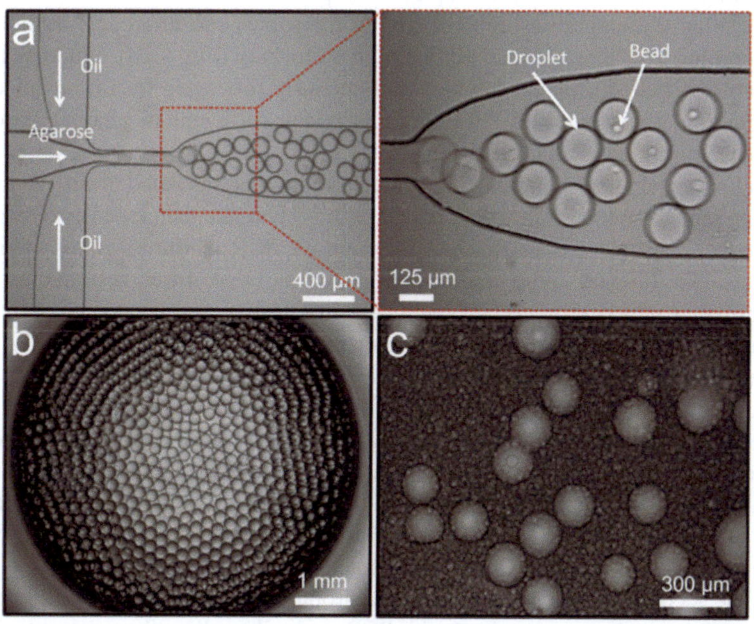

Fig. 4.2 Microfluidic droplet generator design and agarose droplet generation. (**a**) Droplets are generated in a flow-focusing channel geometry with oil flowing into two side channels and agarose flowing into a central channel. Cells and beads are compartmentalized within the droplets. (**b**) Gelled agarose droplets are uniformly suspended in oil. (**c**) Droplets are intact after 32 cycles of PCR in oil. Reproduced from Ref. [64] with permission from the American Chemical Society

Sepharose (a trade name for a cross-linked form of agarose) microbeads are statistically dispersed in the agarose solution and encapsulated within the droplets according to Poisson distribution, thereby isolating individual cells into droplet reactors. The narrow distribution of the droplet size enables precise control over cell and bead occupancy by tuning their concentrations in feeding agarose solutions or the solution and oil flow rates. The beads are conjugated with multiplex primers for targeted STR regions via *N*-hydroxysuccinimide chemistry to function as binding substrates for PCR amplicons from single cells while maintaining the monoclonality of the STR information. Beads with an average diameter of 34 μm are used, because their relatively large surface area provides sufficient spaces to link ample (i.e., at least 100 attomoles) PCR product for downstream analysis [60].

The key feature of agarose droplets is their sol-gel switching in response to thermal treatment, overcoming the limitations of traditional droplet techniques in which the contents in a reaction cannot be completely removed. Agarose gel allows flexible transfer of the droplets between the oil phase and the aqueous phase as well as free exchange of low-molecular-weight reagents (e.g., salts, mono- and oligonucleotides, chemicals, and some proteins) between the exterior and interior of the droplet reactors. Type IX ultra-low-gelling temperature agarose with a gel point of 8–17 °C and a remelting point of $\leq$50 °C is employed to eliminate potential agarose gelation during on-chip droplet generation and detrimental effects on droplet uniformity. After collecting the aqueous droplets from the microfluidic chip via plastic tubing, the agarose droplets rapidly transform into microgels by immediately cooling to 4 °C while maintaining high uniformity (Fig. 4.2b). Unlike aqueous droplets, the gelled droplets can be simultaneously processed by various mechanical manipulations, and 1.5 % agarose provides sufficient mechanical strength without limiting molecular diffusion.

Following agarose gelation, the gelled droplets are isolated from the oil by filtering using a nylon mesh with a pore size of 40 μm, and then extensively washed with water. This step will also efficiently remove PCR inhibitors entrapped in the droplets. To liberate tightly packaged genomic DNA from cell nuclei, cell-encapsulated gelled droplets are incubated in cell lysis buffer containing 0.5 % sodium dodecyl sulfate (SDS), 0.1 mg/mL proteinase K, 100 mM EDTA, and 10 mM Tris–HCl, pH 8. Proteinase K is included to digest histones and other DNA-associated nuclear proteins for efficient release of genomic DNA. Lytic reagents diffuse into the droplet through the cross-linked porous matrix structure, whereas the high-molecular-weight DNA polymer chain is retained within the 1.5 nL droplets. After DNA release, the droplet cleanup is conducted by sequentially rinsing with various solvents to remove residual chemicals (e.g., SDS and proteinase K) and cell lysate, which are potent PCR inhibitors.

In the second stage, a number of gelled droplets containing single-cell genomic DNA and primer beads are mixed with PCR cocktail to allow the transport of PCR components into the agarose matrix. The PCR mixture includes 1× *AmpliTaq*® Gold PCR buffer, 1.5 mM $MgCl_2$, 200 μM dNTP, 0.2 U/μL *AmpliTaq*® Gold DNA polymerase, 4 μg/μL heat-inactivated bovine serum albumin (BSA), 0.01 % Tween 80, and primer mixture. BSA forms a capsule layer at the interface between the droplet and oil during thermal cycling to eliminate droplet coalescence, and the addition of Tween 80 further enhances the thermal stability of the droplets. Since only a single copy or two copies of each DNA template is involved in single-cell PCR, *AmpliTaq*® Gold DNA polymerase, which is inactivated unless exposed to a high temperature, is used to improve the yield and specificity of PCR reactions. In addition to fluorescent dye-labeled primers with the reverse direction of the corresponding primers coupled onto beads, the primer mixture is also composed of a small amount of free primers having identical sequences to the bead oligonucleotides. Although the concentrations of these primers are only 1/10 of their corresponding fluorescently labeled reverse-direction primers, they help to initiate the reaction in free solution and increase the efficiency of primer extension by binding more products onto the primer-immobilized beads. The presence of 1.5 % agarose has no impact on the efficiency of PCR amplification.

To perform emulsion PCR, the gelled droplets are re-dispersed in carrier oil (either fluorinated or silicone oil mixture) by mechanical vibration at a high frequency to form compartmentalized nanoliter PCR reactors. Surfactants are also included in the oil mixture to ensure the structural integrity of water-in-oil emulsions. PCR reagents outside of the microgels are formed into extremely small emulsions with diameters of less than 1 μm, termed microfines, which will enhance agarose droplet stability during PCR reactions. Massively parallel multiplex PCR combining both emulsion-based PCR and on-bead solid-phase PCR is then implemented in a standard PCR tube using a conventional thermocycler. The microgels melt at the hot start stage for DNA polymerase activation and remain aqueous throughout the whole thermal cycling process, maximizing the mixing of reagents and amplicons and hence the PCR reaction rates. The information on STR markers is transferred from the single-cell genomic DNA onto the bead co-encapsulated within the same droplet, and one bead corresponds to the STR information from a single cell. With the optimized recipes of PCR reagents and oil as well as the PCR cycling conditions, droplet merging is minimized to eliminate cross-contamination between amplicons from different cells. Immediately after PCR thermal cycling, the droplets are re-gelled and remain intact, as shown in Fig. 4.2c. The microgels are harvested by centrifuging and washing with various reagents (i.e., 100 % isopropanol, 100 % ethanol, and buffer solutions) to completely remove the oil. The amplicon-carrying beads are recovered by melting the agarose to disrupt the droplets, followed by cleanup of the beads through extensive washes.

In the final stage, the genetic information from an individual cell is transferred from the amplicons immobilized on the beads into free solution to facilitate conventional PCR detection. This is achieved by statistically diluting the DNA-carrying beads at appropriate concentrations in standard PCR well plates and using them as the DNA templates for a low-gain (25 or 23 cycles) secondary bulk PCR. The amplicons in free solutions are processed for conventional CE-based fragment size analysis to identify the differences of STR markers among different cells. The efficiency of this secondary PCR process could also be improved by sorting and amplifying only the positive STR-carrying beads.

Translating traditional STR typing to microfluidic droplet format offers several benefits. First, the sensitivity and selectivity of the assay are significantly improved owing to its efficient sample-handling ability. Multiple steps from single-cell isolation to STR amplification can be accomplished in a droplet, and high single-cell PCR efficiency results from the high concentration of starting genetic materials confined into the nanoliter reactions. Second, this superior approach does not involve the utilization of sophisticated and expensive instrumentation, such as those required for single-cell separation methodologies including micromanipulators, fluorescence-activated cell sorters, and laser capture microdissection systems. In contrast, the disposable microfluidic droplet generators can be manufactured with low cost. As such, it could be immediately applied to forensic analysis using commonly available experimental setups in standard forensic laboratories, and in the future the streamlined workflow can be integrated and optimized for automatic processing. Third, the droplet-based strategy, in principle, can be extended to simultaneously analyze multiple types of cells with any ratios simply by re-amplifying more beads in the secondary PCR. Moreover, the ability of the agarose droplet microfluidics to dilute extracellular contaminants (e.g., exogenous DNA fragments, PCR inhibitors, and other environmental elements) and wash them from microgels enables typing of complex forensic samples.

4.3 Method Validation

Our single-cell STR typing method is validated based on a 9-plex STR system composed of 8 core STR loci from the CODIS database including D3S1358, D5S818, D7S820, D8S1179, D13S317, D21S11, vWA, and TH01 and the sex marker Amelogenin. Pure cell samples of two standard cell lines, GM09947 female and GM09948 male human lymphoid cells, are tested to verify the effectiveness of single-cell droplet PCR-based STR typing. To reduce the frequency of more than one cell in a droplet, the cell samples are substantially diluted to 0.15 cells and 0.9 primer beads on average in each droplet. At this cell occupancy, it is expected that 14 % of the beads would be positive with STR amplicons according to Poisson statistics. In the secondary bulk PCR, beads collected from droplet PCR are statistically diluted into 30 wells and amplified at a bead concentration of 0.9 beads/reaction corresponding to 0.126 (0.9×14 %) positive beads per reaction, ensuring that only one positive bead is included in each PCR reaction. As shown in Fig. 4.3a, b, 4 positive results and 26 null results are obtained for GM09947 female cells, while 2 positive results and 28 null results are for GM09948 male cells. The numbers of positive results are in good accordance with the theoretical value of 3.54 (30×11.8 %) following the Poisson distribution. All positive results obtained from the single cells of both cell types exhibit full STR profiles without allelic drop-in and dropout (i.e., 13 and 15 peaks for the 9 STR loci of GM09947 female cells and GM09948 male cells, respectively), and all alleles are correctly called in agreement with the well-established STR locus information. The results confirm the ability of

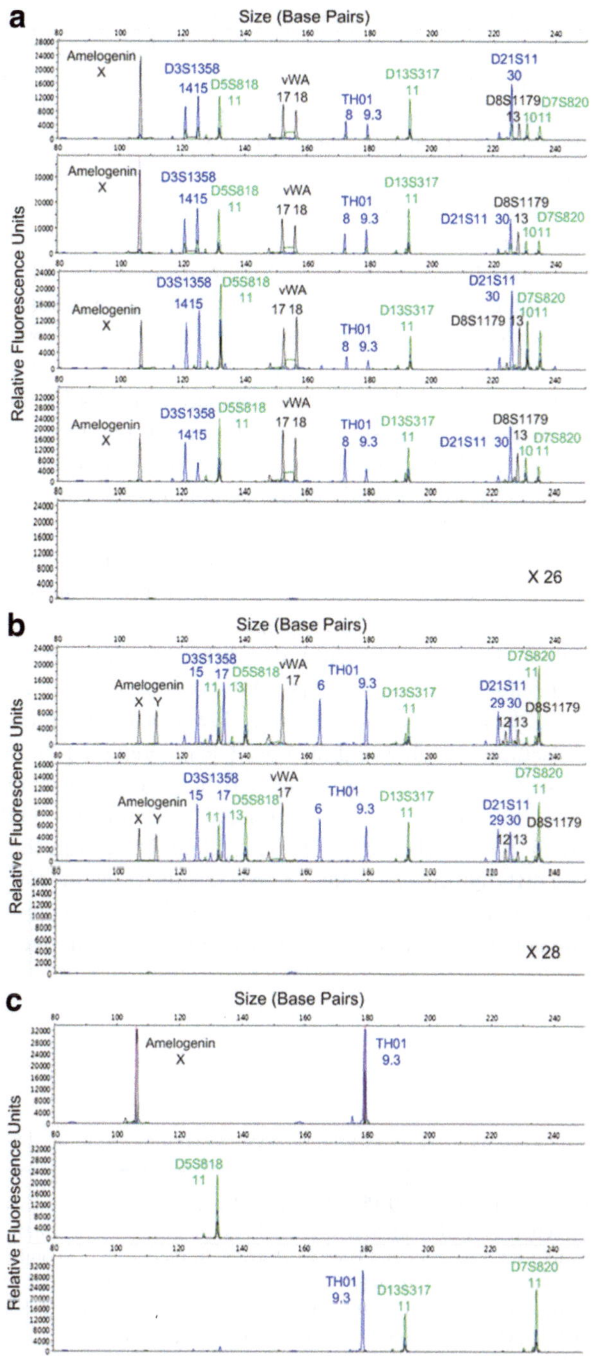

Fig. 4.3 Method validation. (**a**) and (**b**) Single-cell STR typing of pure GM09947 female (**a**) or GM09948 male (**b**) lymphoid cells. The procedure comprises 30 cycles of emulsion PCR using

the agarose droplets to transfer STR information from a single cell to the co-encapsulated bead while preserving single-genome integrity throughout various mechanical manipulations and reactions. In addition, using carefully optimized PCR cocktail and cycling conditions, the STR peaks are generally balanced, and the stutter products for all allelic peaks can be regarded as biological artifact because of their low percentage (below 15 %). It is also worth noting that the optimal amplification protocols overcome the limitations of bead-based solid-phase PCR, as the solid-phase surface may reduce PCR efficiency due to steric hindrance and charge repulsion. In contrast, no peaks are observed in all profiles of the negative samples including samples amplified with negative beads, demonstrating that there is no sample cross-contamination among the array of compartmentalized droplet reactors.

Single-molecule droplet PCR is also conducted with 9947A female genomic DNA to further evaluate the sensitivity of the method. On average 0.2 copies of genomic DNA (i.e., 1.8 alleles) and 0.9 beads are entrapped in a droplet, leading to the probability that 83 % of the beads would be conjugated with one or more STR products. The secondary PCR is carried out at a significantly diluted bead concentration of 0.15 beads/reaction. In this case, 13 % of reactions contain solely one bead, and only 1 % of reactions would amplify from more than one bead. Unlike single-cell typing, DNA molecule-encapsulated droplets exhibit partial STR profiles with different genotypes, because the different allele templates are each statistically distributed into the droplets. Only one to three peaks are profiled from an individual bead due to the extremely low DNA concentration, as shown in the three representative profiles in Fig. 4.3c. The incomplete profiling is expected because the fragments of DNA template are stochastically distributed in the droplets with a probability of 9.5 % for each unique heterozygous STR allele and a possibility of 18 % for each homozygous STR allele. Moreover, the STR markers are all located on different chromosomes. These results reveal the intrinsic nature of digital amplification of single-copy DNA in the droplet-based approach, and demonstrate the advantage of single-cell forensic identification over conventional approaches.

Fig. 4.3 (continued) droplets with 0.15 cells and 0.9 beads per droplet on average, followed by 25 cycles of secondary PCR. (**c**) Single-cell STR typing of 9947A female genomic DNA. The procedure contains 30 cycles of emulsion PCR using droplets with 0.2 copies of DNA and 0.9 beads per droplet on average, followed by 25 cycles of secondary PCR. Different colors represent products labelled with different fluorescent dyes. Blue: 6-FAM (6-carboxy-fluorescein); black: TAMRA (carboxytetramethylrhodamine); and green: JOE (6-carboxy-4′, 5′-dichloro-2′,7′-dimethoxyfluorescein). Reproduced from Ref. [64] with permission from the American Chemical Society

4.4 Single-Cell STR Typing of Cell Mixtures

One benefit of our droplet-based single-cell STR typing method is its good selectivity when processing heterogeneous cell populations, eliminating the subsequent complex statistical mixture interpretation. This is demonstrated with four sets of GM09947 female and GM09948 male human lymphoid cell mixtures at female-to-male cell input ratios of 1:1, 2:1, 5:1, and 10:1. A dramatically diluted total cell concentration is tested at 0.01 cells/droplet on average while keeping the bead concentration at 0.9 beads/droplet. As such, this strategy ensures that no cell aggregates are trapped in the droplets, although PCR products will be bound onto only 1 % of the beads during emulsion PCR and the majority of the beads are empty. To reduce the required reaction number of secondary PCR, the bead concentration is accordingly increased to 20 beads/reaction, resulting in 0.2 (20 × 1 %) positive beads/reaction. 40–100 reactions are carried out according to the cell ratios. As illustrated in Fig. 4.4, a standard curve is established at a constant total cell concentration in the single-cell regime but distinct relative concentrations of the two cell types. The relationship between the observed female-to-male cell number ratios and the corresponding input ratios exhibits a near-perfect linearity ($R^2 = 0.9917$), testifying to the accuracy and reliability of our assay in the range of the tested ratios.

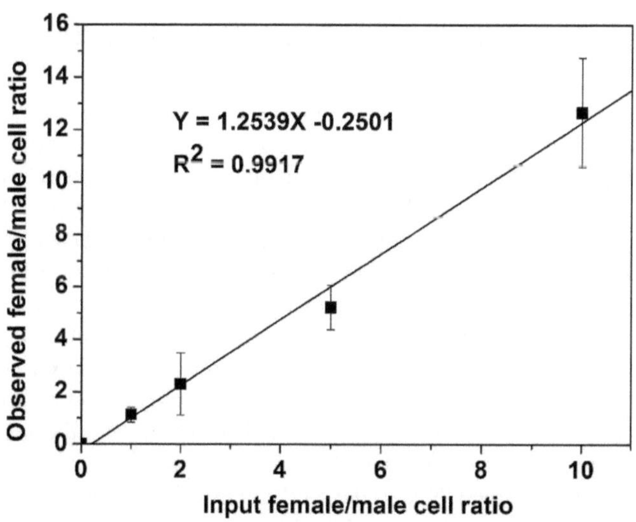

Fig. 4.4 Single-cell STR typing of the GM09947 female and GM09948 male lymphoid cell mixtures. The experimental female-to-male cell number ratio is plotted against input female-to-male cell number ratio. The procedure comprises 30 cycles of emulsion PCR using droplets with 0.01 total cells and 0.9 beads per droplet on average but varying female-to-male cell ratios (1:1, 2:1, 5:1, and 10:1), followed by 25 cycles of secondary PCR. Each data point is based on three independent assays. Reproduced from Ref. [64] with permission from the American Chemical Society

When the mixed sample has equal amounts of two cell types, 40 secondary PCR reactions yield 9 positive results and 31 null results (Fig. 4.5), which aligns perfectly with the theoretical prediction of 18 %. Of the nine positive results, we identify the presence of full STR profiles of four single GM09947 female cells and five single GM09948 male cells, and the absence of mixed genotypes. The experimental female-to-male ratio of 4:5 is close to that expected. Although the heterozygous peaks for some STR loci such as Amelogenin and TH01 are not perfectly balanced due to the stochastic effects in single-copy DNA molecule amplification, the identified STR profiles from this extensively diluted cell mixture do not show significantly compromised quality with respect to stutter formation (below 15 %) and allelic drop-in and dropout compared with those obtained from a pure cell population. When the female-to-male cell input ratio increases, more secondary PCR reactions should be carried out to identify the rarer cells. For instance, 100 PCR reactions are conducted at the ratio of 5:1, and 19 positive STR profiles are discovered including 16 from GM09947 female cells and 3 from single GM09948 male cells. The best performance in cell identification is achieved from the cell mixtures with female cells ten times more prevalent than male cells using 100 PCR reactions. In principle, the single-cell method enables the target cells to be detected in the presence of even more background cells by simply running more secondary PCR reactions.

4.5 Single-Cell STR Typing of Cell/DNA Mixtures

Extracellular cell-free DNA has been reported to be present in diverse biological fluids such as blood, sweat, urine, semen, and saliva which are important sources of evidence samples in crime scenes [66, 67]. In other cases, cell-free DNA from other contributors may contaminate the target cells in a forensic sample. Profiling the cellular STR is challenging, especially when the extracellular DNA is much more prevalent than the target cellular DNA. Droplet microfluidics enables not only minimal interference with each droplet PCR reaction by contaminating DNA due to statistical dilution of the fragments together with the target cells, but also a simple genotype interpretation procedure. The contaminating DNA peaks can be easily ruled out from the target cellular peaks by aligning and comparing all STR profiles obtained from distinct droplets and then identifying the contaminating DNA peaks which will not appear in all profiles.

The impact of cell-free DNA on the STR typing of the target cells is examined using a model system containing GM09947 female lymphoid cells admixed with 9948 male genomic DNA. Comparisons between the known STR profiles of GM09947 cells (13 peaks) and 9948 genomic DNA (15 peaks) demonstrate that six alleles (Amelogenin-Y, D3S1358-17, D5S818-13, TH01-6, D21S11-29, and D8S1179-12) belong exclusively to 9948 genomic DNA, while the other nine allele peaks (Amelogenin-X, D3S1358-15, D5S818-11, vWA-17, TH01-9.3, D13S317-11, D21S11-30, D8S1179-13, and D7S820-11) overlap with those of the GM09947

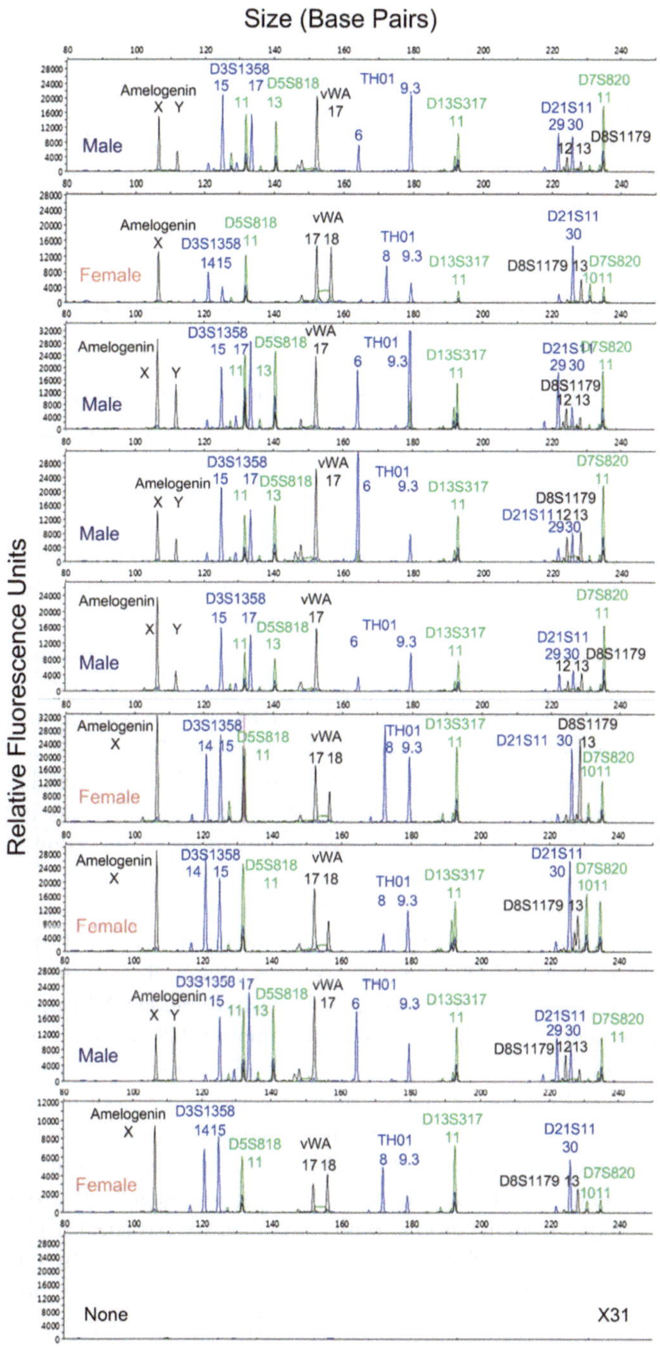

Fig. 4.5 Representative single-cell STR profiles typed from droplets containing a mixture of GM09947 female and GM09948 male lymphoid cells. The procedure comprises 30 cycles of emulsion PCR using droplets with 0.01 total cells and 0.9 beads per droplet on average at a female-to-male cell ratio of 1:1, followed by 25 cycles of secondary PCR. Reproduced from Ref. [64] with permission from the American Chemical Society

cellular profile. Each droplet averages approximately 1 GM09947 female cell, varying copy numbers (0.1, 1, 1.5, 2, and 3) of 9948 genomic DNA, and 0.9 microbeads. In total 70 secondary PCR assays are conducted for each condition at the bead concentration of 0.6 beads/reaction. Consequently, it is theoretically predicted that 63 % of beads will be linked with all 13 STR products from the single genome of GM09947 female cells, and the chances for 9948 male DNA amplicons to be immobilized on beads are determined by the input DNA amount. At a lower concentration of 0.1 copies/droplet, the percentages of beads positive with any of the 15 9948 DNA products and the 6 9948 DNA-specific products are 59 % and 26 %, respectively. Apart from that case, almost all beads should be positive with increasing DNA concentration, and 95, 98.89, 99.75, and 99.99 % of the beads are expected to be bound with at least one of the 9948 DNA-specific STR products for 1, 1.5, 2, and 3 copies/droplet, respectively. However, the number of amplicons on each bead will fluctuate due to the random encapsulation of DNA molecules within the droplets.

As expected, the cell/DNA mixtures yield mixed STR profiles of GM09947 female cells and 9948 male DNA and partial genotypes of 9948 genomic DNA fragments. The percentages of profiles containing all 13 allelic peaks from GM09947 cells and a given number (0, 1, 2, 3, 4, 5, or 6) of 9948-DNA-specific peaks in the total of profiles containing the entire set of GM09947 cellular peaks are plotted as a function of the number of 9948-DNA-specific peaks (Fig. 4.6). A low level of cell-free DNA does not significantly affect the STR genotypes, because the majority (82 %) of the profiles exclusively contain GM09947 cellular peaks and only 18 % of profiles have one additional 9948-DNA-specific peak at 0.1 copies/droplet. When the background DNA concentration increases to 1 copy/droplet, only mixed genotypes could be detected. Nevertheless, still 30 % of the profiles contain solely one additional 9948-DNA-specific peak, in which the contaminating DNA peaks can be easily excluded (Fig. 4.7). When the DNA concentration jumps to 1.5 copies/droplet, most GM09947 cellular profiles contain three to four 9948-DNA-specific peaks. It is clear that the increase in background DNA loading results in a rise in the percentages of profiles containing more 9948-DNA-specific peaks. At 2 copies/droplet, 32 % of the profiles contain six 9948-DNA-specific peaks, and two additional 9948-DNA-specific peaks were detected in 5.3 % of the profiles. With highly concentrated DNA contamination at 3 copies/droplet, 78 % of the profiles have all six contaminating peaks (Fig. 4.8). If sufficient reactions are performed in the second round of PCR amplification, the target cellular STR profile could be ultimately inferred from the profiles with fewer than six extra peaks. This is because the secondary PCR is performed at low gain (25 cycles) and the contaminating DNA is statistically distributed by allele template site. Therefore, our single-cell method is relatively insensitive to background DNA contamination.

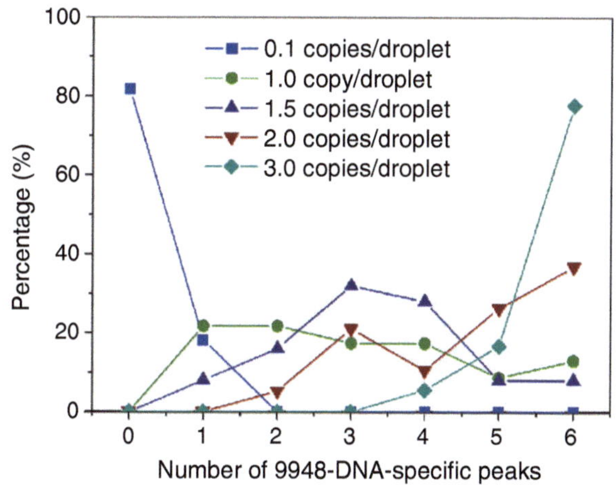

Fig. 4.6 Single-cell STR typing of the mixtures of GM09947 female lymphoid cells and cell-free 9948 male DNA. The percentage of profiles containing all GM09947 cellular peaks and a given number (0, 1, 2, 3, 4, 5, or 6) of 9948-DNA-specific peaks in all profiles containing all GM09947 cellular peaks is plotted against the number of 9948-DNA-specific peaks at varying 9948 DNA concentrations. The procedure comprises 30 cycles of emulsion PCR using droplets with 1 cell and 0.9 beads per droplet on average but varying concentrations (0.1, 1, 1.5, 2, and 3 copies/droplet) of DNA, followed by 25 cycles of secondary PCR. Reproduced from Ref. [64] with permission from the American Chemical Society

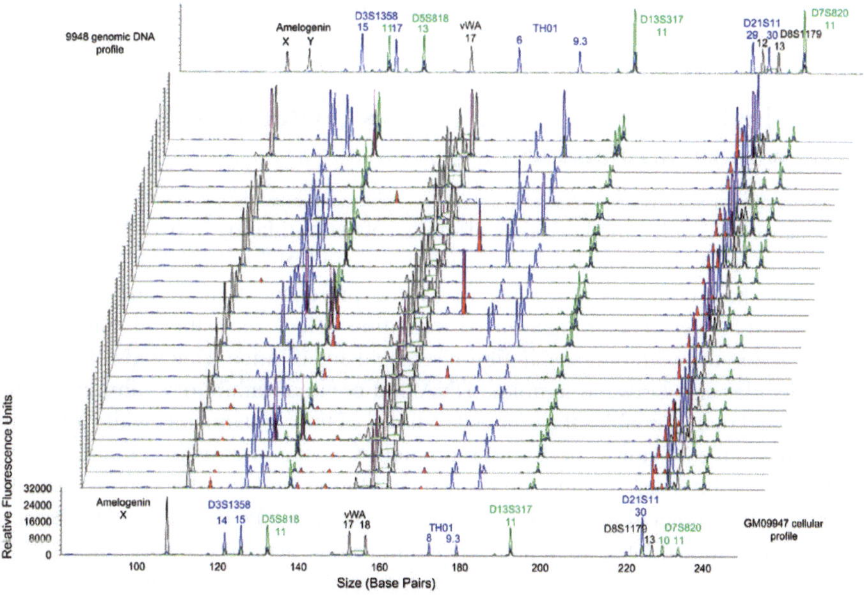

Fig. 4.7 Representative STR profiles typed from droplets containing a mixture of GM09947 female lymphoid cells and 9948 male DNA. The procedure comprises 30 cycles of emulsion PCR using droplets with one cell, one copy DNA, and 0.9 beads per droplet on average, followed by 25 cycles of secondary PCR. *Red*-filled peaks indicate randomly distributed 9948 DNA-specific peaks. Reproduced from Ref. [64] with permission from the American Chemical Society

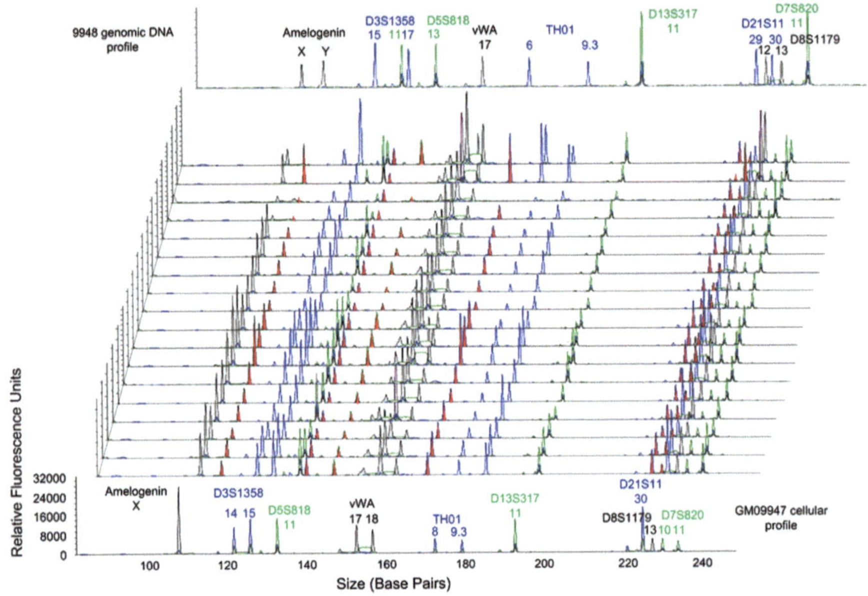

Fig. 4.8 Representative STR profiles typed from droplets containing a mixture of GM09947 female lymphoid cells and 9948 male DNA. The procedure comprises 30 cycles of emulsion PCR using droplets with one cell, three copies DNA, and 0.9 beads per droplet on average, followed by 25 cycles of secondary PCR. *Red*-filled peaks indicate randomly distributed 9948 DNA-specific peaks. Reproduced from Ref. [64] with permission from the American Chemical Society

4.6 Single-Cell STR Typing of Cells Contaminated with PCR Inhibitors

The agarose droplet method also offers a robust technology to overcome PCR inhibition resulting from interference with template DNA, primers, DNA polymerase, and their cofactors (e.g., Mg^{2+}) by the presence of inhibitors in forensic samples [68, 69]. The mitigation of inhibition effects is achieved through the combination of two processes: inhibitor removal and dilution. Due to the difference in molecular size between template DNA and common inhibitory substances, the agarose gel matrix enables the efficient removal of the contaminants while retaining DNA molecules. Prior to performing droplet PCR, the microgels containing both cells and inhibitors are stringently washed both before and after the cell lysis step to remove small molecules from the agarose. Moreover, the PCR inhibitory molecules are significantly diluted to a statistical limit along with the target cells during droplet generation, thereby minimizing the inhibitory effects for each droplet PCR reaction. Unlike dilution of bulk samples, the template DNA concentration in a droplet is not reduced, while only the inhibitor concentration is decreased.

Furthermore, the droplet format ensures that the inhibitors that are not removed will only be able to inhibit single droplet amplification. Thus, this method is practical to analyze low-abundance samples.

The ability of this technology to circumvent PCR inhibition is demonstrated with three commonly encountered PCR inhibitors (i.e., urea, tannic acid, and humic acid). Urea is a soluble small molecule (66.06 g/mol) and the major element of urine. Tannic acid is a larger polyphenol (1701.20 g/mol) commonly found in leather and some plant materials. Humic acid, primarily existing in soil, is a mixture of phenolic compounds with a broad molecular weight distribution from a few hundred to tens of thousands [70]. Droplet PCR is performed with droplets containing 0.6 GM09948 male lymphoid cells and 1 bead per droplet for 32 cycles of emulsion PCR, followed by 23 cycles of secondary PCR at 1 bead/reaction. Two tube-based bulk PCR controls are conducted for comparison using the identical PCR cycling condition. "Bulk PCR-1" is accomplished with 32 cycles of PCR from 0.5 ng 9948 genomic DNA prepared with a commercial kit. Bulk PCR-2 is a biphasic PCR process in which 0.5 μL of PCR products generated from 32 PCR cycles (i.e., bulk PCR-1) is used as the template for the second round of 23 PCR cycles in a new 12.5 μL reaction, eliminating the impact of total PCR cycle number.

The inhibitory level is quantified by the percentage of alleles called for each cell sample (not a single cell) (Fig. 4.9). For all three inhibitors, droplet PCR successfully yields full STR profiles in the range of tested concentrations. However, the bulk PCR reactions are completely inhibited in the presence of high concentrations of inhibitors, although full genotypes can be acquired at lower amounts of inhibitors. Partial profiles are observed in both bulk PCR methods with moderate inhibitor contamination. The introduction of the second-stage bulk PCR enhances the performance to some extent. With 100 mM of urea, the allele recovery rate of bulk PCR-2 (93 %) is higher than that of bulk PCR-1 (67 %). By increasing the urea concentration to 200 mM, biphasic bulk PCR recovers 60 % of the alleles, whereas no STR loci are detected with only one round of bulk PCR. With 200 ng/μL of tannic acid, bulk PCR-1 and bulk PCR-2 recover only 37 and 55 % of the alleles. With 50 ng/μL of humic acid, the allele recovery rates are 47 and 87 % for bulk PCR-1 and bulk PCR-2, respectively. When humic acid is increased to 100 ng/μL, only 23 % of the alleles are called with bulk PCR-2, and bulk PCR-1 is completely inhibited. Therefore, compared to standard bulk PCR, our agarose droplet-based method has 20, 10, and 16 times improved tolerance for PCR inhibition by urea, tannic acid, and humic acid, respectively.

In our agarose droplet-based method, all single-cell STR profiles obtained from samples contaminated with urea and tannic acid exhibit full genotypes, possibly due to their small size and solubility in wash buffers. However, allelic dropout occurs with humic acid concentrations higher than 200 ng/μL. A representative full profile and partial profiles acquired in the presence of 500 ng/μL humic acid are shown in Fig. 4.10. It is observed that larger STR loci more easily drop out. Additionally, the percentage of complete STR profiles among all obtained profiles decreases with

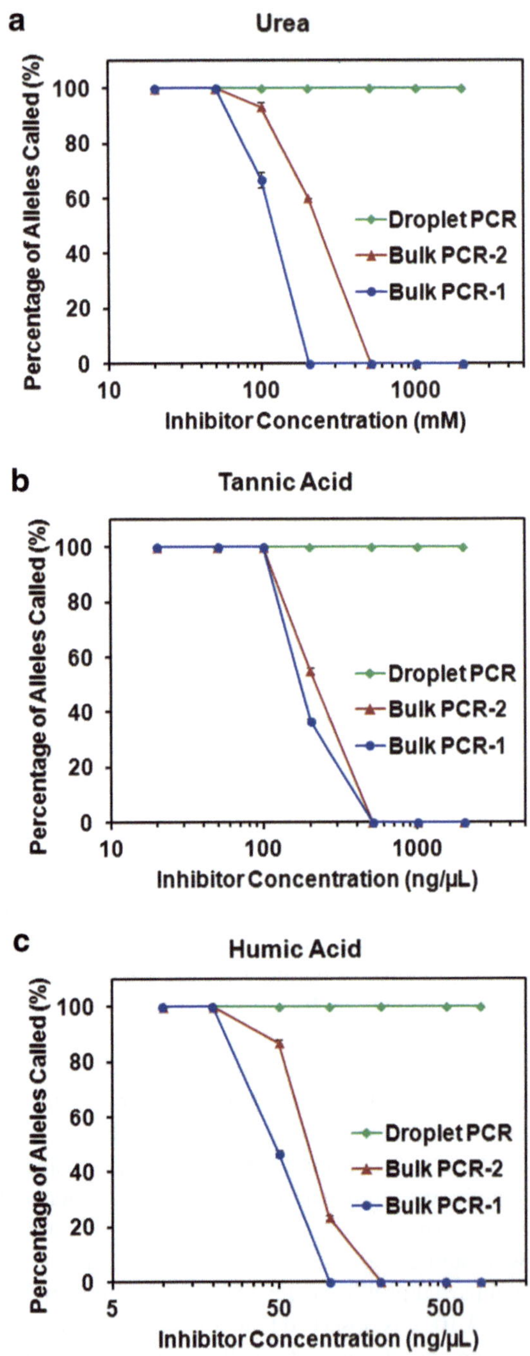

Fig. 4.9 Single-cell STR typing of GM09948 male lymphoid cells contaminated with (**a**) urea, (**b**) tannic acid, and (**c**) humic acid. The percentage of alleles called in a sample is plotted against the concentrations of PCR inhibitors under three different amplification conditions. In droplet PCR, the procedure comprises 32 cycles of emulsion PCR using droplets with 0.6 cells and 1 bead per droplet on average but varying concentrations of PCR inhibitors, followed by 23 cycles of secondary PCR. In bulk PCR-1, 32 cycles of bulk PCR are conducted with 0.5 ng 9948 male genomic DNA in 12.5 μL reactions. In bulk PCR-2, 0.5 μL PCR products from bulk PCR-1 serve as the DNA template for the second round of 23 cycles of bulk PCR in 12.5 μL reactions. Each data point is based on three independent assays. Reproduced from Ref. [65] with permission from Elsevier

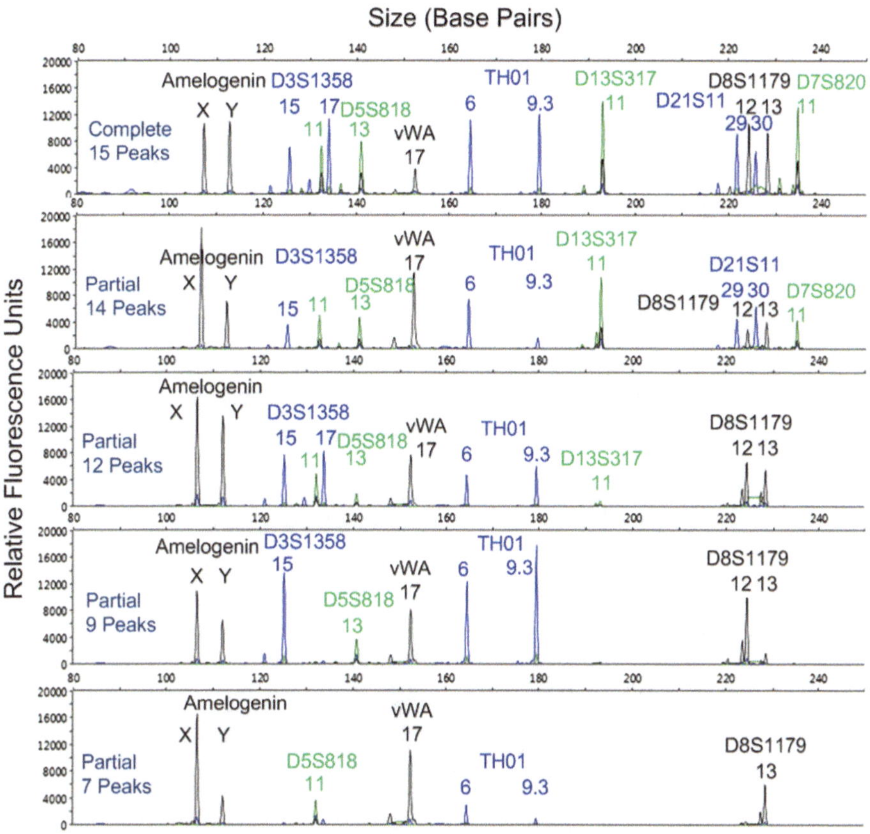

Fig. 4.10 Representative single-cell STR profiles typed from droplets containing GM09948 male lymphoid cells contaminated with humic acid. The procedure comprises 32 cycles of emulsion PCR using droplets with 0.6 cells and one bead per droplet on average at 500 ng/μL humic acid, followed by 23 cycles of secondary PCR. Reproduced from Ref. [65] with permission from Elsevier

more input inhibitor (Fig. 4.11a). As an example, the percentage changes from 72 to 57 % when the concentration increases from 200 to 500 ng/μL. Nevertheless, the percentage of alleles called at the single-cell level did not change significantly (91 %/70 % for 200 ng/μL and 91 %/81 % for 500 ng/μL) (Fig. 4.11a). In contrast, increasing the concentration to 800 ng/μL leads to a drastic drop in both rates (Fig. 4.11). The different effects of humic acid from other inhibitors are attributed to its complex nature of heterogeneous and polydisperse chemicals, and possible direct binding to template DNA molecules.

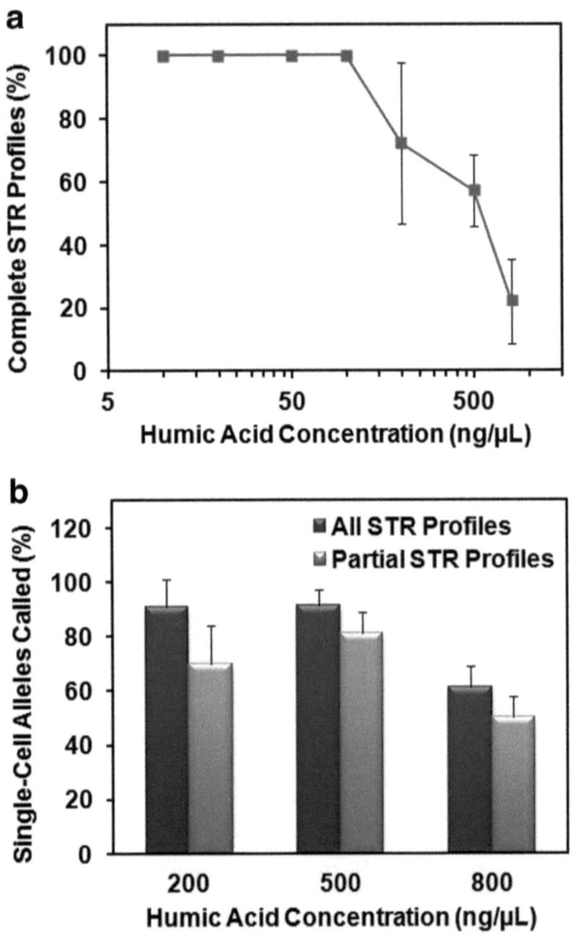

Fig. 4.11 Single-cell STR typing of GM09948 male lymphoid cells contaminated with humic acid. (**a**) The percentage of complete STR profiles. in all positive profiles is plotted against the concentration of humic acid. (**b**) The percentage of alleles called at the single-cell level is plotted against the concentration of humic acid. It is obtained by dividing the number of peaks in all (including full and partial) STR profiles (*black bars*) or in partial STR profiles (*grey bars*) by the total peak number (15) for each genotype, and then calculating the mean. The procedure comprises 32 cycles of droplet PCR using droplets with 0.6 cells and one bead per droplet on average but varying concentrations of humic acid, followed by 23 cycles of secondary PCR. Each data point is based on (**a**) three independent assays and (**b**) 20–30 single-cell STR profiles. Reproduced from Ref. [65] with permission from Elsevier

4.7 Conclusions

In conclusion, our technology combines the advantages of droplet microfluidics, agarose chemistry, and emulsion PCR for forensic STR typing at the single-cell level. The sensitivity and throughput are greatly improved owing to the ultralow

PCR reaction volume and the ability to manipulate the droplets in a highly parallel fashion. The permeability of agarose allows the free introduction of reagents for multiple biochemical steps and efficient removal of PCR inhibitors while entrapping the macromolecules (e.g., genomic DNA). Compared to other droplet-based PCR methods typically yielding 1–2 products, the PCR multiplexing level is increased to nine due to the utilization of beads functionalized with multiplexed primers. Mixed cell populations, mixtures of cells and genomic DNA from different individuals, and PCR inhibitor-contaminated cell samples have been successfully typed, and the digital amplification simplifies the mixture interpretation process. The technology paves the way for dealing with the casework involving low-abundance forensic samples with multiple contributors. The ability to detect multiple genetic loci in parallel from single cells will also be valuable in exploring cellular heterogeneity in phenotypically identical cells and probing rare cells from complex biological mixtures in biotechnological and medical applications.

Acknowledgements This work was supported by the US Department of Justice Office of Justice (2009-DN-BX-K180) and by the Mathies Royalty Fund (UCB-007).

Conflict of Interest R.A.M. has financial interest in IntegenX Inc. that is developing forensic identification technologies.

References

1. Edwards A et al (1991) DNA typing and genetic mapping with trimeric and tetrameric tandem repeats. Am J Hum Genet 49(4):746–756
2. Hammond HA et al (1994) Evaluation of 13 short tandem repeat loci for use in personal identification applications. Am J Hum Genet 55(1):175–189
3. Jobling MA, Gill P (2004) Encoded evidence: DNA in forensic analysis. Nat Rev Genet 5 (10):739–751
4. Butler JM (2006) Genetics and genomics of core short tandem repeat loci used in human identity testing. J Forensic Sci 51(2):253–265
5. Horsman-Hall KM et al (2009) Development of STR profiles from firearms and fired cartridge cases. Forensic Sci Int Genet 3(4):242–250
6. van Oorschot RA, Ballantyne KN, Mitchell RJ (2010) Forensic trace DNA: a review. Investig Genet 1(1):14
7. Torres Y et al (2003) DNA mixtures in forensic casework: a 4-year retrospective study. Forensic Sci Int 134(2–3):180–186
8. Alaeddini R (2012) Forensic implications of PCR inhibition—a review. Forensic Sci Int Genet 6(3):297–305
9. Ladd C et al (2001) Interpretation of complex forensic DNA mixtures. Croat Med J 42 (3):244–246
10. Budowle B et al (2009) Mixture interpretation: defining the relevant features for guidelines for the assessment of mixed DNA profiles in forensic casework. J Forensic Sci 54(4):810–821
11. Shutler GG et al (1999) Removal of a PCR inhibitor and resolution of DNA STR types in mixed human-canine stains from a five year old case. J Forensic Sci 44(3):623–626
12. Greenspoon SA et al (2004) Application of the BioMek 2000 Laboratory Automation Workstation and the DNA IQ System to the extraction of forensic casework samples. J Forensic Sci 49(1):29–39

13. Ye J et al (2004) A simple and efficient method for extracting DNA from old and burned bone. J Forensic Sci 49(4):754–759
14. Kemp BM, Monroe C, Smith DG (2006) Repeat silica extraction: a simple technique for the removal of PCR inhibitors from DNA extracts. J Archaeol Sci 33(12):1680–1689
15. Maher N et al (2001) Magnetic bead capture eliminates PCR inhibitors in samples collected from the airborne environment, permitting detection of Pneumocystis carinii DNA. Appl Environ Microbiol 67(1):449–452
16. Wang J, McCord B (2011) The application of magnetic bead hybridization for the recovery and STR amplification of degraded and inhibited forensic DNA. Electrophoresis 32 (13):1631–1638
17. Broemeling DJ et al (2008) An instrument for automated purification of nucleic acids from contaminated forensic samples. JALA 13(1):40–48
18. Matheson CD et al (2009) Technical note: removal of metal ion inhibition encountered during DNA extraction and amplification of copper-preserved archaeological bone using size exclusion chromatography. Am J Phys Anthropol 140(2):384–391
19. Grubb JC et al (2010) Implementation and validation of the Teleshake unit for DNA IQ robotic extraction and development of a large volume DNA IQ method. J Forensic Sci 55(3):706–714
20. Hudlow WR et al (2011) The NucleoSpin(R) DNA clean-up XS kit for the concentration and purification of genomic DNA extracts: an alternative to microdialysis filtration. Forensic Sci Int Genet 5(3):226–230
21. Schmedes S et al (2013) Effective removal of co-purified inhibitors from extracted DNA samples using synchronous coefficient of drag alteration (SCODA) technology. Int J Legal Med 127(4):749–755
22. Faber KL, Person EC, Hudlow WR (2013) PCR inhibitor removal using the NucleoSpin (R) DNA clean-up XS kit. Forensic Sci Int Genet 7(1):209–213
23. Eckhart L et al (2000) Melanin binds reversibly to thermostable DNA polymerase and inhibits its activity. Biochem Biophys Res Commun 271(3):726–730
24. Scipioni A et al (2008) Detection and quantification of human and bovine noroviruses by a TaqMan RT-PCR assay with a control for inhibition. Mol Cell Probes 22(4):215–222
25. Moretti T, Koons B, Budowle B (1998) Enhancement of PCR amplification yield and specificity using AmpliTaq Gold DNA polymerase. Biotechniques 25(4):716–722
26. Kermekchiev MB et al (2009) Mutants of Taq DNA polymerase resistant to PCR inhibitors allow DNA amplification from whole blood and crude soil samples. Nucleic Acids Res 37(5): e40
27. Eilert KD, Foran DR (2009) Polymerase resistance to polymerase chain reaction inhibitors in bone. J Forensic Sci 54(5):1001–1007
28. Hedman J et al (2011) Applying a PCR inhibitor tolerant DNA polymerase blend in forensic DNA profiling. Forensic Sci Int Genet Suppl 3(1):e349–e350
29. Barbaro A, Cormaci P, Teatino A (2008) Use of "AnyDirect PCR buffer" for PCR amplification of washed bloodstains: a case report. Forensic Sci Int Genet Suppl 1(1):11–12
30. Kreader CA (1996) Relief of amplification inhibition in PCR with bovine serum albumin or T4 gene 32 protein. Appl Environ Microbiol 62(3):1102–1106
31. Abu Al-Soud W, Radstrom P (2000) Effects of amplification facilitators on diagnostic PCR in the presence of blood, feces, and meat. J Clin Microbiol 38(12):4463–4470
32. Gill P, Jeffreys AJ, Werrett DJ (1985) Forensic application of DNA "fingerprints". Nature 318 (6046):577–579
33. Yoshida K et al (1995) The modified method of two-step differential extraction of sperm and vaginal epithelial cell DNA from vaginal fluid mixed with semen. Forensic Sci Int 72(1):25–33
34. Chen J et al (1998) A physical method for separating spermatozoa from epithelial cells in sexual assault evidence. J Forensic Sci 43(1):114–118
35. Schoell WM et al (1999) Separation of sperm and vaginal cells with flow cytometry for DNA typing after sexual assault. Obstet Gynecol 94(4):623–627

36. Di Nunno N et al (2003) DNA identification of sperm cells collected and sorted by flow cytometry. Am J Forensic Med Pathol 24(3):254–270
37. Bennett MR, Hasty J (2009) Microfluidic devices for measuring gene network dynamics in single cells. Nat Rev Genet 10(9):628–638
38. Zare RN, Kim S (2010) Microfluidic platforms for single-cell analysis. Annu Rev Biomed Eng 12:187–201
39. Joensson HN, Andersson Svahn H (2012) Droplet microfluidics—a tool for single-cell analysis. Angew Chem Int Ed 51(49):12176–12192
40. Findlay I et al (1997) DNA fingerprinting from single cells. Nature 389(6651):555–556
41. Bruck S et al (2011) Single cells for forensic DNA analysis—from evidence material to test tube. J Forensic Sci 56(1):176–180
42. Pereira J et al (2012) MtDNA typing of single-sperm cells isolated by micromanipulation. Forensic Sci Int Genet 6(2):228–235
43. Di Martino D et al (2004) Laser microdissection and DNA typing of cells from single hair follicles. Forensic Sci Int 146(Suppl):S155–S157
44. Sanders CT et al (2006) Laser microdissection separation of pure spermatozoa from epithelial cells for short tandem repeat analysis. J Forensic Sci 51(4):748–757
45. Li CX et al (2011) DNA profiling of spermatozoa by laser capture microdissection and low volume-PCR. PLoS One 6(8):e22316
46. Unger MA et al (2000) Monolithic microfabricated valves and pumps by multilayer soft lithography. Science 288(5463):113–116
47. Song H, Chen DL, Ismagilov RF (2006) Reactions in droplets in microfluidic channels. Angew Chem Int Ed 45(44):7336–7356
48. Theberge AB et al (2010) Microdroplets in microfluidics: an evolving platform for discoveries in chemistry and biology. Angew Chem Int Ed 49(34):5846–5868
49. He M et al (2005) Selective encapsulation of single cells and subcellular organelles into picoliter- and femtoliter-volume droplets. Anal Chem 77(6):1539–1544
50. Huebner A et al (2007) Quantitative detection of protein expression in single cells using droplet microfluidics. Chem Commun 12:1218–1220
51. Clausell-Tormos J et al (2008) Droplet-based microfluidic platforms for the encapsulation and screening of Mammalian cells and multicellular organisms. Chem Biol 15(5):427–437
52. Brouzes E et al (2009) Droplet microfluidic technology for single-cell high-throughput screening. Proc Natl Acad Sci U S A 106(34):14195–14200
53. Baret JC et al (2010) Quantitative cell-based reporter gene assays using droplet-based microfluidics. Chem Biol 17(5):528–536
54. Juul S et al (2011) Detection of single enzymatic events in rare or single cells using microfluidics. ACS Nano 5(10):8305–8310
55. Gu SQ et al (2011) Multifunctional picoliter droplet manipulation platform and its application in single cell analysis. Anal Chem 83(19):7570–7576
56. Shi X et al (2011) Real-time PCR of single bacterial cells on an array of adhering droplets. Lab Chip 11(13):2276–2281
57. Rane TD et al (2012) Droplet microfluidics for amplification-free genetic detection of single cells. Lab Chip 12(18):3341–3347
58. Zhang H et al (2012) Massively parallel single-molecule and single-cell emulsion reverse transcription polymerase chain reaction using agarose droplet microfluidics. Anal Chem 84 (8):3599–3606
59. Zhu Z et al (2012) Highly sensitive and quantitative detection of rare pathogens through agarose droplet microfluidic emulsion PCR at the single-cell level. Lab Chip 12 (20):3907–3913
60. Kumaresan P et al (2008) High-throughput single copy DNA amplification and cell analysis in engineered nanoliter droplets. Anal Chem 80(10):3522–3529
61. Zeng Y et al (2010) High-performance single cell genetic analysis using microfluidic emulsion generator arrays. Anal Chem 82(8):3183–3190

62. Novak R et al (2011) Single-cell multiplex gene detection and sequencing with microfluidically generated agarose emulsions. Angew Chem Int Ed 50(2):390–395
63. Shuga J et al (2013) Single molecule quantitation and sequencing of rare translocations using microfluidic nested digital PCR. Nucleic Acids Res 41(16):e159
64. Geng T, Novak R, Mathies RA (2014) Single-cell forensic short tandem repeat typing within microfluidic droplets. Anal Chem 86(1):703–712
65. Geng T, Mathies RA (2015) Minimizing inhibition of PCR-STR typing using digital agarose droplet microfluidics. Forensic Sci Int Genet 14:203–209
66. Vandewoestyne M et al (2013) Presence and potential of cell free DNA in different types of forensic samples. Forensic Sci Int Genet 7(2):316–320
67. Quinones I, Daniel B (2012) Cell free DNA as a component of forensic evidence recovered from touched surfaces. Forensic Sci Int Genet 6(1):26–30
68. Sutlovic D et al (2008) Interaction of humic acids with human DNA: proposed mechanisms and kinetics. Electrophoresis 29(7):1467–1472
69. Opel KL, Chung D, McCord BR (2010) A study of PCR inhibition mechanisms using real time PCR. J Forensic Sci 55(1):25–33
70. Shin HS, Monsallier JM, Choppin GR (1999) Spectroscopic and chemical characterizations of molecular size fractionated humic acid. Talanta 50(3):641–647

Chapter 5
Microfluidic Approaches to Fluorescence In Situ Hybridization (FISH) for Detecting RNA Targets in Single Cells

Robert J. Meagher and Meiye Wu

Abstract Fluorescence in situ hybridization (FISH) is a powerful molecular technique in cell biology and microbiology for detection and localization of a nucleic acid target within an intact cell or chromosome spread, based on hybridization of a fluorescently labeled nucleic acid "probe" to its complementary target. In some instances, FISH analysis is performed on intact samples—whether thin tissue sections, or environmental samples, allowing the nucleic acid target to be localized in context with other cells. FISH evolved from in situ hybridization (ISH) techniques utilizing radiolabeled probes. By comparison, FISH typically utilizes small-molecule fluorophores. This labeling approach eliminates the hazards associated with radioactivity, and allows analysis with common laboratory instrumentation, including epifluorescence or laser-scanning confocal microscopes, or flow cytometers. Unlike PCR, sequencing, or most other nucleic acid analysis methods, FISH is fundamentally a single-cell measurement technique. Whether using imaging or flow cytometry as a readout, the signals from FISH are detected and analyzed on a cell-by-cell basis, affording a unique capability for studying rare events or heterogeneity within a population. Coupling of FISH with flow sorters also affords a unique capability for enriching cell or chromosome populations or even isolating single cells based on presence of specific nucleic acid targets [7, 10, 43].

Keywords Lab on a chip • Molecular diagnostics • Gene expression • Genetic analysis • Single-cell analysis • Nucleic acid hybridization • Flow cytometry

R.J. Meagher, Ph.D. (✉) • M. Wu, M.A., Ph.D.
Biotechnology and Bioengineering Department, Sandia National Laboratories, MS 9291, P.O. Box 969, Livermore, CA 94551, USA
e-mail: rmeaghe@sandia.gov; meiwu@sandia.gov

© Springer International Publishing Switzerland 2016

C. Lu, S.S. Verbridge (eds.), *Microfluidic Methods for Molecular Biology*,
DOI 10.1007/978-3-319-30019-1_5

5.1 Introduction

Fluorescence in situ hybridization (FISH) is a powerful molecular technique in cell biology and microbiology for detection and localization of a nucleic acid target within an intact cell or chromosome spread, based on hybridization of a fluorescently labeled nucleic acid "probe" to its complementary target. In some instances, FISH analysis is performed on intact samples—whether thin tissue sections, or environmental samples, allowing the nucleic acid target to be localized in context with other cells. FISH evolved from in situ hybridization (ISH) techniques utilizing radiolabeled probes. By comparison, FISH typically utilizes small-molecule fluorophores. This labeling approach eliminates the hazards associated with radioactivity, and allows analysis with common laboratory instrumentation, including epifluorescence or laser-scanning confocal microscopes, or flow cytometers. Unlike PCR, sequencing, or most other nucleic acid analysis methods, FISH is fundamentally a single-cell measurement technique. Whether using imaging or flow cytometry as a readout, the signals from FISH are detected and analyzed on a cell-by-cell basis, affording a unique capability for studying rare events or heterogeneity within a population. Coupling of FISH with flow sorters also affords a unique capability for enriching cell or chromosome populations or even isolating single cells based on presence of specific nucleic acid targets [7, 10, 43].

FISH can be used to detect both DNA and RNA targets within cells. FISH for DNA targets is used both for research and clinical applications, including creation of physical maps of chromosomes, copy number analysis of genes, and other applications in cytogenetics. FISH for RNA targets is more typically used as a research technique, e.g., to study gene expression at the single-cell level, or detection of small RNA within cells. FISH for ribosomal RNA (rRNA) as well as functional gene (mRNA) targets is a long-standing technique in microbial ecology for analyzing uncultured microbial communities. Bacterial rRNA and mRNA FISH has also found application in clinical microbiology, for rapid identification of bacteria from positive blood cultures and other clinical specimens [21, 28].

Labeled probes for FISH may be long (>100 bp) double-stranded DNA probes (sometimes referred to as polynucleotide probes), or short oligonucleotides (typically <30 bases) of DNA or RNA. Nucleic acid analogs such as peptide nucleic acids (PNA), locked nucleic acids (LNA), or morpholino oligomers can also be incorporated into FISH probes. Polynucleotide probes often bear multiple fluorescent labels along the length of the probe, with labels incorporated enzymatically (e.g., by PCR or nick translation using dye-modified dNTPs), whereas oligonucleotide probes are typically labeled at one or both ends, with dye labels incorporated during solid-phase synthesis. Polynucleotide probes are most typically used for labeling DNA targets, whereas oligonucleotide probes can be used for either DNA or RNA targets. A variant approach to dye-labeled probes is to use probes labeled with a hapten (e.g., biotin) or an enzyme (e.g., horseradish peroxidase), allowing post-hybridization labeling with a variety of signal amplification techniques, as will be discussed subsequently.

Whether for DNA or RNA targets within eukaryotic or prokaryotic cells, FISH protocols share certain steps in common. These generally include fixation and permeabilization of cells, hybridization with probes, multiple wash steps with increased stringency to remove unbound or nonspecifically bound probes, in some cases a post-hybridization signal amplification with additional wash steps, and finally analysis by microscopy or flow cytometry. The specific protocols for sample prep, hybridization, signal amplification, and analysis are often quite different for DNA or RNA targets. Regardless of the specific protocol, FISH is inherently a complex, multi-step procedure with expensive reagents. FISH is thus a natural protocol to miniaturize and automate onto microfluidic devices. Even conventional FISH (on a microscope slide) is "microfluidic" to the extent that protocols frequently utilize simple hybridization chambers, gaskets, or frame seals adhered to microscope slides. These simple devices reduce the volume of reagents used for hybridization, although the protocol is still manual to the extent that reagent addition and removal are performed using pipettes. Microfluidic flow control offers the possibility of automating multiple steps of the protocol, as well as potentially enhanced mass transfer to improve the efficiency of the experiment. Our review will focus primarily on the use of microfluidics in RNA-targeted FISH, along with a review of techniques used for RNA FISH that may be amenable to microfluidic automation. Microfluidic approaches to DNA FISH will be reviewed to the extent that similar techniques might be amenable to use with RNA FISH. For a more general treatment of RNA FISH in eukaryotic cells, we recommend the reviews by Itzkovits and van Oudenaarden [9] and Crosetto et al. [4], as well as the review by Urbanek et al. [36] specifically for small RNA FISH. For a general treatment of advanced FISH techniques in microbiology we recommend the review by Amman and Fuchs [1].

5.2 FISH Methods and Translation to Microfluidics

FISH labeling has been applied to a wide variety of cell types including both eukaryotes and prokaryotes, for both DNA and RNA targets. We review below some of the methods most applicable to labeling RNA targets within mammalian cells as well as bacteria, and provide examples where these protocols have been translated to microfluidic devices.

5.2.1 Microfluidic FISH and Sample Preparation for Mammalian mRNA and smRNA FISH

Fluorescent in situ hybridization for RNAs in mammalian cells can be performed on either solid tissue sections mounted on slides for microscopy analysis, or in suspension cells for flow cytometry analysis (flow-FISH). The flow cytometry-based RNA-FISH sample preparation methods are the most amenable to adaptation to microfluidic miniaturization due to the fact that suspension cells are easier to manipulate in narrow fluidic channels.

The key to successfully preparing cells for flow-FISH analysis is the optimization of cellular integrity, probe design, probe stability, and signal-to-noise ratio. Cellular integrity can be difficult to maintain in conventional bench-scale protocols due to the cell lysis that result from repeated centrifugation and aspiration steps required for lengthy FISH protocols. Cell morphology can also be affected due to fixation and permeabilization steps, and the accessibility of target RNAs can be blocked by cellular components such as protein complexes and organelles. There are several recent advances in probe chemistry and commercially available flow-FISH kits that address the technical issues that surround RNA-FISH in mammalian cells. A typical RNA flow-FISH protocol contains several steps: first, fresh cells are stained with fixable viability dyes before fixation and permeabilization. The optimal fixation and permeabilization agent and duration need to be determined empirically for each cell type. Typically fixation is performed using 1–8 % paraformaldehyde, for at least 5–10 min. Methanol can be used to both fix and permeabilize cells in a single step, but methanol treatment will destroy cell surface antigens. Methanol treatment should thus be avoided if multiplexed detection of protein and RNA is desired. For multiplexing of cell surface and intracellular protein staining and RNA FISH, it is advisable to perform cell surface protein immunostaining prior to fixation, and use mild permeabilization agents such as Triton X-100 or saponin after cell surface immunostaining. After the fix and permeabilization step, the cells go through FISH hybridization and signal amplification steps, and the design of FISH probes along with the method of amplification varies greatly depending on the type of RNA target as well as the abundance of RNA signal. Triton X-100 is the permeabilization agent of choice when performing immunostaining or FISH in adherent cells or mounted tissue slices, but with suspension cells, permeabilization with Triton X-100 followed by centrifugation will lead to cell lysis. Saponin as a permeabilization agent is widely used for sample preparation for flow cytometry analysis, but saponin permeabilization is reversible. Once saponin is introduced in the sample preparation protocol, it must be included in every subsequent step to ensure that the cells remain permeable throughout the protocol. One very practical advantage of adapting flow-FISH into microfluidic systems is the elimination of centrifugation steps, which allows the use of a single Triton X-100 treatment step to gently and effectively permeabilize the cells while preserving the cell surface antigens for multiplexed immunostaining and FISH detection.

Aside from preserving cell integrity and optimizing permeabilization conditions, the specificity of the oligonucleotide probe used for flow-FISH is crucial in the success of FISH-based detection. Advancements in the novel probe and signal amplification chemistries have greatly enhanced the functionality of RNA FISH assays. Modified nucleic acids, specifically locked nucleic acid (LNA), have become increasingly popular in flow-FISH applications [31]. LNAs are synthetic nucleic acid analogs that contain a methylene bridge between the 2′-oxygen and 4′-carbon, resulting in a strained or "locked" C3′-*endo* conformation [29]. The constrained structure of the LNA provides stability to Watson-Crick pairs by either pre-organization or increased base stacking. Each LNA incorporated into an oligonucleotide will increase the melting temperature of the probe/target duplex by 1–8 °C to DNA and 2–10 °C to RNA [38], and therefore drastically increasing the signal-to-noise ratio of FISH assays. LNA probes have been used to detect difficult-to-access targets such as microRNAs or miRNAs [19], single-nucleotide polymorphisms [13], and viral RNAs [27]. LNA probes have further been combined with rolling circle amplification (RCA) to achieve single-transcript detection sensitivity [12]. The combination of LNA containing probes and RCA has been demonstrated in microfluidic system to detect microRNAs and proteins in single cells [40], making it possible to study the spatiotemporal resolution of miRNA expression in its native cellular context, as illustrated in Fig. 5.1. While technical advances in probe and detection chemistries have greatly enhanced the specificity and sensitivity of RNA-FISH assays, the cost of proprietary FISH reagents can quickly amount to unsustainable levels. The savings offered by microfluidic scale RNA-FISH sample preparation methods become very substantial, in most cases over 90–95 % cost reduction, making microfluidic based RNA-FISH analysis an attractive avenue for further development.

5.2.2 *Microfluidic FISH for Microbial RNA*

FISH has long been utilized as a culture-independent technique for analyzing microbial communities. The 16S rRNA has been widely adopted as a target molecule for FISH, usually with dye-labeled oligonucleotide probes, for two reasons. Firstly, the 16S rRNA has been widely adopted for understanding microbial phylogeny, and extensive databases of 16S rRNA sequences have been collected, allowing design of FISH probes targeting groups of bacteria at the genus or species level. Secondly, the rRNA is the most abundant nucleic acid species within most cells, with several hundred up to tens of thousands of copies, meaning that there is a large multiplicity of labeling sites within the small confines of a microbial cell (on the order of 1 fL volume). The large number of rRNA targets allows for a bright signal, if a high efficiency of labeling can be achieved. rRNA-directed microbial FISH is summarized schematically in Fig. 5.2.

Environmental microbes may have low copy numbers of rRNA, particularly for slowly growing microbes in nutrient-poor environments, which presents a challenge for achieving bright labeling when using simple dye-labeled oligonucleotide probes. Labeling mRNA from functional genes in bacteria is also highly challenging, due to the inherently low copy number of most bacterial mRNA, which are transcribed in bursts and degrade quickly. mRNA copy number for many genes is <10 per cell, and may be zero in the majority of cells at any given time, for many genes. Both for the case of bacterial mRNA and for bacterial species with low copy number of rRNA, a variety of signal amplification techniques have been employed. The most popular of these is catalyzed reporter deposition (CARD), using

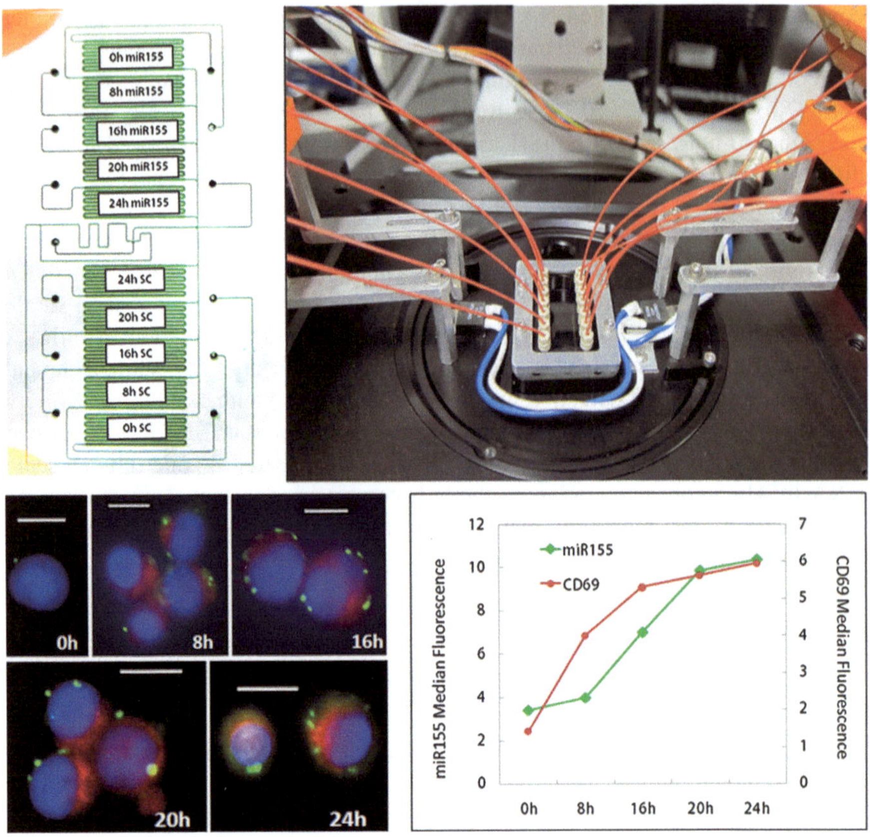

Fig. 5.1 miRNA flow-FISH was performed in a 10-chamber microfluidic chip (*top left*), where T cells were stimulated for 0, 8, 16, 20, and 24 h with phorbol myristate acetate (PMA) and ionomycin. The resultant upregulation of miR155 was detected using LNA-FISH coupled to rolling circle amplification. The cells and reagents were moved on and off by the chip by positive pressure through tubing connected to electronic valves (*top right*). The multiplexed detection of miR155 (*green*) with early T cell activation cell surface marker CD69 (*red*) was demonstrated by imaging as well as flow cytometry (*bottom*). Scale bar 5 µm. Adapted from [40] with permission of the authors in accordance with the Creative Commons Attribution (CC BY) license

fluorescently labeled tyramide signal amplification (TSA), in which an enzyme (horseradish peroxidase, HRP) results in precipitation or covalent linkage of dye labels in the vicinity of the enzyme [14, 23]. In the microbial FISH literature, the technique is frequently called "CARD-FISH," and less frequently "TSA-FISH." CARD-FISH requires an HRP-labeled FISH probe, which can be achieved either by direct covalent conjugation of HRP to the FISH probe or by use of a biotinylated probe followed by addition of streptavidin-HRP conjugate. In either event, the microbial cell membrane must be sufficiently permeabilized to allow passage of the bulky enzyme, or even bulkier streptavidin-enzyme conjugate without completely lysing the cell and releasing its contents or losing the cell morphology.

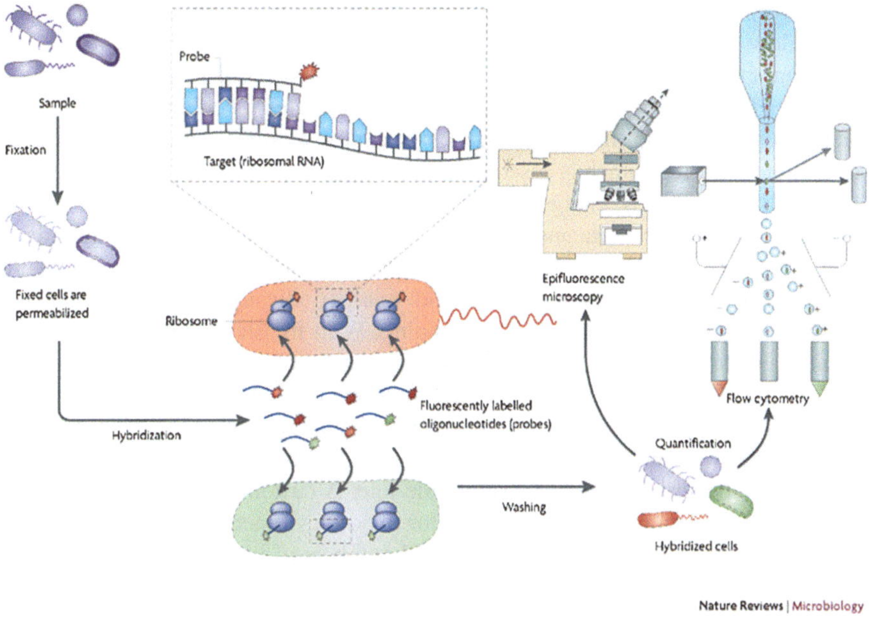

Fig. 5.2 Conceptual illustration of rRNA-directed microbial FISH, illustrating detection by either flow cytometry (flow-FISH) or epifluorescence microscopy. Adapted by permission from Macmillan Publishers Ltd, from Ref. [1], copyright 2008

One common approach is to embed the microbial cell in a thin layer of agarose. CARD-FISH is difficult to perform in suspension, being almost exclusively performed on microscope slides or filter membranes. The large number of steps involved (with additional signal amplification, washes, and blocking steps, beyond the conventional FISH hybridization) makes CARD-FISH a more challenging protocol than conventional FISH. We are unaware of reports demonstrating automation of CARD-FISH on a microfluidic platform, but such an approach could conceivably save both labor and costly reagents on this procedure.

Other approaches to achieve bright labeling for bacterial mRNA include the use of multiple probes per target, such as Stellaris FISH which uses up to 48 probes hybridizing along the length of the target RNA [25, 33], or use of multiply labeled polynucleotide probes [45]. In situ PCR or in situ LAMP, followed by hybridization of labeled probes, has also been reported with bacterial cells [18]. However, we are aware of relatively few applications of microfluidic technologies to microbial FISH, either mRNA or rRNA. The earliest example we are aware of is the use of the Agilent Bioanalyzer 2100 flow cytometry chip, to perform flow analysis of FISH labeled marine microbes [6], or FISH-labeled *Pseudomonas* cells in milk [42]. However, in these examples, FISH hybridization was performed in a conventional protocol using microbes in suspension, or enriched on a filter membrane, followed by resuspension of microbes and introduction into the Bioanalyzer chip.

In this respect, the Bioanalyzer chip simply functions as a surrogate for a conventional flow cytometer.

In an example of a more fully integrated analysis, Liu et al. demonstrated on-chip FISH labeling and analysis of defined mixtures of cultured microbes, as well as environmental microbes, with on-chip flow cytometry analysis [15]. This microfluidic device, which is described in Fig. 5.3, relied upon electrokinetic forces to mobilize negatively charged bacterial cells and oligonucleotide FISH probes within the device. Bacteria and probes were concentrated together in a "hybridization chamber" bounded by two photopatterned semipermeable hydrogel membranes. The direction of the electric field was periodically reversed to maintain the cells in suspension and prevent clumping. With the short length scales of the device, a diffusion-based washing protocol was utilized, relying upon the relatively fast diffusion of excess probe out of the hybridization chamber, while the much slower diffusing bacteria remained essentially stationary. Following this wash step, the bacteria were mobilized to a detection zone, where a "pinching" electric field focused the bacterial cells into a single-file line. A custom-built detector detected both forward scatter and fluorescence, for a flow cytometry-like detection. This device was capable of analyzing very small numbers of input cells (100–1000 per sample), which is well suited to environmental samples with low microbial density, presumably following a concentration step (e.g., by centrifugation or filtration). The device was used to detect *Pseudomonas* species bacteria in groundwater collected from a chemically contaminated waste site, as part of a bioremediation study.

FISH detection on a microfluidic device using an 18S rRNA-directed probe was combined with immunostaining for detection of cysts of *Giardia lamblia*, a protozoan parasite [44]. These cysts have characteristic dimension of approximately 10 μm, an order of magnitude larger than the microbes in the examples described above, and presumably with correspondingly larger rRNA content. An array of micropillars acts as a filter to trap the cysts from a flowing suspension, while a coarse pre-filter upstream of the micropillars prevents clogging of the device with debris. Dual-mode labeling was performed by flushing the device successively with solutions containing the Cy3-labeled FISH probe and a quantum dot-labeled antibody, with brief intermediate wash steps. The on-chip labeling required less than 30 min, with washing steps significantly accelerated relative to conventional tube-based processing which requires multiple centrifugation and pipetting steps. The FISH probe and quantum dot-labeled antibody were detected by imaging with different fluorescent filters, allowing identification of cysts based on both RNA content and surface markers.

Besides these examples of FISH utilizing cytometry or imaging for single-cell detection of microbes, several examples of FISH or FISH-like protocols have been demonstrated in microfluidic devices, with detection of an integrated signal to detect the presence and/or concentration of specific microbes. Another example of electrokinetic-assisted concentration and on-chip labeling of bacterial nucleic acid was demonstrated by Packard et al. [22]. In this device, bacteria from a dilute suspension are labeled with a nonspecific nucleic acid stain (SYTO 9), and concentrated greater than 400-fold using dielectrophoresis in a microchip with

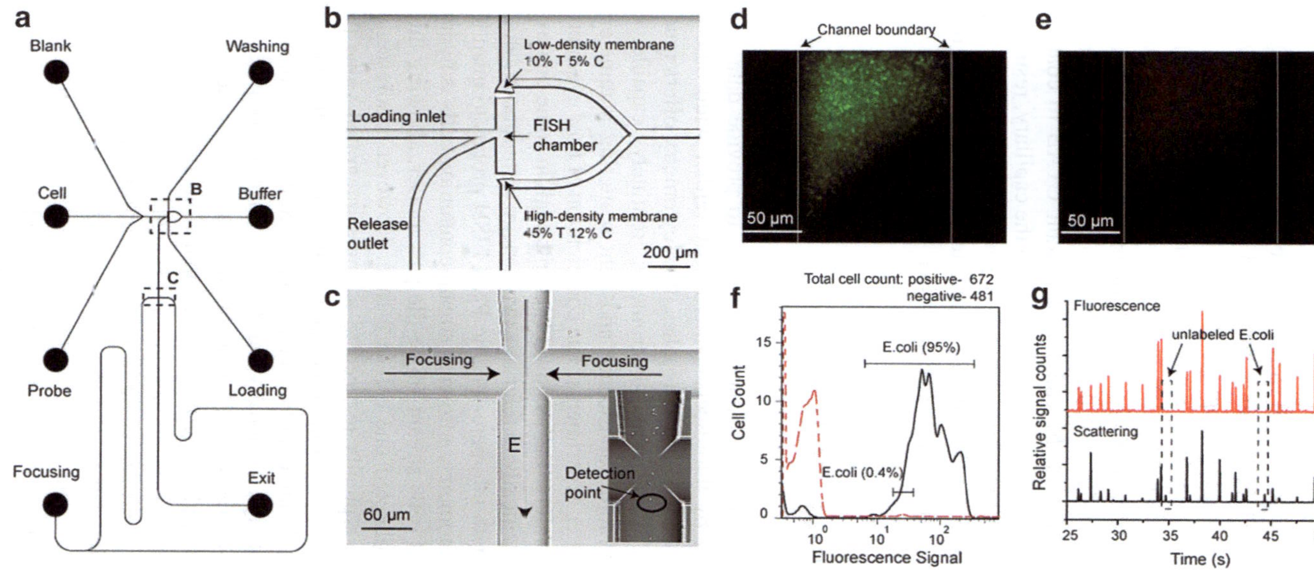

Fig. 5.3 μFlowFISH device for integrated FISH labeling and flow cytometry detection of bacteria, applied here to cultured *E. coli* K12. The panels at left show a schematic and micrographs of the microchip design for fluorescence in situ hybridization (FISH) and flow cytometry (μFlowFISH). (**a**) The mask design of the μFlowFISH chip. (**b**) An image of the FISH chamber formed by two photopolymerized membrane in the channel. (**c**) The cross-channel structure for electrokinetically focusing microbial cells into a single stream line along the center of the vertical channel for flow cytometry. The enlarged image of the channel cross shows *Escherichia coli* being focused in the center of the channels. (**d**) Fluorescence image of *E. coli* hybridized with Alexa488-labeled Eco681 probes in the FISH chamber. Two *vertical lines* indicate the boundaries of the chamber. (**e**) Very weak fluorescence signals in the negative control using Cy3-labeled NON338. (**f**) On-chip flow cytometry results of *E. coli*. The *black line* is the positive control experiment using Eco681 probes, showing that about 95 % of the cells were successfully labeled with probes. The *red dashed line* indicates that only 0.4 % of the cells have strong fluorescence signals when incubated with the negative control NON338 probes. (**g**) Typical raw data traces obtained from the flow cytometry detection system. Unlabeled cells, as indicated in the *dashed rectangles*, produce only scattering peaks (*bottom trace*) without corresponding fluorescence peaks (*top trace*), while labeled *E. coli* cells have aligned peaks in both channels. Adapted from [15] by the author and reproduced by permission of the Royal Society of Chemistry

patterned microelectrodes. The cells are then lysed, and nucleic acid is labeled by hybridization using a target-specific oligonucleotide probe, labeled with HEX, which accepts as an acceptor dye for FRET, using the SYTO 9 as a donor. This assay, termed "FRET-ISH," is not performed with a single-cell counting or imaging readout would be typical for FISH, but rather an integrated signal representative of the fraction of target bacteria in a mixed population.

Although not performed on a chip, FISH analysis of bacteria in a glass capillary was reported by Lantz et al. [11]. Rapid hybridization and wash steps were performed on suspensions of cultured *Salmonella* Typhimurium, either alone or in mixed culture, followed by analysis of the resulting microbial suspension by capillary electrophoresis. Addition of the cationic surfactant CTAB to coat the cells, in combination with a discontinuous buffer system in the capillary, results in focusing of a wide variety of microbial cells into a single narrow zone. The presence of the target bacteria (*S.* Typhimurium) is detected by the presence of a bright fluorescent signal, using on-capillary LIF detection.

5.3 Detection Modalities in FISH

Traditionally FISH has relied on two distinct approaches to provide data with single-cell resolution: imaging by fluorescence microscopy (either wide field or confocal), or flow cytometry. Of the two approaches, flow cytometry offers much higher analysis rates, with commercial instruments routinely capable of >100,000 cells/s, with 10 or more fluorescence and light scattering parameters, allowing complex multiparameter experiments. The combination of flow cytometry with RNA FISH is sometimes referred to as flow-FISH, although in some contexts the term FlowFISH is used specifically to refer to a cytogenetic technique in which flow cytometry is combined with telomere-specific PNA FISH probes to quantify telomere length [2]. In flow cytometry, a flowing stream containing cells is hydrodynamically focused such that cells stream single-file past a laser, allowing fluorescence information to be collected one cell at a time. However, fluorescence information from each cell is typically reduced to a single integrated value per channel, which means there is no information about subcellular localization of fluorescence. The logarithmic signal amplifiers commonly used in flow cytometers offer a wide dynamic range of signal intensity, but lack the quantitative precision required for counting individual transcripts. Some commercial flow cytometers are also configured for flow sorting, allowing unique subpopulations of cells, or even individual cells, to be enriched or isolated based on FISH signals.

Fluorescence imaging, by contrast, yields a wealth of information about the subcellular localization of fluorescent probes, as well as spatial relationships between labeled cells within heterogenous tissues or microbial communities. Individual probe-labeled RNA transcripts can be identified and counted as bright, diffraction-limited spots with high-resolution fluorescence imaging. The location of fluorescently tagged molecules in relation to other cellular components or structures can also be revealed. Conventional epifluorescence microscopy based upon band-pass filters and dichroic mirrors to separate light into spectral components is typically limited to 3–4 fluorescence channels, whereas multispectral approaches (increasingly available with laser scanning confocal microscopes) can offer a much greater degree of spectral multiplexing. Combinatorial labeling and spectral imaging FISH (CLASI-FISH) has been demonstrated, for example allowing identification of 15 pylotypes of bacteria in a biofilm from the human oral microbiome [37]. Although fluorescence microscopy has traditionally been a low-throughput, labor-intensive technique, a wide variety of laboratory automation is now available, ranging from motorized stages and filter changers to fully automated laser scanning image cytometers, enabling higher throughput, high-content imaging, although automated imaging does not yet approach the analysis rates of flow cytometers.

Both imaging and flow cytometry modalities have been adapted for use with FISH in microfluidic chips. Within academic research groups, the stage of an inverted microscope equipped with a CCD camera is frequently the "platform" for testing and development of microfluidic devices of all sorts, and fluorescence imaging of chip-based FISH is thus a natural development. High-resolution imaging of cells (whether by FISH or otherwise) is best performed with high-numerical-aperture (NA) oil immersion objective lenses, which are almost exclusively designed to work optimally with #1.5 (0.17 mm thick) glass cover slips. These objectives can have $NA > 1.3$ and achieve lateral resolution of <0.3 μm. Thicker glass substrates necessitate using longer working distance dry objectives, often fitted with correction collars for different thicknesses of glass; these objectives typically have $NA < 0.7$ which limits resolution to approximately 0.5 μm. Optically clear plastics such as PMMA, polycarbonate, or COC can be used to construct microfluidic chips designed for imaging, and these materials may be available in coverslip thickness, allowing oil immersion lenses to be used. However, the refractive index of plastic materials typically differs from that of glass. Modern objective lenses are designed to correct for aberrations, based on the assumption that only materials between the objective and the specimen are glass and the immersion medium (i.e., index-matched immersion oil or air). Using materials of differing refractive index may introduce aberrations into the image.

Microfluidic devices constructed from PDMS can readily be bonded directly to glass cover slips, using an oxygen plasma treatment to oxidize the PDMS surface prior to bonding. This offers a straightforward approach to constructing microfluidic chips compatible with FISH with high-resolution imaging. Direct fabrication of glass microfluidic devices with coverslip-thickness substrates is less common. Micronit Microfluidics BV offers custom fabrication of glass chips with coverslip-thickness bottom substrates. Another approach that we have employed is to manufacture or purchase glass chips fabricated from thicker substrates, and then grinding and polishing the bottom substrate to coverslip thickness, although the resulting thinned chips are more fragile than the original (thicker) devices.

A variety of flow cytometry-like detectors have also been described for microfluidic chips, either for flow-FISH or other types of analysis. Detectors described in academic publishing literature tend to be custom-built laboratory prototypes, which are frequently much more limited in throughput and spectral channels than the detectors available in commercial flow cytometers. The cylindrical sheath flow geometry of most commercial flow cytometers has enabled unique arrangements of filters, dichroic beam splitters, and PMTs fully surrounding the sheath flow. For example, eight fluorescence channels may be interrogated from a single 488 nm laser, with detectors placed in an octagonal arrangement around the sheath flow. Multiple lasers are used to interrogate cells at separate positions in the sheath, with multiple "rings" of detectors. Such a compact arrangement is impossible with planar microfluidic chips, particularly for laboratory prototypes built around an inverted microscope with a single objective lens used in an epifluorescence configuration, and using commercial off-the-shelf optics. Coupling lasers and detection optics with chips can also be achieved by carefully embedding optical fibers directly into the chip [16], which necessarily increases the complexity of device fabrication. Most examples of chip-based flow cytometry detectors that we are aware of are limited to one or two lasers, and 2–3 fluorescence and scatter channels. An alternative approach that enables more spectral channels is to replace the cascade of filters and photomultiplier tubes with a spectrograph (e.g., a diffraction grating and a multianode PMT or CCD detector), enabling collection of an entire fluorescence spectrum for each cell, with multivariate curve analysis to resolve individual spectral components [8, 20]. A variety of cell sorting modalities based on optical trapping [24], valving [5], deflection with a piezoelectric transducer [3], or electrokinetics [34] have been coupled to chip-based flow cytometry, and these are expected to be compatible with FISH detection. In general chip-based cytometers and sorters operate at much lower flow rates than commercial flow sorters, with sort rates of a few to tens of cells per second, as opposed to tens of thousands of cells per second. The low throughput of chip-based systems may actually become an advantage when dealing with very small populations of cells that may get lost in the noise of a high-throughput sorter, or when dimly labeled cells require longer interrogation times that are afforded by low flow rates.

Besides construction of an LIF point detector (with the associated challenges of fast data acquisition), the other main challenge for a microfluidic flow cytometry detector is focusing cells into a single-file line. This is accomplished in commercial flow cytometers using a sheath flow cuvette, which focuses the cell-containing "core" into a narrow cylindrical stream, with average flow velocities on the order of 10 m/s [30]. In planar microfluidic chips, using either hydrodynamic flow or electrokinetic forces, it is much more straightforward to achieve a one-dimensional (1D) focusing, in which a flowing stream is "pinched" into a narrow plane between two flowing streams from the side. Although cells are not confined within two dimensions along the center of the flow channel, good optical performance can still be achieved if the depth of the microfluidic channel is roughly the same as the depth of focus of the objective lens.

A separate concern is presented by the nonuniform velocity profile within the microchannel when using hydrodynamic forces to achieve one-dimensional (1D) focusing: the velocity at the centerline of the channel is higher than velocity at either the top or the bottom. If cells are focused in two dimensions only, cells traveling near the center are traveling faster, and thus interrogated for a shorter period of time, than cells traveling near the top or bottom of the channel. This may confound interpretation of results, as more signal will be collected from cells that move slowly through the illuminated region, compared to cells at the center of the channel that travel quickly through. One approach would be to normalize the integrated intensity of each event by the length of the duration (measured, for example, by the duration or peak width of a coincident light scattering signal). A second approach would be to combine 1D focusing with a spatial filter (i.e., a confocal pinhole) to reject signals arising out of the plane of focus of the laser, although this would result in a fraction of the flowing cells being ignored by the detector. Although numerous papers have described flow cytometry based upon 1D hydrodynamic focusing, it is not clear whether the effect of nonuniform flow velocity is typically considered. We note that a purely electrokinetic flow (either electrophoresis or electroosmosis) gives rise to a much more nearly plug-flow profile, with no velocity gradients across the channel. A 1D electrokinetic focusing should not be problematic from the point of view of velocity gradients, although heterogeneity in surface charge of cells may still result in variations in transit time through the beam.

Several approaches to achieve a two-dimensional hydrodynamic focusing in microchannels have also been reported, with the intent of approaching the cylindrical focusing achieved in a sheath flow cuvette. We note that examples cited below of flow focusing in microfluidics are not exclusively limited to flow cytometry applications. One approach has been to design channel structures with multiple inlets that pinch the cell stream in two dimensions (i.e., from the top and bottom as well as the sides) [26, 32], resulting in a roughly rectangular stream at the center of the channel. A second approach has been to simply introduce cells through a narrow capillary or needle, threaded into a microfluidic channel network, which is similar to a conventional sheath flow cuvette. A third approach is to use a technique called "microfluidic drifting," relying upon inertial effects within curved microchannels in laminar flow at high Reynolds number [16, 17].

5.4 Comparison to Other RNA-Based Detection Technologies

Of the microfluidic methods that profile RNA expression, single-cell RNA-FISH has some distinct advantages as well as drawbacks. The most high-profile methods in single-cell genomic analysis are single-cell RT-PCR and single-cell RNA sequencing (RNA-seq) using next-generation sequencing (NGS). Single-cell whole RNA-seq holds an advantage over FISH due to its comprehensive nature. Single cells can be sorted and placed into microfluidic reservoirs and automatically lysed, and made into sequencing libraries for comprehensive gene expression analysis of the entire genome or transcriptome. No prior information about gene targets is required, and the RNA-seq is capable of producing sensitive and theoretically unbiased profile of "omic" information of any cell. We note that, in principle, FISH labeling can be used as a selection technique for FACS with single-cell deposition, to allow targeted single-cell RNA-seq. As of the writing of this chapter, there are two limitations to single-cell RNA-seq: 1) the degree to which cDNA must be amplified in order to yield enough material for NGS. The degree of amplification required to generate enough cDNA for NGS library preparation will introduce significant bias from trace contamination of genome amplification reagents and bias related to different rates of degradation of mRNA transcripts, along with inherently high variability due to stochastic transcription of some genes. The large experimental error from cell to cell will necessitate sequencing hundreds perhaps thousands of single cells to generate statistically meaningful results, and this may be prohibitively expensive even as sequencing costs per base continue to drop. 2). The vast quantity of RNA-seq data generated from each cell, times several hundreds or thousands of cells, presents a significant challenge in bioinformatics expertise and computing infrastructure.

In contrast to sequencing, both single-cell RT-PCR and single-cell RNA-FISH are target-driven techniques where prior knowledge of the gene target must be available in order to design primers for RT-PCR and FISH, and therefore provide more specific information regarding the transcriptome, but not "omics"-level interrogation afforded by single-cell RNA-seq. Dynamic arrays for microfluidic single-cell RT-PCR commonly use matrices with fluid channels containing individual DNA samples of interest with gene expression assay channels [35, 39]. The throughput of single-cell RT-qPCR is limited by the number of unique sample channels that can be fabricated onto a single-chip setup—typically tens to hundreds, and each microfluidic sample channel can perform as many as 96 PCR assays. Single-cell RNA-FISH is limited to much fewer RNA targets labeled with fluorophores of different colors. Conventional RNA-FISH is limited to three or less RNA targets/cell, and with spectral barcoding techniques there can be up to dozens of RNA targets interrogated in the same cell. Though single-cell RNA-FISH has limited bandwidth in terms of number of targets/cell, there are two key advantages to performing single-cell RNA-FISH. (1) There is no need to sort and separate single cells from a mixture in order to perform single-cell RNA-FISH, as the subsequent analysis (imaging or flow) is inherently single cell in nature; (2) - single-cell RNA-FISH does not require the lysis and extraction of RNA, and therefore the RNA-FISH can be multiplexed with protein immunostaining, with the RNA and protein targets unified within the same cell. Elimination of the cell sorting step prior to RNA analysis is a great benefit because not all cell types are amenable to sorting. Some cell types, such as primary cells, exhibit adverse response to the sorting process, and functional assays using sorted cells can be a challenge. Also, cells from solid tissue samples are often incompatible with cell sorting, and laser microdissection of solid tissue will yield samples with hundreds of cells, but not single cells. Hence, the elimination of sample sorting prior to

RNA-FISH means that all cell types, including tissue slices, can be used for RNA-FISH. In addition, if a functional assay is to be performed prior to RNA-FISH to monitor the upregulation of several known RNA targets in response to stimulation, RNA-FISH is the method of choice.

The multiplexing capability presented by RNA-FISH is significant especially for small noncoding RNAs such as miRNAs, whose expression and function are tissue specific. The same miRNA can exert opposite modulatory effects on signaling pathways, modulate entirely different signaling pathways, or have no effect, all depending on the tissue type and cellular context. Without the ability to monitor mRNAs, proteins, and transient signaling events along with miRNAs in the same cell, it is very difficult to ascertain what relationship, if any, miRNA expression has to health or disease of that cell. Bulk profiling methods such as RNA-seq and RT-PCR can provide measurements of numerous RNA and DNA species from single cells, but the relationship between miRNA expression levels in relation to their mRNA targets and protein indicators of the cellular physiological state is lost in the sample preparation process. FISH allows for multiplexing of different categories of targets in their native context, including proteins, and posttranslational modifications, with RNA and DNA detection [41], which makes FISH unique in its ability to preserve the cellular context and single-cell resolution.

5.5 Future Outlook for Microfluidic FISH

FISH remains a common and useful technique in cytogenetics, biomedical research, and microbial ecology. RNA-directed FISH specifically is proving useful for understanding the dynamics of gene expression at the single-cell level. While there have been relatively few examples to date of fully integrated RNA-directed FISH on microfluidic devices, we believe that microfluidics offers several advantages. First, the inherently small volume of microfluidic devices allows experiments to be performed on small numbers of cells, without loss of cells associated with handling at large volumes. This is particularly valuable for unique, rare, or limited cell types, such as primary cells or patient samples, or low-cell-density samples such as environmental samples. Second, FISH protocols, particularly those involving signal amplification such as mRNA or miRNA FISH, tend to involve several steps, with multiple intermediate washes. These protocols are tedious and time consuming when performed manually, but are amenable to automation using microfluidic devices with programmable pumps and valves, or programmable power supplies (enabling electrokinetic FISH protocols). Further, the combination of RNA-directed FISH and protein labeling affords the ability to study both gene expression and protein expression simultaneously; such complex multi-step labeling protocols are greatly simplified when using a unified microfluidic device. Further, microfluidic chips are uniquely capable of offering both common detection modalities for FISH—high-content fluorescence imaging and flow cytometry—on a single platform.

Despite its great potential, microfluidic FISH for RNA targets faces dual challenges. RNA-directed FISH itself is a complex task, with multiple possible labeling strategies, requiring extensive optimization and validation for each target. New labeling and signal amplification approaches are still reported periodically, suggesting that existing approaches are sometimes inadequate, whether in sensitivity, simplicity, or reliability. The second challenge is that many microfluidic systems for working with cells are relatively primitive. "Proof-of-concept" devices have been demonstrated mostly in academic research labs, and these custom-built devices require significant engineering expertise to fabricate, set up, and operate. Microfluidic FISH has thus far lacked the extensive development and optimization (and private-sector investment) that are required to move an engineered device from the proof-of-concept stage to a commonly accepted, user-friendly tool that can be used by (and taken for granted by) biological or clinical researchers. Given the vast array of different protocols and applications for FISH in research labs, the research laboratory market is likely too diverse to be served by any one commercial device or kit for microfluidic FISH. Clinical diagnostic applications may thus be a more attractive market for commercial investment in microfluidic FISH. Clinical diagnostics based on FISH currently exist for cytogenetics as well as microbial identification. If these applications can be simplified or made more reliable using microfluidic automation, then these may represent sufficient incentive for commercial investment in microfluidic FISH. For clinical applications specifically targeting RNA, such as quantifying gene expression, qRT-PCR is currently the technique of choice, and is characterized by exquisite sensitivity and good reproducibility, but faces challenges with certain sample types such as formalin-fixed paraffin-embedded (FFPE) tissues. Any clinical diagnostic for gene expression based upon FISH in general, or microfluidic FISH in particular, must show compelling advantages over qRT-PCR in terms of speed, simplicity, and predictive value, or otherwise improve patient care over and above current techniques. As academic research groups continue to push the limits of FISH for RNA detection, it is possible that we are only a single killer app away from "FISH and Chips" in every clinical laboratory.

Acknowledgments The authors acknowledge support from Sandia National Laboratories' Laboratory Directed Research and Development (LDRD) program. Sandia National Laboratories is a multi-program laboratory managed and operated by Sandia Corporation, a wholly owned subsidiary of Lockheed Martin Corporation, for the US Department of Energy's National Nuclear Security Administration under Contract DE-AC04-94AL85000.

References

1. Amann R, Fuchs BM (2008) Single-cell identification in microbial communities by improved fluorescence in situ hybridization techniques. Nat Rev Micro 6:339–348
2. Baerlocher GM, Vulto I, de Jong G, Lansdorp PM (2006) Flow cytometry and FISH to measure the average length of telomeres (flow FISH). Nat Protocols 1:2365–2376

3. Chen CH, Cho SH, Tsai F, Erten A, Lo Y-H (2009) Microfluidic cell sorter with integrated piezoelectric actuator. Biomed Microdevices 11:1223–1231
4. Crosetto N, Bienko M, van Oudenaarden A (2015) Spatially resolved transcriptomics and beyond. Nat Rev Genet 16:57–66
5. Fu AY, Chou HP, Spence C, Arnold FH, Quake SR (2002) An integrated microfabricated cell sorter. Anal Chem 74:2451–2457
6. Gerdts G, Luedke G (2006) FISH and chips: marine bacterial communities analyzed by flow cytometry based on microfluidics. J Microbiol Methods 64:232–240
7. Giorgi D, Farina A, Grosso V, Gennaro A, Ceoloni C, Lucretti S (2013) FISHIS: fluorescence in situ hybridization in suspension and chromosome flow sorting made easy. PLoS One 8: e57994
8. Gregori G, Patsekin V, Rajwa B, Jones J, Ragheb K, Holdman C, Robinson JP (2012) Hyperspectral cytometry at the single-cell level using a 32-channel photodetector. Cytometry A 81:35–44
9. Itzkovitz S, van Oudenaarden A (2011) Validating transcripts with probes and imaging technology. Nat Methods 8:S12–S19
10. Kalyuzhnaya MG, Zabinsky R, Bowerman S, Baker DR, Lidstrom ME, Chistoserdova L (2006) Fluorescence in situ hybridization-flow cytometry-cell sorting-based method for separation and enrichment of type I and type II methanotroph populations. Appl Environ Microbiol 72:4293–4301
11. Lantz AW, Brehm-Stecher BF, Armstrong DW (2008) Combined capillary electrophoresis and DNA-fluorescence in situ hybridization for rapid molecular identification of Salmonella Typhimurium in mixed culture. Electrophoresis 29:2477–2484
12. Larsson C, Grundberg I, Soderberg O, Nilsson M (2010) In situ detection and genotyping of individual mRNA molecules. Nat Meth 7:395–397
13. Latorra D, Campbell K, Wolter A, Hurley JM (2003) Enhanced allele-specific PCR discrimination in SNP genotyping using 3′ locked nucleic acid (LNA) primers. Hum Mutat 22:79–85
14. Lebaron P, Catala P, Fajon C, Joux F, Baudart J, Bernard L (1997) A new sensitive, whole-cell hybridization technique for detection of bacteria involving a biotinylated oligonucleotide probe targeting rRNA and tyramide signal amplification. Appl Environ Microbiol 63:3274–3278
15. Liu P, Meagher RJ, Light YK, Yilmaz S, Chakraborty R, Arkin AP, Hazen TC, Singh AK (2011) Microfluidic fluorescence in situ hybridization and flow cytometry (mu FlowFISH). Lab Chip 11:2673–2679
16. Mao X, Nawaz AA, Lin S-CS, Lapsley MI, Zhao Y, McCoy JP, El-Deiry WS, Huang TJ (2012) An integrated, multiparametric flow cytometry chip using "microfluidic drifting" based three-dimensional hydrodynamic focusing. Biomicrofluidics 6:024113-024113-024119
17. Martel JM, Toner M (2013) Particle focusing in curved microfluidic channels. Sci Rep 3:3340
18. Maruyama F, Kenzaka T, Yamaguchi N, Tani K, Nasu M (2003) Detection of bacteria carrying the stx(2) gene by in situ loop-mediated isothermal amplification. Appl Environ Microbiol 69:5023–5028
19. Nelson PT, Baldwin DA, Kloosterman WP, Kauppinen S, Plasterk RH, Mourelatos Z (2006) RAKE and LNA-ISH reveal microRNA expression and localization in archival human brain. RNA 12:187–191
20. Nolan JP, Condello D, Duggan E, Naivar M, Novo D (2013) Visible and near infrared fluorescence spectral flow cytometry. Cytometry A 83:253–264
21. Oliveira K, Procop GW, Wilson D, Coull J, Stender H (2002) Rapid identification of Staphylococcus aureus directly from blood cultures by fluorescence in situ hybridization with peptide nucleic acid probes. J Clin Microbiol 40:247–251
22. Packard MM, Shusteff M, Alocilja EC (2012) Microfluidic-based amplification-free bacterial DNA detection by dielectrophoretic concentration and fluorescent resonance energy transfer assisted in situ hybridization (FRET-ISH) (†,‡). Biosensors 2:405–416
23. Pernthaler A, Amann R (2004) Simultaneous fluorescence in situ hybridization of mRNA and rRNA in environmental bacteria. Appl Environ Microbiol 70:5426–5433
24. Perroud TD, Kaiser JN, Sy JC, Lane TW, Branda CS, Singh AK, Patel KD (2008) Microfluidic-based cell sorting of Francisella tularensis infected macrophages using optical forces. Anal Chem 80:6365–6372

25. Raj A, van den Bogaard P, Rifkin SA, van Oudenaarden A, Tyagi S (2008) Imaging individual mRNA molecules using multiple singly labeled probes. Nat Methods 5:877–879
26. Rhee M, Valencia PM, Rodriguez MI, Langer R, Farokhzad OC, Karnik R (2011) Synthesis of size-tunable polymeric nanoparticles enabled by 3D hydrodynamic flow focusing in single-layer microchannels. Adv Mater 23:H79–H83
27. Robertson KL, Verhoeven AB, Thach DC, Chang EL (2010) Monitoring viral RNA in infected cells with LNA flow-FISH. RNA 16:1679–1685
28. Salimnia H, Fairfax MR, Lephart P, Morgan M, Gilbreath JJ, Butler-Wu SM, Templeton KE, Hamilton FJ, Wu F, Buckner R, Fuller D, Davis TE, Abdelhamed AM, Jacobs MR, Miller A, Pfrommer B, Carroll KC (2014) An international, prospective, multicenter evaluation of the combination of AdvanDx Staphylococcus QuickFISH BC with mecA XpressFISH for detection of methicillin-resistant Staphylococcus aureus isolates from positive blood cultures. J Clin Microbiol 52:3928–3932
29. Schmidt KS, Borkowski S, Kurreck J, Stephens AW, Bald R, Hecht M, Friebe M, Dinkelborg L, Erdmann VA (2004) Application of locked nucleic acids to improve aptamer in vivo stability and targeting function. Nucleic Acids Res 32:5757–5765
30. Shapiro HM (2003) Practical flow cytometry, 4th edn. Wiley, Hoboken, NJ
31. Silahtaroglu A, Pfundheller H, Koshkin A, Tommerup N, Kauppinen S (2004) LNA-modified oligonucleotides are highly efficient as FISH probes. Cytogenet Genome Res 107:32–37
32. Simonnet C, Groisman A (2005) Two-dimensional hydrodynamic focusing in a simple microfluidic device. Appl Phys Lett 87:114104
33. Skinner SO, Sepúlveda LA, Xu H, Golding I (2013) Measuring mRNA copy-number in individual Escherichia coli cells using single-molecule fluorescent in situ hybridization (smFISH). Nat Protoc 8:1100–1113
34. Song Y, Peng R, Wang J, Pan X, Sun Y, Li D (2013) Automatic particle detection and sorting in an electrokinetic microfluidic chip. Electrophoresis 34:684–690
35. Spurgeon SL, Jones RC, Ramakrishnan R (2008) High throughput gene expression measurement with real time PCR in a microfluidic dynamic array. PLoS One 3:e1662
36. Urbanek M, Nawrocka A, Krzyzosiak W (2015) Small RNA detection by in situ hybridization methods. Int J Mol Sci 16:13259
37. Valm AM, Welch JLM, Rieken CW, Hasegawa Y, Sogin ML, Oldenbourg R, Dewhirst FE, Borisy GG (2011) Systems-level analysis of microbial community organization through combinatorial labeling and spectral imaging. Proc Natl Acad Sci 108:4152–4157
38. Vester B, Wengel J (2004) LNA (locked nucleic acid): high-affinity targeting of complementary RNA and DNA. Biochemistry 43:13233–13241
39. Warren L, Bryder D, Weissman IL, Quake SR (2006) Transcription factor profiling in individual hematopoietic progenitors by digital RT-PCR. Proc Natl Acad Sci U S A 103:17807–17812
40. Wu M, Piccini M, Koh CY, Lam KS, Singh AK (2013) Single cell microRNA analysis using microfluidic flow cytometry. PLoS One 8:e55044
41. Wu M, Singh AK (2014) Microfluidic molecular assay platform for the detection of miRNAs, mRNAs, proteins, and posttranslational modifications at single-cell resolution. J Lab Autom 19:587–592
42. Yamaguchi N, Ohba H, Nasu M (2006) Simple detection of small amounts of Pseudomonas cells in milk by using a microfluidic device. Lett Appl Microbiol 43:631–636
43. Yilmaz S, Haroon MF, Rabkin BA, Tyson GW, Hugenholtz P (2010) Fixation-free fluorescence in situ hybridization for targeted enrichment of microbial populations. ISME J 4:1352–1356
44. Zhang Q, Zhu L, Feng H, Ang S, Chau FS, Liu W-T (2006) Microbial detection in microfluidic devices through dual staining of quantum dots-labeled immunoassay and RNA hybridization. Anal Chim Acta 556:171–177
45. Zwirglmaier K (2005) Fluorescence in situ hybridisation (FISH)—the next generation. FEMS Microbiol Lett 246:151–158

Chapter 6
Microfluidic Multistage Integration for Analysis of Circulating Exosomes

Mei He, Andrew Godwin, and Yong Zeng

Abstract The study of circulating exosomes has received growing interest in fields of biology and medicine. Exosomes, small (50–150 nm) membrane vesicles of endocytic origin, have emerged as new mechanisms for regulation of a wide range of biological processes, including intercellular communications, immune response, and cancer metastasis. Molecular analysis of exosomes is an emerging paradigm for noninvasive cancer diagnosis and monitoring of disease treatment; however, exosome analysis is challenging due to current low-efficient isolation and molecular measurement approaches. In this chapter, a novel microfluidic strategy is described for rapid, quantitative isolation and analysis of exosomes, as well as intravesicular markers directly from human blood. This microfluidic platform enables multistage integration of immunomagnetic isolation and enrichment, chemical lysis, and immune-sandwich chemifluorescence probing in one sequential process. The developed microfluidic exosome profiling platform has been employed to measure expression levels of total and phosphorylated IGF-1R in non-small-cell lung cancer patient plasma-derived exosomes, thereby opening a new avenue for monitoring IGF-1R and its activation status in a noninvasive manner (i.e., liquid biopsy). This developed microfluidic exosome profiling

M. He, Ph.D.
Department of Biological and Agricultural Engineering, Johnson Cancer Research Center, College of Engineering, Kansas State University, Seaton Hall 129, 920N. 17th Street, Manhattan, KS 66506, USA
e-mail: meih@ksu.edu

A. Godwin
Pathology and Laboratory Medicine, University of Kansas Medical Center, Kansas City, KS, USA

University of Kansas Cancer Center, Kansas City, KS, USA
e-mail: agodwin@kumc.edu

Y. Zeng, Ph.D. (✉)
University of Kansas Cancer Center, Kansas City, KS, USA

Department of Chemistry, University of Kansas, 1251 Wescoe Hall Dr., Lawrence, KS 66045, USA
e-mail: yongz@ku.edu

© Springer International Publishing Switzerland 2016
C. Lu, S.S. Verbridge (eds.), *Microfluidic Methods for Molecular Biology*,
DOI 10.1007/978-3-319-30019-1_6

platform shows great potential for forming the basis of a critically needed infrastructure to advance the biological and clinical utilization of exosomes.

Keywords Exosomes • Biomarker • Cancer diagnosis • Microfluidic multistage integration • Intravesicular marker

6.1 Introduction

6.1.1 Microfluidic Exosome Analysis

Microfluidic technology has shown tremendous capability in applications of molecular biology, medicine, and bioanalytical chemistry, including single molecule/cell, high-sensitive detection, rapid speed for efficient analysis, and automated operation with microliter consumption [1–6]. Multistage functional integration in microfluidic technology has been of a long-standing interest which is essentially needed for processing complex biological samples. Although recent advances in microfluidic technology have significantly impacted many areas of biology, only several initial attempts have been made to accelerate exosome analysis. Chen et al. first reported a microfluidic isolation system for affinity capture of serum microvesicles using CD63 antibody immobilized on the microchannel surface [7]. In contrast to targeting surface biomarkers of microvesicles, a physical method has been developed to purify nano-sized extracellular vesicles by employing a microfluidic filtration system integrated with a monolithic porous polymer membrane [8]. Although microfluidic systems demonstrated superior isolation performance compared to bench-top centrifugation methods, most of these microfluidic platforms still rely on conventional bench-top techniques for downstream characterization of isolated microvesicles. Recently, Shao et al. presented the first demonstration of microfluidic molecular microvesicle analysis using a miniaturized nuclear magnetic resonance (μNMR) sensor to detect immunomagnetic bead-labeled microvesicles [9]. Im et al. developed a label-free detection approach called nano-plasmonic exosome (nPLEX) assay (nPLEX) to analyze ascites-derived exosomes in ovarian cancer patients [10]. In order to gain accurate definition of microvesicles in particular exosomes and a deeper understanding of exosome biological functions, comprehensive characterization of molecular composition of microvesicles must be enabled by advantageously utilizing microfluidic technology. Multistage integration in microfluidic technology significantly contributes to efficient processing and analysis of exosome samples.

Cells secrete numerous membrane-derived vesicles, which are mainly composed of 10–1000 nm exosomes and other microvesicles [11]. Exosomes differ from other microvesicles in their cellular origin, abundance, and biogenesis. Exosomes commonly recognized as membrane vesicles derived from the endolysosomal pathway have a size range of 30–150 nm and are abundantly found in the plasma and malignant effusions derived from cancer patients [12–14]. Exosome populations share certain common characteristics, including shape, size, density, and general protein composition [15]. Exosomes have been found to mediate cellular communications via transfer of cargo consisting of an array of proteins, selected functional cellular RNAs [16, 17], and mitochondrial DNA [18]. Due to the constitutive release and enrichment of certain proteins in exosome populations, exosomes have been considered novel and promising biomarker resources for cancer diagnosis and prognosis [19, 20]. However, despite the potential for cancer research, progress in clinical utilization of exosomes is still slow [11]. Conventional exosome isolation and analysis approaches are of low throughput and they are time consuming, consequently posing a significant technical barrier for processing such nanoscale and molecularly diverse vesicles. Improvement and standardization of approaches for exosome isolation and analysis are urgently needed. Current exosome purification methods, including ultracentrifugation and sucrose gradient ultracentrifugation, are tedious, time consuming (>10 h), and unable to completely discriminate exosomes from other membrane vesicles [21, 22]. Most importantly, differential ultracentrifugation co-purifies multiple extracellular vesicles that are secreted by variable intracellular mechanisms, thereby potentially masking disease-related biosignatures [23]. Conventional filtration has been used to isolate microvesicles in uniform sizes of less than 150 nm [24]. However, forced filtration and shearing force may cause membrane fusion and integrity loss in vesicles. Molecular analysis of isolated microvesicles is primarily performed using Western blot, ELISA, and mass spectrometry, which require lengthy processes and concentrated exosome samples (purified from more than 3 mL plasma or 300 mL cell culture media), thereby limiting progress in clinical investigation and utilities of exosomes [21, 25]. For precise quantitative measurement of exosome size and concentration, nanoparticle tracking analysis (NTA) requires a large amount (on the order of mL) of purified and concentrated vesicles (~10^9 mL^{-1}) and is less reliable for analyzing polydisperse system [26]. Flow cytometry analyzes small particles, but it is limited for quantitative measurement of particles below 300 nm [27]. Transmission electron microscopy (TEM) and atomic force microscopy (AFM) have recently been utilized in order to investigate exosome morphology and size distribution [28]. However, microscopic protocols do not lend themselves to the high throughput and rapid measurement. In this chapter, a "sample-in-answer-out" microfluidic exosome profiling approach is described that integrates immunoaffinity isolation of plasma exosomes with quantitative characterization of exosomal phenotypes. A cascading microfluidic circuit is introduced for streamlining multistep assay, including exosome isolation and enrichment, online chemical lysis and target capture, and sandwich immunoassay assisted by chemifluorescence detection.

6.1.2 Exosomes in Blood-Based Cancer Diagnosis and Prognosis

Development of noninvasive blood-based tests is extremely appealing, especially for diseases in which biopsy is difficult, costly, and sometimes impossible.

Probing circulating exosomes is an emerging paradigm for noninvasive cancer diagnosis and monitoring of disease treatment. Lung cancer is the leading cause of cancer-related deaths worldwide [29]. Non-small-cell lung cancer (NSCLC) accounts for approximately 85 % of lung cancer cases, with an overall 5-year survival (stage IIIA) rate of only 15 % [30]. This fact highlights the need for advanced diagnostic and prognostic biomarkers, therapeutic targets, and molecular targeted anticancer agents [31]. Type 1 insulin growth factor receptor (IGF-1R) has attracted increasing interest as a promising therapeutic target in NSCLC and other cancers because of its implication in promoting oncogenic transformation, growth, and survival of cancer cells [32]. However, current clinical assessment of IGF-1R expression primarily relies on immunohistochemistry (IHC) of tumor tissue sections [33], which is invasive and contains several disadvantages. Because a majority of NSCLC patients present with unresectable advanced disease, acquisition of adequate tissue for diagnosis can be challenging. Furthermore, it is extremely difficult to obtain tissue biopsies prior to each therapy and therefore histologic and molecular information is substantially limited [34]. Although the insulin-like growth factor-1 (IGF-1), insulin-like growth factor-2 (IGF-2), and insulin-like growth factor-binding protein-3 (IGFBP-3) are circulating in blood, their plasma levels do not always correlate well with cancer diagnosis and prognosis [35]. The expression of specific circulating exosomal miRNAs in lung cancer has been emerging as a potential surrogate of tumor miRNA expression, suggesting a new paradigm that includes use of exosomes for noninvasive early diagnosis, prognostics, and therapeutics [36, 37]. The diagnostic and prognostic value of exosomal proteins has recently been increasing dramatically in cancer research [38–40]. In this chapter, exosomal IGF-1R and its prognostic value in NSCLC patients are discussed. The developed microfluidic platform can be used to measure expression levels of total and phosphorylated IGF-1R in NSCLC plasma-derived circulating exosomes, opening a new avenue for noninvasive monitoring of IGF-1R and its activation status. Overall, the ability of microfluidic technology to detect specific circulating microvesicular markers directly from patient plasma samples has been demonstrated. The possibility of using circulating microvesicles to monitor the status of IGF-1R during targeted therapy in NSCLC patients is also discussed in this chapter.

6.2 Methods

6.2.1 Microfluidic Device Fabrication

The microfluidic exosome profiling platform was a hybrid device composed of a glass slide bound with a layer of poly (dimethylsiloxane) (PDMS) membrane. The PDMS layer with microfluidic channel network was fabricated using standard soft lithography. A master was the patterns of SU8 photoresist on a 4-in. silicon wafer and was silanized to facilitate removal of PDMS. Negative PDMS replicas were made by pouring a 10:1 mixture of PDMS base with the curing agent over the wafer, followed by curing at 60 °C overnight. After removing the replicas from the master, fluidic interfacing via holes were drilled on the PDMS layer at the desired location. The PDMS layer was bonded to a clean glass substrate to form enclosed microchannels immediately after exposing to UV ozone for 3 min. The surface of the PDMS microfluidic channel was treated using (3-aminopropyl) trimethoxy silane. The PDMS microfluidic channel was continuously flushed with a mixture of $H_2O/H_2O_2/HCl$ (in a volume ratio of 5:1:1) for 30 min to obtain the hydrophilic silanol-covered surface. After purging the microchannel with deionized water and dry Ar, the neat (3-aminopropyl) trimethoxy silane (Sigma-Aldrich) was pumped into the microchannel sequentially to perform a silanization reaction at room temperature for 10 min. Unreacted silane was flushed out using deionized water mixed with ethanol (1:1) in order to generate the hydrophilic surface. The nonspecific binding was further minimized by employing the blocking buffer (1 % BSA containing 0.1 % Tween 20) and washing buffer (1× PBS buffer, 0.01 % Tween 20).

6.2.2 Immunomagnetic Beads and Antibody Conjugation

Exosome capture beads (Dynal beads M-270 Streptavidin, 2.8 μm in diameter) were purchased from Invitrogen. Antibodies against α-IGF-1R, EpCAM, CD9, CD81, and CD63 were conjugated to the Dynal beads through biotin-streptavidin linkage per vendor's instructions, generating typical binding capacity of ~10 μg biotinylated antibody per 1 mg of beads. Relative lower concentrations of capture beads demonstrated higher capture efficiency. The 20 μL of antibody-coated beads (1 mg/mL) were premixed with 150 μL human plasma for 30 min and then mixture was introduced into microfluidic device. The local magnetic field was generated using a disk magnet (Licensed NdFeB, Grade N40, 2 mm in diameter, poles on flat face). Although the diameter of the magnet was 2 mm, which is smaller than the capture chamber, the extended magnetic field right covered the full chamber region as empirically optimized. IGF-1R capture beads were generated by coupling mouse monoclonal IGF-1R antibody to surface-activated Dynabeads (M-270 Epoxy, 2.8 μm in diameter) through epoxy-amine covalent bonds. The epoxy coupling reaction was performed at 37 °C overnight with gentle shaking. Typical binding capacity is ~5–8 μg antibody per 1 mg of beads, according to the vendor's instruction. The specificity of antibodies to IGF-1R/pIGF-1R was tested using ELISA.

6.2.3 Transmission Electron Microscopy and Exosome Characterization

Agar and resin embedding protocols were employed to ensure that exosome morphology was maximally maintained under TEM imaging. Exosome-bead complexes were fixed in 2 % (v/v) glutaraldehyde in PBS for 3 h, followed by two 15-min washes in PBS. Beads were resuspended in 4 % agar. After cutting the agar into 1 mm pieces, the specimen was fixed overnight in 2 % (w/v) osmium tetroxide in 0.1 M cacodylate buffer kept in the refrigerator. The specimen was rinsed two times for 15 min each with distilled water and then dehydrated in a graded series of ethanol: 30 % for 10 min, 70 % for 10 min, 95 % for 10 min, 100 % for 30 min, and 100 % for 30 min. The specimen was then placed into a graded series of mix of pure L.R. white (hard grade) and 100 % ethanol: 1:2 for 30 min, 1:1 for 30 min, and 2:1 for 30 min. Then three changes of ethanol-free pure L.R. white were conducted for 30 min and kept overnight. The resin-embedded specimen was polymerized at 60 °C for 24 h. Ultrathin sections (80 nm) were cut on Leica Ultracut-S Ultramicrotome and counterstained with 4 % (w/v) aqueous uranyl acetate for 5 min, followed by 3 min with a solution of lead acetate. Sections were viewed after counterstaining in a JEOL JEM-1400 Transmission Electron Microscope (equipped with a Lab6 gun) operating at 80 kV. Micrographs were prepared to a known scale, and diameters of vesicles bound to the bead surface were measured. Exosome sizes were measured and calculated using JEM-1400 TEM software with ruler function at 20 K magnification and normalized to the scale bar.

Other methods and protocols, including exosome isolation using differential ultracentrifugation and nanoparticle tracking analysis (NTA), are detailed below. Exosome microvesicles were prepared from fresh frozen blood plasma (2 mL). Differential centrifugations were carried out of the collected plasma or cell culture media in order to obtain exosomes. The plasma was centrifuged for 45 min at 10,800 rpm using ultracentrifugation at 4 °C (Thermo Scientific SORVALL WX ULTRA series centrifuge). Supernatants were purified by two successive centrifugations for 2 h each at 35,800 rpm. Pellets were resuspended in 20 μL PBS and washed once in a second 20 μL PBS. The amount of pellet proteins recovered was measured by Bradford assay (BioRad). Exosome microvesicles were conserved at −80 °C until use. Exosome microvesicles purified by ultracentrifugation were collected and diluted in 1:50, 1:125, 1:250, and 1:500 in 1× PBS buffer. The size and concentrations of diluted microvesicles were counted using NanoSight following standard protocols. By monitoring the trajectory of microvesicle movement, particle numbers within the size range of 0–500 nm were estimated under serial dilutions. Concentrations were calibrated back to the plasma concentration. For consistent reading, measurement settings were optimized and five replicas were performed to obtain average measurements.

Western blotting was performed following standard protocol using Tris-glycine pH 8.3, 4–12 % precast polyacrylamide slab mini-gels with Blot Module (BioRad). Samples were prepared by adding running buffer (0.1 % SDS) and heating at 65 °C for 5 min. After electrophoresis at 125 V for 2 h, gels were electrotransferred to cellulose membranes (0.2 μm) at 25 V for 2.5 h. After twice washing (1× PBS, 0.5 % Tween 20, pH 7.4), the membranes were blocked with 5 % dry milk overnight at 4 °C with shaking. The solution of primary antibody (1:1000) was added into blocking buffer for 2-h incubation with shaking at room temperature. After incubation, the membrane was washed three times for 10 min each. The secondary antibody streptavidin-HRP (Invitrogen, ELISA grade, 1.1 mg/mL) diluted 1:2500 in the blocking solution was added for 1-h incubation at room temperature with agitation, and then the washing step was repeated three times. The membrane was subsequently developed with Chromogenic Substrate Reagent (BioRad) until desired band intensity was achieved. Imaging was performed using FluorChem E (ProteinSimple) with a proprietary filter.

6.3 Microfluidic Multistage Integration

A PDMS microfluidic device was fabricated that contained a cascading microchannel circuit for streamlining multistage analysis of exosomes directly from human plasma, including exosome isolation and enrichment (1st stage capture), exosome lysis, fluidic mixing and immunomagnetic capture of intravesicular targets (2nd stage capture), and protein immuno-assays. The device architecture and multistage procedures are detailed in Fig. 6.1. The two cascading magnetic-bead capture chambers were 4 mm in diameter, capable of capturing up to 10^9 2.8 μm microbeads each. The plasma sample premixed with antibody-labeled magnetic beads was introduced through sample Inlet #1 into the first capture chamber in which the magnetic beads with bound exosomes were isolated and washed by PBS washing buffer. The lysis buffer was introduced through Inlet #2 and incubated with captured exosomes in the first chamber by stopping the flow. The lysate then flowed into a serpentine microchannel along with antibody-labeled magnetic beads introduced from two side channels in order to specifically capture released intravesicular proteins (Fig. 6.1). The protein capture beads were magnetically retained in the second chamber in which detection antibodies (Inlet #3) and chemifluorescence reagents (Inlet #4) were introduced to perform sandwich immunodetection of captured intravesicular protein markers. The complete analysis, including two-stage immunomagnetic capture, was completed in ~2 h (0.5 h off-chip incubation and ~1.5 h on-chip assay) with as low as 30 μL plasma samples.

Multiple fluidic introductions were precisely controlled at a certain flow rate using a 4-syringe programmable pump system (HARVARD Pump11 Elite) with pico-litter resolution. Plasma samples (150 μL) premixed with immunomagnetic microbeads were introduced through Inlet #1 into the first-stage magnetic capture chamber at a speed of 2 μL/min for exosome immunomagnetic isolation. The lysis buffer was passed into the first-stage magnetic chamber after exosome capture and stopped in order to complete lysis. The lysate was then washed out from the first-stage magnetic capture chamber and subsequently flowed into the serpentine

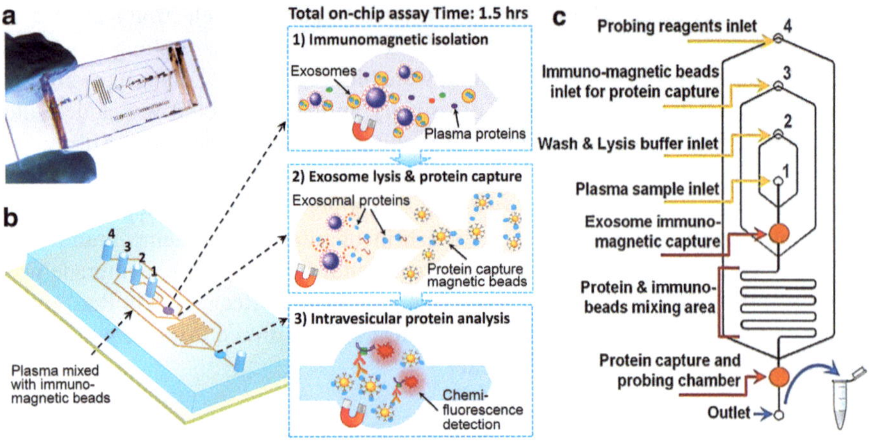

Fig. 6.1 Multistage integrated microfluidic exosome profiling platform. (**a**) Image of the prototype PDMS chip containing a microchannel network for cascading exosome profiling. (**b**) Schematic illustration of streamlined on-chip immunomagnetic isolation of exosomes, chemical lysis, and intravesicular protein analysis. (**c**) Workflow and operation procedures of microfluidic exosome profiling (reproduced from He et al. 2014)

mixing microchannels designed for immunomagnetic binding of potential exosomal protein markers. After capture of target exosomal protein, the microbeads were collected at the second-stage magnetic chamber for total IGF-1R antibody (1:100 dilution) probing. The secondary antibody anti-rabbit IgG-AP (1:100 dilution) was introduced for chemifluorescence detection. The fluorogenic substrate DiFMUP was introduced through Inlet #4 for quantitative and sensitive detection of target protein marker. The washing buffer (1× PBS, 0.01 % Tween 20) was employed for 5 min between each step. Fluidic exchange was performed carefully to avoid any unwanted air bubbles.

6.3.1 Flow Conditions for Exosome Capture and Processing

The magnetic capture of beads has been investigated, because it dictates the overall performance of exosome isolation and molecular analysis. Beads suspended in a buffer solution were infused into the chamber and retained by the magnet placed underneath the centerExosomes:flow conditions of of the chamber (Fig. 6.2a, inset), forming a field-induced aggregate due to dipolar interactions between the beads. Previous research reported that the amount of magnetically captured beads in a microchannel can be represented by the size of bead aggregate that increases linearly with time at a constant flow rate [41, 42]. This approach was adopted in order to estimate the amount of beads captured in the chamber as a function of total sample infusion volumes, as summarized in Fig. 6.2a. In this case, flow rates used to reach total volumes were randomly varied during bead capture

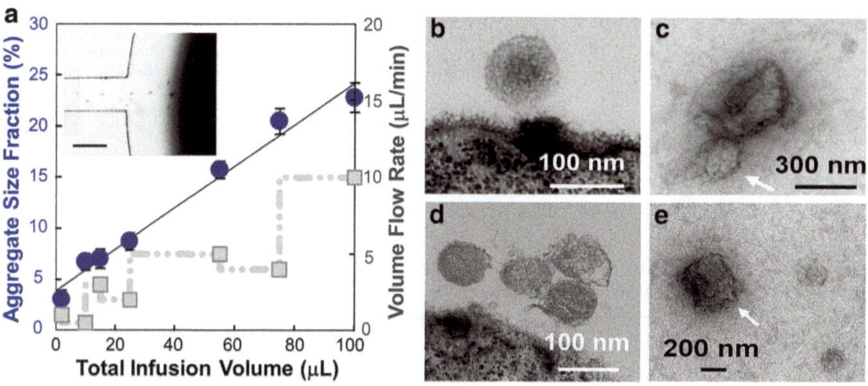

Fig. 6.2 Optimization of flow conditions for microfluidic immunomagnetic capture of circulating exosomes. (**a**) Plot of the amount of beads captured in the chamber represented by the aggregate area fraction as a function of total infusion volume and flow rate. The error bars are the standard deviation of three replicates. *Inset*: Bright-field microscopic photograph of magnetic bead capture in the first magnetic capture chamber. Scale bar is 100 μm. (**b, c**) Representative TEM images of exosomes from NSCLC patient plasma on-chip isolated by immunomagnetic beads and ultracentrifugation. (**d, e**) Large aggregates (**d**) and other membranous particles (**e**) presented in ultracentrifugation-purified exosomes indicated by *white arrows*

(Fig. 6.2a, dash line). Results showed that the size of aggregate normalized to the chamber size was linearly dependent on the total sample infusion volume regardless of applied flow rates (1–10 μL/min). This observation was consistent with the reported simulation and experimental results [43]. Linear volume dependence indicated the well-maintained capacity and efficiency for bead capture using developed microfluidic platform, thereby ensuring quantitative measurement of exosomes from biological samples. The very low flow rates were chosen for exosome capture and intravesicular protein capture in both chambers (2 and 1 μL/min, respectively). The bead recovery rate was determined to be >99 % by counting residual beads in the elute volume. In order to investigate the effects of flow conditions on exosome immunomagnetic capture, the morphology of bead-captured exosomes has been examined using TEM imaging. Several representative TEM images are shown in Fig. 6.1b, c for on-chip immunomagnetically isolated plasma exosomes as compared to ultracentrifugation-purified plasma microvesicles (Fig. 6.2d, e). Immunomagnetic isolation revealed a population of vesicles in a typical round, homogenous morphology with a major size distribution of 40–150 nm, which is consistent with reported exosome morphology [44]. The cup-shaped morphology is often erroneously considered to be a typical appearance of microvesicles resulting from membrane collapse during sample drying for TEM without embedding and sectioning. Relatively large aggregates were often observed, as well as other membranous particles (arrows in Fig. 6.2d, e) in a heterogeneous population of vesicles purified by ultracentrifugation, although ultracentrifugation is the most commonly used technique to isolate exosomes.

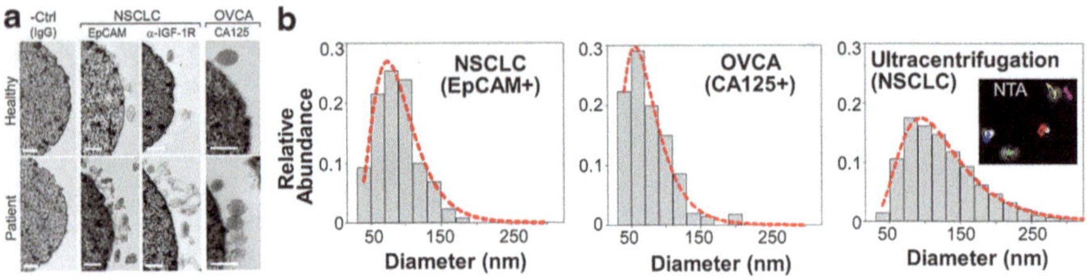

Fig. 6.3 (**a**) Representative TEM images showing enriched exosomes on the surface of antibody-conjugated beads from NSCLC and ovarian cancer (OVCA) samples. Significantly fewer vesicles from healthy plasma and almost no vesicles on the negative control beads without specific antibodies were observed. (**b**) Representative size histograms of on-chip isolated exosomes from NSCLC (EpCAM+, $n = 130$) and OVCA (CA125+, $n = 130$) compared to that of ultracentrifugation-purified NSCLC vesicles measured by NTA using NanoSight (*insets*). Sizes were obtained by averaging five measurements. Red dot plots are log-normal fitting ($R^2 > 0.98$). Scale bars: 100 nm

On-chip purification of exospores was performed from NSCLC and healthy plasma using beads labeled with EpCAM, α-IGF-1R, or CA125 antibodies. As demonstrated in Fig. 6.3, TEM examination of collected microbeads showed that bead surfaces were densely coated with numerous round vesicles, approximately 50–150 nm in diameter, in patient samples, while significantly less vesicles from healthy plasma and no vesicles for the negative control beads without antibody conjugation were observed. In addition, most bead-bound vesicles remained round in shape and only very few appeared irregular, indicating the high integrity of exosomes [13, 45]. These results suggest that flow conditions used effectively minimized nonspecific binding while preventing specifically captured exosomes from being eluted out by shearing force. On-chip capture performance was further characterized by size distribution of individual exosome subpopulations isolated by targeting tumor-associated markers (EpCAM, α-IGF-1R, and CA125) and common exosomal markers (CD9, CD81, and CD63). Size is the most acceptable criterion for exosome identification and differentiation from other extracellular vesicle types. The current consensus is that exosomes originated from multivesicular endosome fusion are typically smaller than 150 nm, while a majority of microvesicles derived from plasma membranes are relatively larger (150–2000 nm). Compared to nano-particle tracking analysis (NTA) using NanoSight, which requires ~1 mL of concentrated vesicles (~10^9 mL^{-1}) for accurate size determination, TEM provides a robust means of sizing and counting exosomes in small volumes collected from microfluidic isolation without significant dilution (~30 μL). Most immunocaptured exosomes were found to be smaller than 150 nm, with a notably smaller size range (e.g., 97 % of EpCAM+ and CA125+ vesicles <150 nm) than those exosomes obtained by ultracentrifugation (72.1 %) (Fig. 6.3). Current "gold standard" approaches based on ultracentrifugation yield a mixed population of various extracellular vesicle types with a wide size distribution. NTA analysis of ultracentrifugation-purified vesicles yielded a broader size variation and no distinct profiles between healthy and NSCLC cases (Fig. 6.3). These findings suggest that this microfluidic immunocapture method provides a more specific means of purifying exosomes than ultracentrifugation.

6.3.2 Microfluidic Streamlined Chemical Lysis of Exosomes

In order to chemically lyse captured exosomes, lysis buffer was pumped to fill the first capture chamber and then stopped to incubate with the immobilized beads. A mild nonionic detergent, Triton X-100, was used because it is one of the most widely used surfactants to lyse cells but retain activity of released proteins. The lysis conditions, including Triton X-100 concentrations and incubation times, were characterized. As shown in Fig. 6.4, TEM examination suggested that incubation with 5 % Triton X-100 for 5 min is sufficient to completely lyse the exosome cluster bound on the bead surface in comparison to control beads collected before lysis.

After lysis of exosomes, the lysate was flushed into a serpentine microchannel to mix with magnetic beads conjugated with specific antibody in order to capture

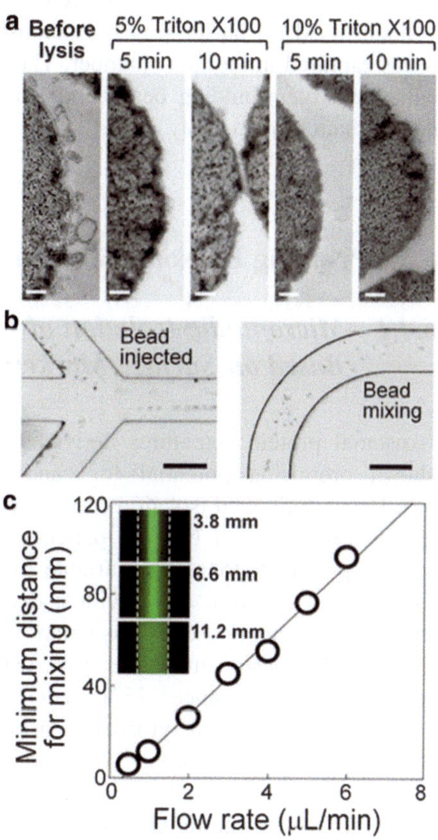

Fig. 6.4 (**a**) Exosome-bound microbeads before and after lysis were observed by TEM. Scale bar: 100 nm. (**b**) Bright-field images showing the process of introducing protein capture beads (*left*) and mixing in the serpentine microchannel (*right*). Scale bar: 200 μm. (**c**) Plot of the minimum flow travel distance required for uniform mixing in the serpentine channel over a flow rate range of 0.5–6 μL/min. *Inset*: Fluorescence images of mixing a stream of 0.1 μM FITC-BSA solution with the bead suspension at the flow rate of 1 μL/min

released intravesicular proteins of interest. In order to facilitate fluidic mixing, a microchannel configuration was employed such that bead suspension was injected (Inlet #3) from the two side channels in order to flank the main stream of exosome lysate (Fig. 6.4b, left), thereby reducing the cross-sectional distance for mass transfer [46]. The effect of flow rate on mixing was characterized in the serpentine channel by co-flowing a 0.1 μM solution of FITC-labeled BSA and the bead solution. The injected beads were well mixed to randomly distribute across the channel width (200 μm) within a short flow travel distance of ~10 mm (Fig. 6.4b, right). Fluorescence imaging showed that uniform BSA concentration across the channel can be obtained within a flow travel distance of 11 mm at a flow rate of 1 μL/min (Fig. 6.4c, inset). The minimum travel distance was measured for complete mixing at a flow rate range of 0.5–6 μL/min (Fig. 6.4c), which exhibited a linear response. This result provided a guidance for determining suitable flow rate under given microfluidic dimensions for efficient mixing and protein capture. A low flow rate of 1 μL/min was chosen, resulting in an estimated incubation time of 3.8 min when flowing through a 25 cm long mixing channel and the second magnetic capture chamber. This extensive incubation aided by convective mixing allowed efficient mixing and solid-phase affinity capture inside microchannels [47, 48]. In order to maximize recovery of released exosomal proteins, a total volume of 30 μL buffer was continuously infused to wash the exosome-binding beads captured in the first chamber. The exosome-binding beads were then mixed with the protein-binding beads and finally collected in the second chamber for downstream protein assay.

6.4 Profiling Exosome Subpopulations

6.4.1 Microfluidic Isolation of Exosome Subpopulations Based on Surface Markers

Exosomal protein signatures derived from the plasma of cancer patients have shown promising potential for cancer research and clinical diagnosis [49–52]. Using optimized isolation conditions, plasma-derived exosomes were characterized from stage II NSCLC patients and healthy controls. Based on exosome surface protein topography, variable exosome subpopulations can be identified. Microfluidic isolation of individual exosome subpopulations was conducted by targeting commonly used surface markers (e.g., CD9, CD81, and CD63) [15]. Epithelial cell adhesion molecule (EpCAM)-positive exosomes have been recently reported in lung cancer [53]. α-IGF-1R, the extracellular alpha-subunit of IGF-1R, is involved in tyrosine kinase signal transduction pathways in NSCLC [54]. TEM visualization of the bead-exosome complex allows morphological identification and abundance analysis of exosome surface protein expression. Exosome size is a commonly used, yet arguable, criterion for differentiating various components in extracellular vesicles and for defining exosomes [55]. Current consensus is that exosomes originated from multivesicular endosome fusion are typically smaller than 150 nm, while a majority of microvesicles derived from plasma membrane are relatively larger. The log-normal fitted size distributions ($R^2 > 98$ %) were obtained for individual exosome subpopulations (EpCAM+, α-IGF-1R+, CD9+, CD81+), as summarized in Fig. 6.5. Results showed that targeting various surface markers leads to considerable variation in exosome subpopulation isolation: ~97 % of EpCAM+ and CD9+ exosomes are smaller than 150 nm (Fig. 6.5a, c) as compared to 80 % in α-IGF-1R+ and CD81+ sub-populations, respectively (Fig. 6.5b, d). For comparison, NTA analysis of ultracentrifugation-purified vesicles was performed and found no distinct size distribution profiles between healthy and NSCLC samples (Fig. 6.5e, f). Compared to ultracentrifugation, microfluidic immunoaffinity isolation of EpCAM+ and CD9 + exosome subpopulations yielded notably narrower size distributions with a higher percentage of exosomes smaller than 150 nm (~97 % vs. 72.1 %). Based on general size criterion to define exosomes, the results suggested that the EpCAM- or CD9-based microfluidic immunoaffinity capture provides a more specific approach to identify and purify exosomal subpopulations in NSCLC compared to centrifugation-based methods. This conclusion is consistent with the report by Tauro et al. which observed better homogeneity and more than twofold higher exosomal proteins associated with exosome biogenesis from immuno-captured exosomes than ultracentrifugation and density gradient-purified exosomes [56].

In order to demonstrate the ability to detect exosomal expression patterns associated with cancer, the relative quantification of five exosome subpopulations defined by individual surface markers was conducted using TEM. Figure 6.6 illustrates tested representative patients and controls. NSCLC patients typically exhibit distinct exosome expression profiles compared to healthy controls, with a three- to fivefold increase in exosome abundance except when captured with CD63. CD63, commonly thought to be expressed in exosomes, has been a universal surface marker for exosome isolation and profiling in cell lines and biopsies of various cancers [57, 58]; however, observations developed here suggested differently. The number of CD63+ exosomes in the plasma from NSCLC patients compared to other surface markers was universally low, indicating inefficiency of CD63 capture for exosome purification. The adaptability of developed microfluidic approach was further demonstrated for other cancers by testing OVCA with the tumor markers (EpCAM and CA125) and exosomal markers (CD9, CD81, and CD63). OVCA samples also provided a positive control for NSCLC studies because CD63 was found to be highly expressed in OVCA exosomes. As expected, high CD63 expression was observed, thereby validating developed method and supporting the observation of low CD63 expression in NSCLC cases. The ability to discriminate disease from healthy subjects and detect differential expression of markers (e.g., CD63) in cancers indicates high specificity of the microfluidic method.

In order to validate microfluidic results, parallel measurements of ultracentrifugation-purified vesicles were performed using standard analytical

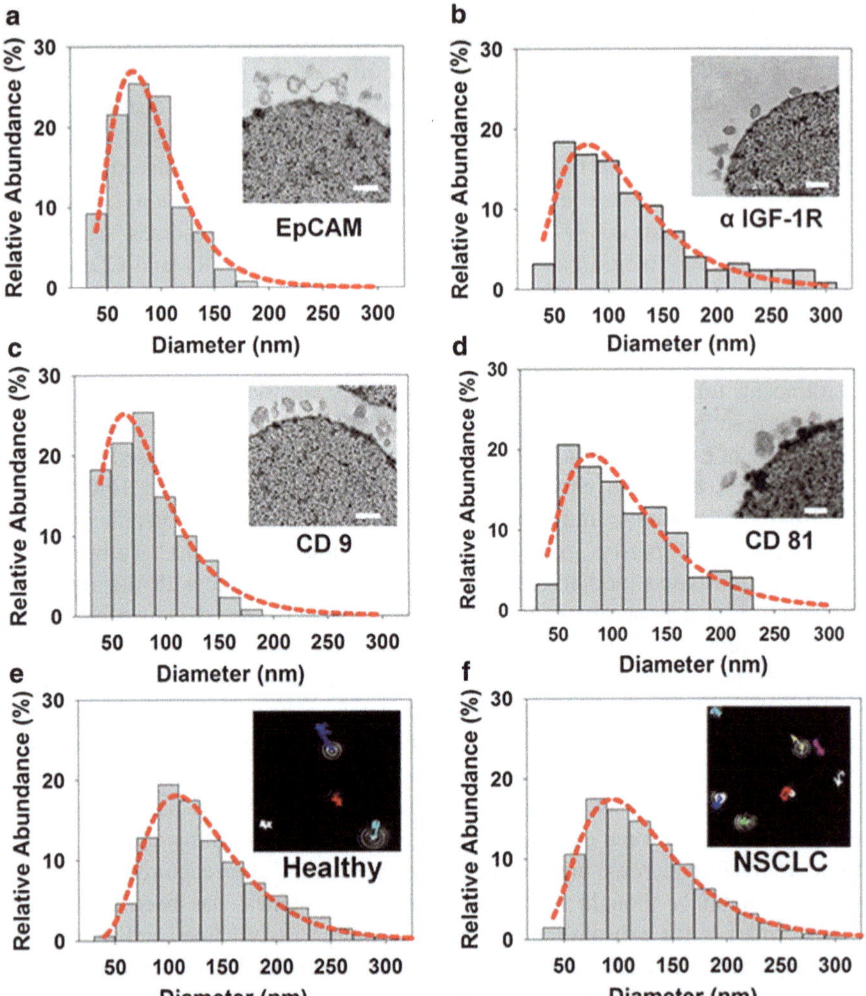

Fig. 6.5 Size distributions of exosomes isolated from NSCLC plasma by microfluidic immunomagnetic capture and the ultracentrifugation-based method. (**a–d**) Size histograms of exosome subpopulations isolated by targeting four exosomal surface proteins: EpCAM, α-IGF-1R, CD9, and CD81, respectively. A total of 130 captured vesicles were measured for each marker. *Insets*: Representative TEM images of exosomes from each subpopulation. Scale bar is 100 nm. (**e**, **f**) Size histograms of ultracentrifugation-purified vesicles from (**e**) healthy and (**f**) NSCLC plasma. NTA by NanoSight was employed to analyze microvesicle sizes. *Insets* show the movement trajectory of microvesicles (in *color*). Size was obtained by averaging five NTA measurements. The *dotted red lines* are log-normal fitted plot ($R^2 > 0.98$)

methods. Exosome abundance in patient plasma measured by NTA showed a ~4-fold increase on average compared to healthy controls ($p = 0.0001$, Fig. 6.6b), thereby agreeing with total protein levels estimated by the Bradford assay. Results

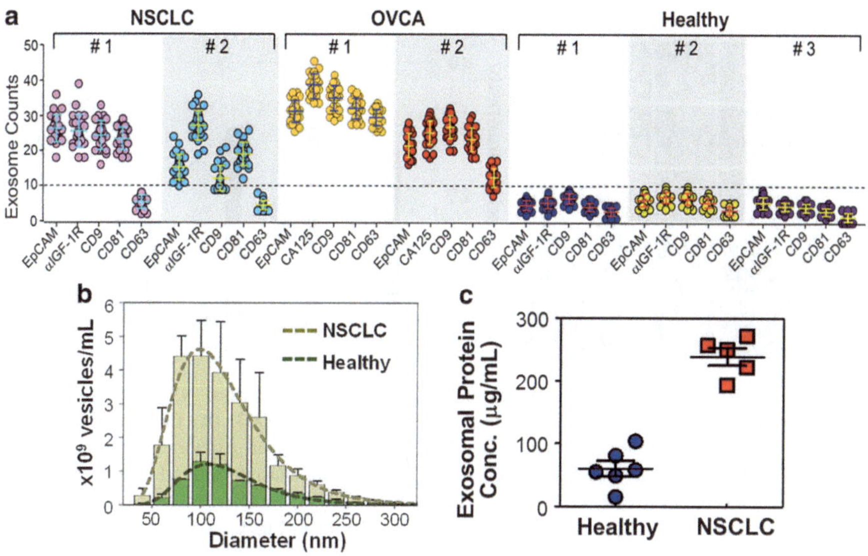

Fig. 6.6 Microfluidic isolation and surface phenotyping of circulating exosomes in cancer. (**a**) The scattered dot plot of abundance of bead-bound exosomes from NSCLC, OVCA, and healthy plasma obtained by TEM ($n = 25$). A panel of surface markers (EpCAM, α-IGF-1R, CA125, CD9, CD81, and CD63) were used for exosome isolation. The *dashed line* indicates the highest exosome counts observed for healthy controls. (**b**) NTA analysis of size distribution and abundance of vesicles purified from NSCLC and healthy controls by ultracentrifugation. The error bars are the standard deviations. The *dashed lines* are log-normal fitting ($R^2 > 0.98$). (**c**) Bradford assay of total proteins in ultracentrifugation-purified exosomes from NSCLC patients (stage II) and healthy subjects ($p = 0.0007$)

of the Bradford protein assay (Fig. 6.6c) showed significantly higher total exosomal protein levels in NSCLC patients than the healthy controls (3.9-fold increase in the averaged concentration, $p = 0.0007$). The change of total exosomal protein associated with disease status correlated well with the microvesicle quantity determined by NTA, collaboratively validating the ability of developed microfluidic method to detect disease-caused variations in exosome expression.

6.4.2 Exosomal Surface Protein Expressions

In order to determine if plasma-derived exosome subpopulations show similar protein profiles as the tumor origin in NSCLC, the three-color immunofluorescence histological (IFH) study of patient-matched lung tumor tissues was performed, as presented in Fig. 6.7. EpCAM, α-IGF-1R, CD9, and CD81 were found to be highly expressed in tumor tissues from patients, supporting the corresponding plasma exosomal subpopulation profiles generated by microfluidic technology (Fig. 6.6a). A high expression of four markers was also found in the tumor tissue collected from

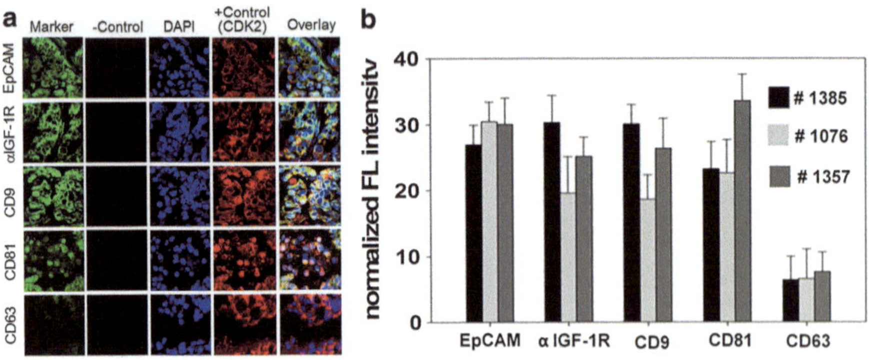

Fig. 6.7 (**a**) Representative IFH images of matched tumor tissue from NSCLC patient #1 in Fig. 6.6a, showing high expression of biomarkers except for CD63. (**b**) Immunofluorescence analysis of NSCLC tumors (patient ID 1385, 1357, 1076) for exosome-specific markers (EpCAM, α-IGF-1R, CD9, CD81, CD63) found in matching plasma-derived exosomes. CD63 protein levels were not readily detected in the examined NSCLC tumors, which is consistent with the on-chip observation of patient-matched plasma exosomes in Fig. 6.6a

Patient #1076 before treatment, in contrast to the low-abundance exosome subsets found in plasma sampled after the completion of surgery and chemotherapy. This result suggests the feasibility of microfluidic method for diagnosis and monitoring residual disease after surgery and tumor progression during subsequent therapy. Finally, the expression level of CD63 was considerably low in tumor tissues from all three NSCLC cases, further confirming results of microfluidic-based analyses. The matched molecular profiles between circulating exosomes and tumor tissues suggested the possibility of exosomal profiling as a noninvasive blood-based test for cancer diagnosis and monitoring of cancer treatment. The studies in this chapter also provided more experimental evidences that support recent questions regarding the use of CD63 as a general surface marker for unbiased isolation of exosomes [59].

In order to confirm the microfluidic study of exosomal surface protein expressions, protein composition of plasma vesicles from various NSCLC stages and healthy controls was analyzed by Western blotting, as presented in Fig. 6.8. Differential expression of tested protein markers agreed well with variation in the abundance level of exosome subpopulations detected by the microfluidic approach. In order to verify detection of CD63 expression, an ovarian cancer cell line (C30) known to produce CD63+ exosomes was included as the positive control. All NSCLC cases in various disease stages (e.g., stages I, II, and IV) and healthy cases showed indiscernible or low CD63 levels as opposed to the ovarian cancer cell line (Fig. 6.8). Overall, the microfluidic and centrifugation-based methods showed significant elevation in exosome concentration in NSCLC patients as compared to healthy controls, proving clinical relevance of circulating exosomes in cancers. Matched molecular profiles between circulating exosomes and tumor tissues suggest the possibility of exosomal profiling as a noninvasive blood-based test for cancer diagnosis and monitoring of cancer treatment. The studies in this

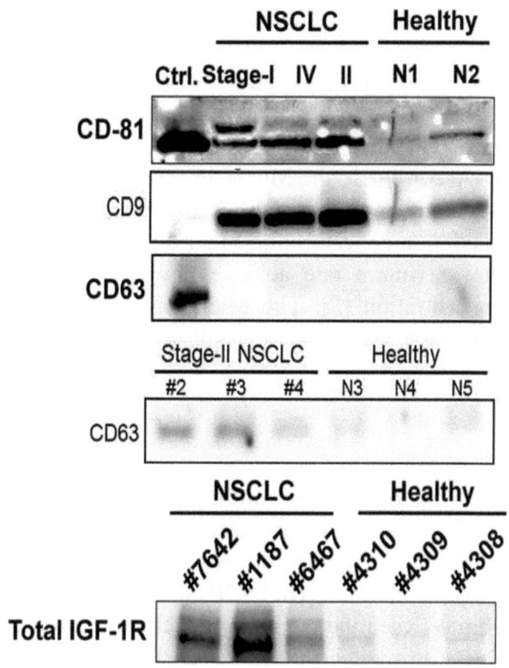

Fig. 6.8 Western blot analysis of ultracentrifugation-purified exosomes derived from human plasmas. C30, an ovarian cancer cell line that produces CD63-containing exosomes, served as a positive control. CD9 and CD81 expressions were readily detectable in NSCLC patient plasma-purified exosomes as compared to CD63. Total IGF-1R derived from NSCLC plasma exosomes showed higher expression compared to healthy controls, thereby confirming the on-chip assay using the same samples

chapter also provided experimental evidences that supported recent questions regarding the use of CD63 as a general surface marker for unbiased isolation of exospores. Microfluidic protein topography profiling technology revealed more detailed molecular signatures associated with cancer, which is inaccessible to centrifugation-based methods.

6.5 Noninvasive Analysis of IGF-1R in NSCLC Through Microfluidic Exosome Profiling

6.5.1 Microfluidic Chemifluorescence Detection of Intravesicular Markers

The cascading microfluidic immunomagnetic capture strategy enables integration of intravesicular proteomic assays with upstream exosome subpopulation isolation in order to confer a "sample-in-answer-out" exosome analysis capability. To demonstrate this capability, a complete assay was performed that combined exosome subpopulation isolation and enrichment, chemical lysis, flow mixing for intravesicular protein capture, and chemifluorescence-based sandwich immunoassays in order to quantify total IGF-1R and phosphorylated IGF-1R in EpCAM+ NSCLC exosomes. The on-chip immunomagnetic sandwich assay and chemifluorescence detection were optimized using a matched antibody pair of rabbit anti-total IGF-1R and alkaline phosphatase (AP)-conjugated anti-rabbit IgG, as well as the substrate DiFMUP. The incubation time was an important factor for enzymatic chemifluorescence detection in microfluidic system because the second capture chamber was not completely closed after the flow stopped.

According to the Michaelis-Menten equation below [60], the turnover rate of an enzyme was fastest and became independent of the substrate concentration at the maximum concentration of substrate (saturation). In the equation, $V_{\max}$ represents the maximum rate achieved by the system at maximum (saturating) substrate concentration $[S]$. The Michaelis constant K_{M} is the substrate concentration at which the reaction rate is half of $V_{\max}$:

$$\upsilon = \frac{d[P]}{dt} = \frac{V_{\max}[S]}{K_{\mathrm{M}} + [S]}$$

Because of enzyme-limiting conditions, the turnover rate rose linearly with increased enzyme concentration, as reflected by the conversion of more DiFMUP substrate to strongly fluorescent DiFMU product. Therefore, application of approximately saturating DiFMUP concentration allowed acquisition of a linear relationship between the rates of product conversion and the amount of enzyme presented in the third-stage magnetic chamber. In order to achieve a saturating substrate concentration at the site of the reaction chamber, a 1000 μM DiFMUP solution diluted from 10 mM stock solution of DiFMUP in dimethyl sulfoxide (DMSO) was introduced through a syringe pump for 5 min and then stopped for a quiescent flow condition. Product DiFMU accumulation during the assay occurred linearly after the introduction of DiFMUP and then reached maximum as the equilibrium was achieved (Fig. 6.9). Product DiFMU was imaged using a UV filter cube (excitation 325–375 nm, emission 435–485 nm). The fluorescent signal reached maximum after 6-min incubation (Fig. 6.9) as a possible result of the balance between

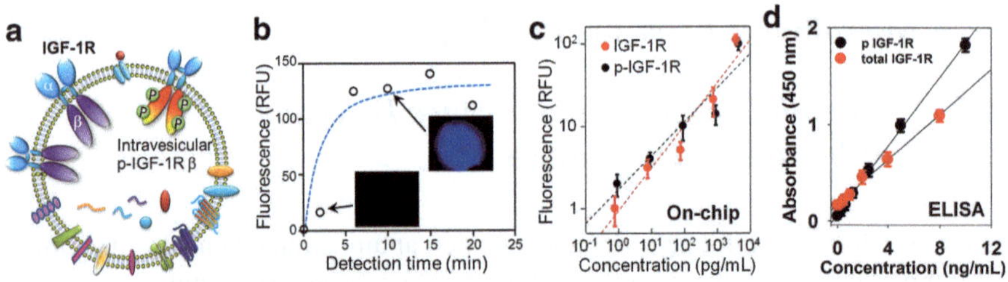

Fig. 6.9 Integrated microfluidic platform analysis of exosome intravesicular marker. (**a**) Schematic of transmembrane IGF-1R in exosomes. The extravesicular IGF-1R α unit and phosphorylated β domain inside the vesicle were targeted for surface phenotyping and intravesicular protein analysis of exosomes. (**b**) The effect of incubation time on chemifluorescence detection using alkaline phosphatase and substrate DiFMU. (**c**) Calibration of on-chip capture and detection of IGF-1R and p-IGF-1R. (**d**) Calibration of quantitative detection of total IGF-1R and phospho-IGF-1R using the standard 96-well microplate ELISA kit

enzymatic conversion and diffusion of the fluorescent product out of the chamber. The generated blue fluorescence throughout the entire reaction chamber was collected to eliminate spatial nonuniformity in the system and normalize to background. Image analysis was conducted using ImageJ in order to obtain average intensity. The calibration curve was developed for on-chip IGF-1R and pIGF-1R assays by conducting the entire process, except the lysis step, using standard human IGF-IR and pIGF-1R. As plotted in Fig. 6.9 (linear range over employed concentrations, $R^2 > 0.98$), the microfluidic assay achieved quantitative detection of total IGF-1R and pIGF-1R with a limit of detection of 8.43 fg and 11.5 fg, respectively (S/N = 3), equivalent to plasma concentrations of 0.281 pg/mL and 0.383 pg/mL. Such sensitivity was at least 100-fold higher than commercial ELISA kits [3].

6.5.2 *Microfluidic Profiling of Exosomal IGF-1R for Noninvasive Prognosis*

The integrated microfluidic profiling has been demonstrated directly using human plasma for noninvasive measurement of IGF-1R in pretreatment stage II NSCLC patients. Therefore, we examined total IGF-1R and pIGF-1R levels of EpCAM+ exosomes in each sample (30 μL) and compared to parallel standard ELISA measurement of ultracentrifugation-purified exosomes from the same patient plasma (2 mL). As shown in Fig. 6.10, microfluidic analysis showed a significantly higher average total IGF-1R expression (EpCAM+) for NSCLC patients compared to healthy individuals ($p = 0.0001$), as verified by ELISA results ($p = 0.0097$) (Fig. 6.10b). This finding of exosomal total IGF-1R level was consistent with observations reported from NSCLC cell lines and tumor tissues [61–63]. Moreover, ELISA analysis of ultracentrifugation-purified vesicles showed significant patient-to-patient heterogeneity in IGF-1R expression with a coefficient of variation (CV) of 56.4 %; microfluidic approach revealed fairly consistent protein profiles (CV = 11.2 %) by confining the exosome target to the EpCAM+ subpopulation. The detected IGF-1R concentration was found to correlate linearly with the total abundance of plasma vesicles determined by NTA, while microfluidic approach measured a fraction of the circulating vesicles (Fig. 6.11). In addition, quantitative detection was achieved for vesicle concentrations much lower than healthy levels. These results validated the method for sensitive and quantitative characterization of circulating exosomes in clinical samples. IGF-1R overexpression was also evident in ELISA results presented in Fig. 6.10 ($p < 0.01$), consistent with previous observations reported for NSCLC cell lines and tumor tissues. ELISA detects the total IGF-1R level in all vesicle types co-purified by ultracentrifugation, while the microfluidic approach enables characterization of molecularly defined subpopulations. Many factors can contribute to the observed patient heterogeneity, including distinct exosome subpopulation composition, differential protein expression in subpopulations, and other types of extracellular vesicles. As demonstrated in

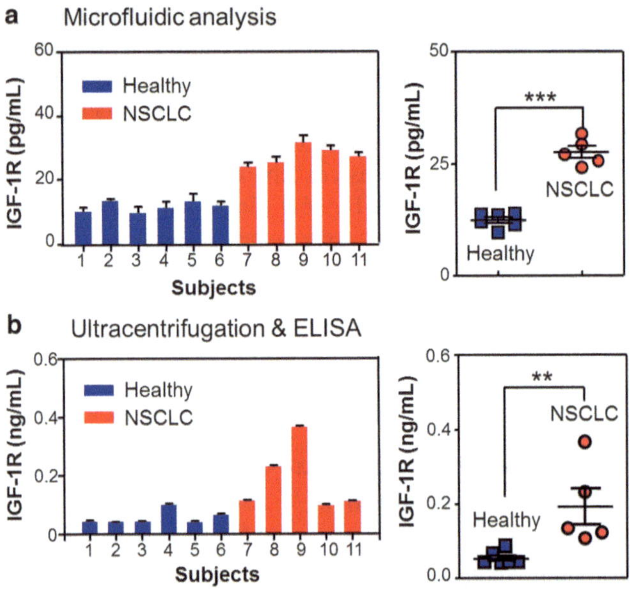

Fig. 6.10 Quantitative detection of total IGF-1R in circulating exospores directly from clinical plasma samples. (**a**) Results of integrated microfluidic analysis presented in the bar (*left*) and scattered dot (*right*) plots show significant overexpression of IGF-1R in EpCAM+ exospores of NSCLC patients compared to healthy controls ($p = 0.0001$, CV = 11.2 %). (**b**) Parallel ELISA analysis confirmed the overexpression of IGF-1R in total exosomes purified from the same subjects by ultracentrifugation ($p = 0.0097$, CV = 56.4 %). The error bars are standard deviations ($n = 3$) in all cases

Fig. 6.11 Quantitative detection of exosomal IGF-1R as a function of total exosome concentration in plasma. Tenfold dilutions of an NSCLC plasma sample (patient #11 in Fig. 6.10) were assayed to obtain the calibration curve (*solid triangles*). Measurements of the subjects in Fig. 6.10 were superimposed in the plot (*inset*). Error bars are standard deviations ($n = 5$ for x-axis and $n = 3$ for y-axis)

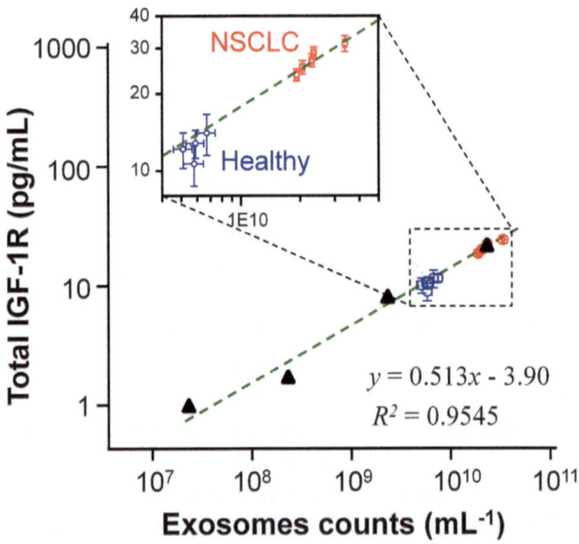

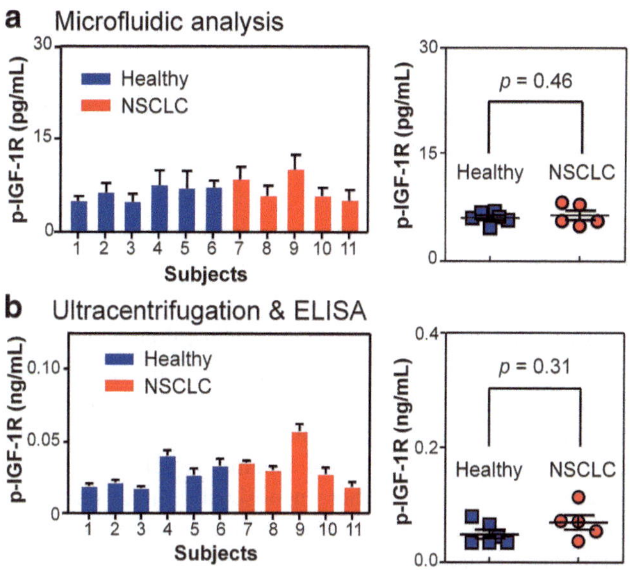

Fig. 6.12 Quantitative analysis of intravesicular p-IGF-1R in circulating exospores using the same clinical samples as in Fig. 6.10. (**a**) Bar (*left*) and scattered dot (*right*) plots of microfluidic results show no significant difference of the p-IGF-1R level in the EpCAM+ exospores between NSCLC and healthy subjects ($p = 0.46$). (**b**) No correlation between the p-IGF-1R level and the disease state was verified by ELISA analysis ($p = 0.31$). The error bars are standard deviations ($n = 3$) in all cases

Figs. 6.10 and 6.11, the microfluidic approach allows interrogating these factors in well-defined conditions.

In addition to total IGF-1R, the phosphorylation profile of exosomal IGF-1R was evaluated from corresponding patient plasma samples. No correlation was found between pIGF-1R and total IGF-1R levels in EpCAM+ exosomes and ultracentrifugation-purified vesicles (Fig. 6.12). Previous studies have also shown lack of correlation between the pIGF-1R level, the total IGF-1R level, and the response to treatment in NSCLC [64–66] and other cancers. One proposed explanation was the cross-reactivity of the pIGF-1R antibody with the phosphorylated insulin receptor (IR) that has close homology to IGF-1R [67]. In this study, the cross-reactivity of anti-IGF-1R and anti-pIGF-1R antibodies from different sources has been investigated using a commercial ELISA kit for IR (regardless of phosphorylation state). Results showed no discernible cross-reactivity between IGF-1R antibodies with IR, while considerable cross-reactivity between pIGF-1R antibodies with IR was observed. Because beads conjugated with a specific IGF-1R antibody, in order to capture the total IGF-1R released from exosomes, microfluidic approach developed in this chapter should rule out the possibility of IR interference with the detection of pIGF-1R. However, understanding the regulation of IGF-1R and its phosphorylation level in cancer requires mechanistic studies of IGF-1R signaling pathways, which is beyond the scope of this work. Overall, microfluidic technology demonstrated in this chapter enables sensitive and specific proteomic profiling of exosomes in well-defined conditions to facilitate elucidation of biological functions and clinical implications of circulating exosomes.

6.6 Discussion

Exosome analysis is an emerging paradigm in clinical care, especially for noninvasive diagnosis and monitoring of patient response to disease treatment, due to the fact that exosomes are abundant in plasma and malignant effusions and contain a selective repertoire of enriched proteins that reflect their cellular origin [68]. Currently, lack of technologies for efficient isolation, molecular classification, and comprehensive characterization of these nanosized vesicles remains a major hindrance in the field. Microfluidic technology offers a unique solution to surmount this technical obstacle because of its inherent merits in manipulation of small-scale particles and biochemical analysis. However, limited efforts have been made to improve isolation of microvesicles or quantitative detection of microvesicle number and surface protein expression. The developed microfluidic system in this chapter has a threefold distinction: (1) it offers a "sample-in-answer-out" capability for high-performance total analysis of exosomes by integrating cascading steps for isolation and molecular profiling of circulating exosomes in one microfluidic device (Fig. 2.1); (2) it enables profiling of subpopulation landscapes in patient-derived exosomes defined by surface protein expression patterns (Fig. 6.6); and (3) it allows in-depth proteomic characterization of selectively isolated subpopulations (Fig. 6.10), allowing new opportunities to facilitate molecular definition of extracellular vesicles and identification of biomarkers and therapeutic targets for cancer. Compared to conventional bench-top methods, the microfluidic exosome analysis platform increases assay sensitivity and reduces assay time and sample requirement by at least two orders of magnitude (Fig. 6.10). These advantages are beneficial for large-scale, comprehensive studies of exosomes at the system level.

Size characterization has shown that the microfluidic method that targets certain surface markers was able to purify exosome subpopulations from plasma with notably narrower size distributions than the ultracentrifugation-isolated vesicles (Fig. 6.5). These results suggest that microfluidic technology provides a means for more selective isolation of exosomes than the ultracentrifugation-based approach that is known to exhibit a mixed population of various extracellular microvesicle types with a wide size distribution. By selecting tumor-specific biomarkers, microfluidic approach allows for isolation and phenotyping of tumor-derived exosomes from patient plasma. This study demonstrated the ability to detect distinct landscape profiles of exosome subpopulations from pretreatment and treated NSCLC patients and control subjects by probing a panel of five exosomal and tumor-specific markers. Thus, this surface protein-based subpopulation profiling method should provide a useful approach for identifying exosomal signatures that indicate disease status. Furthermore, comparative analyses confirmed that subpopulation landscapes of NSCLC exosomes reflect protein profiles of patient-matched tumor tissue biopsies (Fig. 6.7). Results in this chapter supported potential use of tumor-derived exosomes as noninvasive surrogate biomarkers for clinical care. Accumulating evidences have shown markedly differential response to anti-IGF-1R therapy depending on molecular characteristics of NSCLC subsets. Advantages of microfluidic approach in differentiation and selective isolation of exosome subpopulations would be beneficial for deconvoluting the complexity of extracellular microvesicles and identifying molecular determinants of sensitivity to IGF-1R-targeted therapy.

Relevant to future clinical indications, microfluidic exosome analysis technology enables investigation of IGF-1R, an emerging pathway for targeted therapy in cancer. The IGF-1R pathway plays important roles in tumorigenesis and metastasis. Phosphorylation of IGF-1R, initiated by binding of ligands such as IGF-1 and IGF-2, is required for activation of MAPK, PI3K, AKT, and other signaling pathways involved in cell proliferation and survival [69]. Consequently, intense interest has arisen regarding the study of IGF-1R and pIGF-1R as predictive markers and therapeutic targets [70]. In lung cancer, expression levels of IGF ligands and IGF-1R have been associated with disease risk, clinical benefit of therapy, prognosis, and resistance to anticancer therapy [71]. However, IHC of tumor tissue currently predominates clinical assessment of IGF-1R expression, which is invasive and problematic for regular monitoring of disease progression and treatment. Microfluidic exosome analysis allows more accurate and noninvasive prognosis.

Rapid and sensitive analyses of total IGF-1R and its phosphorylation level in the EpCAM+ exosome subpopulation in a "sample-in-answer-out" manner have been demonstrated, proving the feasibility of microfluidic exosomal profiling technology for noninvasive monitoring of potential tumor burden in patients with NSCLC. Many more patients collected on clinical trials will be needed in order to evaluate the ultimate utility of this technology. Moreover, the ability to combine selective subpopulation isolation with intravesicular analysis of exosome contents would broadly impact the development of effective biomarkers and therapeutic targets. Clinical value of the IGF-1R family as a predictive biomarker and targeted therapy remains invalidated, and mechanisms underpinning resistance to anti-IGF-1R therapy are still unclear. Significant efforts have been focused on rational combination therapies that target various pathways activated by or interacting with the IGF-1R pathway, such as PI3K-AKT pathway and EGFR pathways. Microfluidic methodology developed in this study could be readily adaptable for implementation in various assays in order to probe various combinations of signaling proteins and their activated forms in exosomes for noninvasive evaluation of combination therapies. Although this work focused on NSCLC as the disease model for the proof-of-concept study, this microfluidic platform should yield a general approach that can be adapted for various cancer types by choosing tumor-associated biomarkers for exosome isolation, subpopulation profiling, and intravesicular protein probing. In order to demonstrate generalizability of the technology, some results from investigation of circulating exosomes in ovarian cancer patients have been shown. The microfluidic platform was scalable in nature in order to integrate simultaneous measurements of multiple exosomal markers for high-throughput screening of cancer as well as non-cancerous diseases. Therefore, this advanced methodology holds the potential to accelerate the study and utilization of exosomes in cancer research and clinical tests.

References

1. Guo MT, Rotem A, Heyman JA, Weitz DA (2012) Droplet microfluidics for high-throughput biological assays. Lab Chip 12:2146–2155
2. Witters D, Knez K, Ceyssens F, Puers R, Lammertyn J (2013) Digital microfluidics-enabled single-molecule detection by printing and sealing single magnetic beads in femtoliter droplets. Lab Chip 13:2047–2054
3. Wang T, Zhang M, Dreher DD, Zeng Y (2013) Ultrasensitive microfluidic solid-phase ELISA using an actuatable microwell-patterned PDMS chip. Lab Chip 13(21):4190–4197
4. Zeng Y, Novak R, Shuga J, Smith MT, Mathies RA (2010) High-performance single cell genetic analysis using microfluidic emulsion generator arrays. Anal Chem 82:3183–3190
5. He M, Herr AE (2010) Automated microfluidic protein immunoblotting. Nat Protoc 5:1844–1856
6. He M, Novak J, Julian BA, Herr AE (2011) Membrane-assisted online renaturation for automated microfluidic lectin blotting. J Am Chem Soc 133:19610–19613
7. Chen C, Skog J, Hsu CH, Lessard RT, Balaj L et al (2010) Microfluidic isolation and transcriptome analysis of serum microvesicles. Lab Chip 10:505–511
8. Davies RT, Kim J, Jang SC, Choi EJ, Gho YS, Park J (2012) Microfluidic filtration system to isolate extracellular vesicles from blood. Lab Chip 12:5202–5210
9. Shao H, Chung J, Balaj L, Charest A, Bigner DD et al (2012) Protein typing of circulating microvesicles allows real-time monitoring of glioblastoma therapy. Nat Med 18:1835–1840
10. Im H, Shao H, Park YI, Peterson VM, Castro CM et al (2014) Label-free detection and molecular profiling of exosomes with a nano-plasmonic sensor. Nat Biotechnol 32:490–495
11. Raposo G, Stoorvogel W (2013) Extracellular vesicles: exosomes, microvesicles, and friends. J Cell Biol 200:373–383
12. Hendrix A, Hume AN (2011) Exosome signaling in mammary gland development and cancer. Int J Dev Biol 55:879–887
13. Staals RH, Pruijn GJ (2011) The human exosome and disease. Adv Exp Med Biol 702:132–142
14. Schaeffer D, Clark A, Klauer AA, Tsanova B, van Hoof A (2011) Functions of the cytoplasmic exosome. Adv Exp Med Biol 702:79–90
15. Thery C, Zitvogel L, Amigorena S (2002) Exosomes: composition, biogenesis and function. Nat Rev Immunol 2:569–579
16. Valadi H, Ekstrom K, Bossios A, Sjostrand M, Lee JJ, Lotvall JO (2007) Exosome-mediated transfer of mRNAs and microRNAs is a novel mechanism of genetic exchange between cells. Nat Cell Biol 9:654–659
17. Skog J, Wurdinger T, van Rijn S, Meijer DH, Gainche L et al (2008) Glioblastoma microvesicles transport RNA and proteins that promote tumour growth and provide diagnostic biomarkers. Nat Cell Biol 10:1470–1476
18. Guescini M, Genedani S, Stocchi V, Agnati LF (2010) Astrocytes and Glioblastoma cells release exosomes carrying mtDNA. J Neural Transm 117:1–4

19. Keller S, Ridinger J, Rupp AK, Janssen JW, Altevogt P (2011) Body fluid derived exosomes as a novel template for clinical diagnostics. J Transl Med 9:86
20. Rountree RB, Mandl SJ, Nachtwey JM, Dalpozzo K, Do L et al (2011) Exosome targeting of tumor antigens expressed by cancer vaccines can improve antigen immunogenicity and therapeutic efficacy. Cancer Res 71:5235–5244
21. Taylor DD, Zacharias W, Gercel-Taylor C (2011) Exosome isolation for proteomic analyses and RNA profiling. Methods Mol Biol 728:235–246
22. Rani S, O'Brien K, Kelleher FC, Corcoran C, Germano S et al (2011) Isolation of exosomes for subsequent mRNA, MicroRNA, and protein profiling. Methods Mol Biol 784:181–195
23. Bobrie A, Colombo M, Krumeich S, Raposo G, Thery C (2012) Diverse subpopulations of vesicles secreted by different intracellular mechanisms are present in exosome preparations obtained by differential ultracentrifugation. J Extracell Vesicles 1:18397
24. Gyorgy B, Modos K, Pallinger E, Paloczi K, Pasztoi M et al (2011) Detection and isolation of cell-derived microparticles are compromised by protein complexes resulting from shared biophysical parameters. Blood 117:e39–e48
25. Kalra H, Adda CG, Liem M, Ang CS, Mechler A et al (2013) Comparative proteomics evaluation of plasma exosome isolation techniques and assessment of the stability of exosomes in normal human blood plasma. Proteomics 13:3354–3364
26. Sokolova V, Ludwig AK, Hornung S, Rotan O, Horn PA et al (2011) Characterisation of exosomes derived from human cells by nanoparticle tracking analysis and scanning electron microscopy. Colloids Surf B Biointerfaces 87:146–150
27. Dragovic RA, Collett GP, Hole P, Ferguson DJ, Redman CW et al (2015) Isolation of syncytiotrophoblast microvesicles and exosomes and their characterisation by multicolour flow cytometry and fluorescence nanoparticle tracking analysis. Methods 87:64–74
28. van der Pol E, Coumans FA, Grootemaat AE, Gardiner C, Sargent IL et al (2014) Particle size distribution of exosomes and microvesicles determined by transmission electron microscopy, flow cytometry, nanoparticle tracking analysis, and resistive pulse sensing. J Thromb Haemost 12:1182–1192
29. Siegel R, Ma J, Zou Z, Jemal A (2014) Cancer statistics, 2014. Cancer J Clin 64:9–29
30. Hirsh V (2012) Future of personalized medicine in non-small-cell lung cancer. Curr Oncol 19: S86
31. Kucherlapati R (2010) Personalized medicine for non-small-cell lung cancer. Oncology (Williston Park) 24:399–400
32. Scagliotti GV, Novello S (2012) The role of the insulin-like growth factor signaling pathway in non-small cell lung cancer and other solid tumors. Cancer Treat Rev 38:292–302
33. Nakagawa M, Uramoto H, Oka S, Chikaishi Y, Iwanami T et al (2012) Clinical significance of IGF1R expression in non-small-cell lung cancer. Clin Lung Cancer 13:136–142
34. Ofiara LM, Navasakulpong A, Ezer N, Gonzalez AV (2012) The importance of a satisfactory biopsy for the diagnosis of lung cancer in the era of personalized treatment. Curr Oncol 19: S16–S23
35. Clyne M (2012) Prostate cancer: reducing IGF-1 levels unlikely to ProtecT against prostate cancer initiation. Nat Rev Urol 9:6
36. Rosell R, Wei J, Taron M (2009) Circulating MicroRNA signatures of tumor-derived exosomes for early diagnosis of non-small-cell lung cancer. Clin Lung Cancer 10:8–9
37. Taylor DD, Gercel-Taylor C (2008) MicroRNA signatures of tumor-derived exosomes as diagnostic biomarkers of ovarian cancer. Gynecol Oncol 110:13–21
38. Pisitkun T, Gandolfo MT, Das S, Knepper MA, Bagnasco SM (2012) Application of systems biology principles to protein biomarker discovery: urinary exosomal proteome in renal transplantation. Proteomics Clin Appl 6:268–278
39. Madhavan B, Yue S, Galli U, Rana S, Gross W et al (2015) Combined evaluation of a panel of protein and miRNA serum-exosome biomarkers for pancreatic cancer diagnosis increases sensitivity and specificity. Int J Cancer 136:2616–2627

40. Alvarez ML, Khosroheidari M, Kanchi Ravi R, DiStefano JK (2012) Comparison of protein, microRNA, and mRNA yields using different methods of urinary exosome isolation for the discovery of kidney disease biomarkers. Kidney Int 82:1024–1032
41. Orita T, Moore LR, Joshi P, Tomita M, Horiuchi T, Zborowski M (2013) A quantitative determination of magnetic nanoparticle separation using on-off field operation of quadrupole magnetic field-flow fractionation (QMgFFF). Anal Sci 29:761–764
42. Minc N, Futterer C, Dorfman KD, Bancaud A, Gosse C et al (2004) Quantitative microfluidic separation of DNA in self-assembled magnetic matrixes. Anal Chem 76:3770–3776
43. Sinha A, Ganguly R, Puri IK (2009) Magnetic separation from superparamagnetic particle suspensions. J Magn Magn Mater 321:2251–2256
44. Atay S, Gercel-Taylor C, Kesimer M, Taylor DD (2011) Morphologic and proteomic characterization of exosomes released by cultured extravillous trophoblast cells. Exp Cell Res 317:1192–1202
45. Conde-Vancells J, Rodriguez-Suarez E, Embade N, Gil D, Matthiesen R et al (2008) Characterization and comprehensive proteome profiling of exosomes secreted by hepatocytes. J Proteome Res 7:5157–5166
46. Sethu P, Anahtar M, Moldawer LL, Tompkins RG, Toner M (2004) Continuous flow microfluidic device for rapid erythrocyte lysis. Anal Chem 76:6247–6253
47. Proczek G, Gassner AL, Busnel JM, Girault HH (2012) Total serum IgE quantification by microfluidic ELISA using magnetic beads. Anal Bioanal Chem 402:2645–2653
48. Herrmann M, Veres T, Tabrizian M (2006) Enzymatically-generated fluorescent detection in micro-channels with internal magnetic mixing for the development of parallel microfluidic ELISA. Lab Chip 6:555–560
49. Kumar D, Gupta D, Shankar S, Srivastava RK (2015) Biomolecular characterization of exosomes released from cancer stem cells: Possible implications for biomarker and treatment of cancer. Oncotarget 6:3280–3291
50. Aliotta JM (2011) Tumor exosomes: a novel biomarker? J Gastrointest Oncol 2:203–205
51. Hosseini-Beheshti E, Pham S, Adomat H, Li N, Tomlinson Guns ES (2012) Exosomes as biomarker enriched microvesicles: characterization of exosomal proteins derived from a panel of prostate cell lines with distinct AR phenotypes. Mol Cell Proteomics 11:863–885
52. Lasser C (2015) Exosomes in diagnostic and therapeutic applications: biomarker, vaccine and RNA interference delivery vehicle. Expert Opin Biol Ther 15:103–117
53. Rabinowits G, Gercel-Taylor C, Day JM, Taylor DD, Kloecker GH (2009) Exosomal microRNA: a diagnostic marker for lung cancer. Clin Lung Cancer 10:42–46
54. Larsson O, Girnita A, Girnita L (2005) Role of insulin-like growth factor 1 receptor signalling in cancer. Br J Cancer 92:2097–2101
55. Akers JC, Gonda D, Kim R, Carter BS, Chen CC (2013) Biogenesis of extracellular vesicles (EV): exosomes, microvesicles, retrovirus-like vesicles, and apoptotic bodies. J Neurooncol 113:1–11
56. Tauro BJ, Greening DW, Mathias RA, Ji H, Mathivanan S et al (2012) Comparison of ultracentrifugation, density gradient separation, and immunoaffinity capture methods for isolating human colon cancer cell line LIM1863-derived exosomes. Methods 56:293–304
57. Cocucci E, Racchetti G, Meldolesi J (2009) Shedding microvesicles: artefacts no more. Trends Cell Biol 19:43–51
58. Poliakov A, Spilman M, Dokland T, Amling CL, Mobley JA (2009) Structural heterogeneity and protein composition of exosome-like vesicles (prostasomes) in human semen. Prostate 69:159–167
59. Jorgensen M, Baek R, Pedersen S, Sondergaard EK, Kristensen SR, Varming K (2013) Extracellular vesicle (EV) array: microarray capturing of exosomes and other extracellular vesicles for multiplexed phenotyping. J Extracell Vesicles 2:20920
60. Cheminant M, Labia R (1991) The Michaelis-Menten equation: computing substrate concentration as a function of time without restrictions on the initial conditions. Comput Appl Biosci 7:99–101

61. Gong Y, Yao E, Shen R, Goel A, Arcila M et al (2009) High expression levels of total IGF-1R and sensitivity of NSCLC cells in vitro to an anti-IGF-1R antibody (R1507). PLoS One 4: e7273
62. Janeway KA, Zhu MJ, Barretina J, Perez-Atayde A, Demetri GD, Fletcher JA (2010) Strong expression of IGF1R in pediatric gastrointestinal stromal tumors without IGF1R genomic amplification. Int J Cancer 127:2718–2722
63. Tarn C, Rink L, Merkel E, Flieder D, Pathak H et al (2008) Insulin-like growth factor 1 receptor is a potential therapeutic target for gastrointestinal stromal tumors. Proc Natl Acad Sci U S A 105:8387–8392
64. Yeo CD, Park KH, Park CK, Lee SH, Kim SJ et al (2015) Expression of insulin-like growth factor 1 receptor (IGF-1R) predicts poor responses to epidermal growth factor receptor (EGFR) tyrosine kinase inhibitors in non-small cell lung cancer patients harboring activating EGFR mutations. Lung Cancer 87:311–317
65. Peled N, Wynes MW, Ikeda N, Ohira T, Yoshida K et al (2013) Insulin-like growth factor-1 receptor (IGF-1R) as a biomarker for resistance to the tyrosine kinase inhibitor gefitinib in non-small cell lung cancer. Cell Oncol 36:277–288
66. Tisi E, Lissoni P, Rovelli F, Mandelli D, Barni S, Tancini G (1991) Blood levels of IGF-I in non-small cell lung cancer: relation to clinical data. Int J Biol Markers 6:99–102
67. Khosravi J, Krishna RG, Bodani U, Diamandi A, Khaja N et al (2007) Immunoassay of serine-phosphorylated isoform of insulin-like growth factor (IGF) binding protein (IGFBP)-1. Clin Biochem 40:86–93
68. Mobergslien A, Sioud M (2014) Exosome-derived miRNAs and cellular miRNAs activate innate immunity. J Innate Immun 6:105–110
69. Pollak M (2012) The insulin and insulin-like growth factor receptor family in neoplasia: an update. Nat Rev Cancer 12:159–169
70. Fidler MJ, Shersher DD, Borgia JA, Bonomi P (2012) Targeting the insulin-like growth factor receptor pathway in lung cancer: problems and pitfalls. Ther Adv Med Oncol 4:51–60
71. Han JY, Choi BG, Choi JY, Lee SY, Ju SY (2006) The prognostic significance of pretreatment plasma levels of insulin-like growth factor (IGF)-1, IGF-2, and IGF binding protein-3 in patients with advanced non-small cell lung cancer. Lung Cancer 54:227–234

Chapter 7
Microfluidic Single-Cell Functional Proteomics

Shay Mailloux, Lisa Ramirez, and Jun Wang

Abstract We review an emerging microfluidic single-cell functional proteomics field and the associated technologies. Functional proteins, such as secreted signaling proteins from immune cells and phosphoproteins in cancer cells, refer to those that play important roles in promoting live cells' physiological activities. Assay of single-cell functional proteins can be carried out by either cytometry or newly developed microfluidic tools, each of which possesses its own advantages and disadvantages. The use of microfluidic chips brings benefits of high multiplexity, low cost, and high flexibility to integrate various cell manipulation strategies into one platform, while such platforms are normally less developed and automated. We focus the discussion specifically on single-cell barcode technology, which has been more mature than others, and is able to quantitate up to 42 different functional proteins from single cells. This platform has also been uniquely extended to the study of cell–cell interactions. Quantitative analysis of the single-cell functional proteins offers new perspectives of biological systems and provides a conduit between biology and the physicochemical laws. And finally we discuss the challenges and future of the microfluidic single-cell functional proteomics field.

Keywords Single-cell proteomics • Signaling networks • Functional proteomics • Microfluidics • Immunoassay

S. Mailloux, Ph.D. • L. Ramirez, B.S.
Multiplex Biotechnology Laboratory, Department of Chemistry, University at Albany, State University of New York, 1400 Washington Avenue, CH 120, Albany, NY 12222, USA
e-mail: smailloux@albany.edu; lsramirez@albany.edu

J. Wang, Ph.D. (✉)
Cancer Research Center, University at Albany, State University of New York, Rensselaer, NY 12144, USA
e-mail: jwang34@albany.edu

© Springer International Publishing Switzerland 2016

C. Lu, S.S. Verbridge (eds.), *Microfluidic Methods for Molecular Biology*,
DOI 10.1007/978-3-319-30019-1_7

7.1 Introduction

Cellular heterogeneity occurs intrinsically and universally at the genetic level, the epigenetic level, and the protein level in otherwise identical cell populations [1, 2]. The protein level's heterogeneity is of particular importance because functional proteins carry out a multitude of cellular activities such as signal transduction, cytokine/chemokine secretion, cell migration and invasion, and other physiological processes [3]. Moreover, disease status is often reflected by aberrant signaling pathways, which can be studied by quantitating levels of signaling proteins. In recent years, functional proteomics has constituted an emerging research area that is focused on monitoring and analyzing the spatial and temporal properties of molecular networks and fluxes involved in living cells [4–7]. The information obtained from those studies overcomes the old paradigm "one gene-one protein" that does not reflect the real nature of the cell system. Since functional proteins are on the frontier of cell response to environmental stimuli (e.g., signaling cues from neighboring cells), studying cellular heterogeneity in the functional proteome would be expected to be crucial for understanding the societal aspects of a cell population. Considering that the single-cell proteomics field has only been in existence for a few years, the new format of data produced by single-cell tools presents a challenge to the available analytical tools for protein quantitation. Of immense value is the analytical ability to exploit single-cell data and measurements for use in generating conclusions about the whole picture of a multicellular system—such an approach is frequently ignored in bulk cell measurements [8].

This chapter offers a review of single-cell proteomics technologies that have emerged recently, focusing on microfluidics and microchip-based methods. Applications to cancer and immune biology are discussed, as well as the approaches for carrying out analyses on the large body of data generated from these technologies. Two-cell heterogeneity is also discussed here as a study that bridges the gap between single-cell proteomics and whole population studies.

Single-cell proteomics challenges the field of proteomics, pushing technology towards innovative solutions for technical and biological problems. There is a constant drive towards miniaturization, and in parallel multiplexing, while increasing sensitivity and maintaining robustness. With advances in these categories, the need for more extensive tools in data collection and analyses becomes more imperative [6]. Such challenges must be addressed and overcome in order to develop useful single-cell technologies. Further, the cost of the method must also be minimized in order for the technology to gain widespread use, encompassing the broadest sectors of research and clinical audiences. Single-cell methods have the opportunity to revolutionize cell studies, by not only providing deeper understanding of intracellular processes and cell–cell interactions, but also serving as important tools for the discovery of new information in research areas like the cancer microenvironment and targeted therapy.

7.2 Single-Cell Functional Protein Detection

Major functional proteins are secreted proteins and intracellular phosphorylated signaling proteins, both of which can be associated with signaling networks stimulated by growth factors. Secreted proteins such as cytokines also have well-studied roles in immune response. The choice of which proteins to detect in single-cell functional proteomics assays is dependent on the specific goals of the proteomic study. For instance, signaling networks within cell populations may be investigated by detecting proteins involved in signaling pathways.

A single-cell functional proteomics assay has to be sensitive enough to detect and measure even very low quantities of proteins; protein copy numbers are determined for isolated cells, for two-cell systems, or for other small groups of *n* cells. Another key challenge lies in recapitulating the physiological conditions under which the protein is activated or produced. The in vivo microenvironment for cell populations is always complicated, and some in vitro artifacts from proteomic techniques, when not accounted for in data analysis, may threaten the validity of the research method. Moreover, the biological heterogeneity across individual cells might be masked by the noise of detection measurement; this is where single-cell statistics is crucial. Finally, it is important to be aware of the kinetics associated with both the generation and degradation of the protein sample—the proteomics assays described here are typically carried out in under 12 h. If these factors are carefully considered, along with more general knowledge like cell phenotypes, and culture conditions, it is possible to carry out single-cell functional proteomics assays for a wide range of applications.

7.2.1 In Immune System Studies

Immune system studies are essential for the understanding of immune defense mechanisms, as well as the exploitation of our own immune functions for disease treatment. Of particular interest in immune studies are leukocytes, or T cells from blood and tissue, specifically the quantities of specific cell subsets, secretions, transcriptional states, and their proliferative profiles [9]. T cells are naturally heterogeneous; that is how they are able to recognize numerous pathogens and other invaders by surface T cell receptors (TCR) [10]. The different T cell types and their functions are conventionally defined by surface markers such as cluster of differentiation 4 (CD4) and CD8 [7]. However, those functions, often indicated by secreted effector proteins that mediate the tasks of target killing, self-renewal, inflammation, and cell communications, are not always consistent with surface markers. Recent studies show that only a small subset of T cells of a particular type can secrete large amounts of cytokines and chemokines, and therefore such a subset serves as the major player in the immune response [11, 12]. In addition, the secretion profiles of different types of T cells defined by surface markers have significant overlaps. It is also found that a portion of hematopoietic stem cells and progenitor cells are the active responders in immune defense against pathogen invasion [13]. Thus, single-cell functional assays on immune cells are highly demanded, which would fundamentally refresh knowledge of our immune system.

7.2.2 *In Cancer Biology Studies*

The hallmarks of cancer include sustaining proliferative signaling, evading growth suppressors, resisting cell death, enabling replicative immortality, inducing angiogenesis, and activating invasion and metastasis [14], all of which involve aberrant signaling pathways where functional proteins are activated. These signaling proteins have been considered as drug target candidates in cancer treatment, ideally to minimize the drug side effects [15]. Although pharmaceutical companies and institutes are investing billions each year to develop targeted therapeutic agents, most of these therapies only minimally improve the overall survival rate. The major challenge here is that cancer cells are highly heterogeneous, and a specific drug may only influence a portion of tumor cells while the rest still sustains the growth of the tumor. In addition, cancer cells possess the ability to employ alternative pathways to survive once a pathway is shut down; in some cases other functional proteins may be activated and the drug targets are shifted. Those factors will ostensibly cause drug resistance, and the cancer treatment will eventually lose efficacy [16, 17]. Therefore, it is urgent to have effective single-cell functional proteomic tools for detecting the tumor-specific drug targets and for tracking target shifting in cancer cell signaling [18, 19]. Proteomics strategies have hitherto been lacking these important features. Conventional proteomic tools, unfortunately, are limited in their capacity to delineate cellular heterogeneity in cancer studies.

In the following content, we discuss the use of single-cell barcode chips as alternatives to conventional proteomic tools and as useful devices for fulfilling the abovementioned purposes.

7.3 Single-Cell Functional Proteomic Tools

7.3.1 *Flow Cytometry Methods*

Flow cytometry methods, including fluorescence flow cytometry (FFC, or fluorescent activated cell sorting—FACS) and time-of-flight mass cytometry (Cytof), have become the standard in single-cell protein detection. In FFC, cellular proteins are detected using fluorophore-labeled antibodies (Fig. 7.1) where each type of antibody corresponds to one type of fluorophore, limiting the technology in terms of multiplexing (<17) due to the overlapping of fluorescence spectra [20, 21].

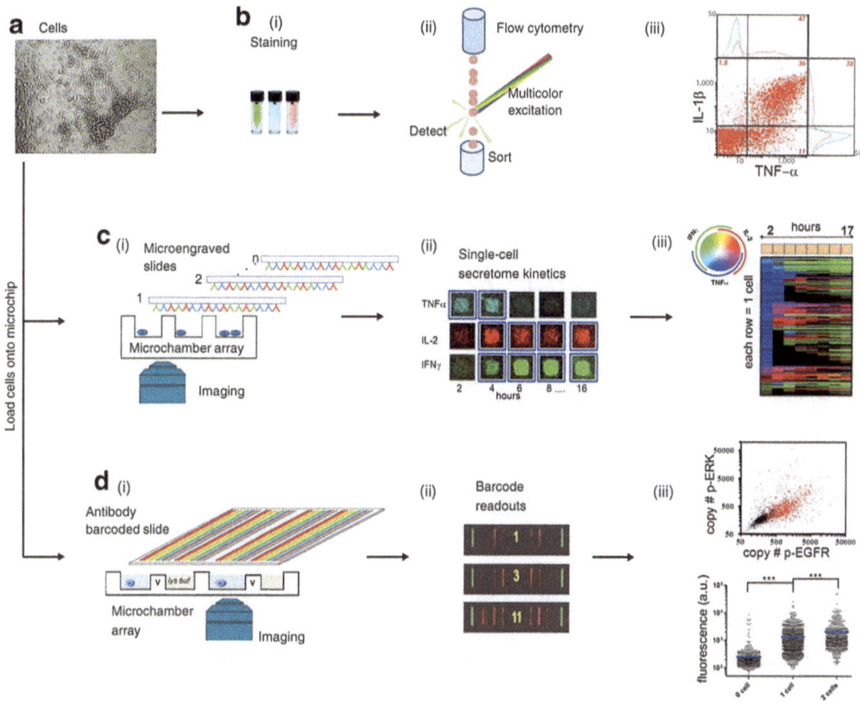

Fig. 7.1 An overview of three technologies used for single-cell functional proteomics. The cell suspension of interest (**a**), which is stained (**b**) (*i*) for flow cytometry (*ii*) and detected by a scanner to generate a flow cytometry scatterplot to show the correlation of two proteins (*iii*). The microengraving method begins by isolating cells into chambers functionalized with capture antibodies (**c**) (*i*) where the secreted protein levels are then read using fluorophore labeled secondary antibodies (*ii*) and the data is analyzed (*iii*). Finally, the SCBC assay involves isolating single cells into respective chambers of the chip (**d**) (*i*), using the barcode method to analyze species of interest (*ii*), and perform statistical analysis (*iii*). Reprinted with permission from "Microchip platforms for multiplex single-cell functional proteomics with applications to immunology and cancer research," by Wei, W. et al. 2013, *Genome Med* 5. Copyright 2013 by BioMed Central

Mass cytometry allows an approximate twofold increase of multiplexing due to its use of transition metal mass labels [22–24]. However, disadvantages of the latter technology include the use of costly equipment and the requirement of extensively trained technicians.

7.3.2 ELISpot Method

Surface methods for single-cell proteomics include the enzyme-linked immunospot (ELISpot) technique, a well-established, inexpensive, and simple method for detecting protein secretions from live cells [25]. This method resembles enzyme-linked immunosorbent assay (ELISA): cells are immobilized on a surface coated with antibodies and sandwich-immuno-assays are performed to detect secreted proteins. Signals are detected near the immobilized cell, resulting in single-cell protein detection. The major limitation is low multiplexity: usually, only one secreted protein is detected for one cell. The sensitivity at the single-cell level is also very low, so most studies are conducted on a few single cells.

7.3.3 Microfluidics Methods

Microfluidics tools offer unique advantages over conventional technologies, such as being inexpensive to manufacture, low consumption of reagents for analysis, as well as versatility and customizability, to name just a few [26]. In regard to single-cell proteomics there are two distinct classes. The first uses antibody staining of the cells of interest—a method similar to flow cytometry—while the second involves surface-based immunoassays to measure proteins released from the cells—a method similar to ELISpot.

The flow-based fluorescence assays on microchips are essentially small-scale versions of FACS [27]. Fu et al. developed a platform that could separate *E. coli* that expressed green fluorescent protein from a larger population of the bacterium. This basic approach and its afforded versatility led to important advances where cells could be spatially separated into specific arrays (Fig. 7.2) [28], or even entrapped in droplets [29]. These methods allow specific control over the cell before the screening process; however, because they are based on staining, the multiplexing capacity is limited. This limitation may be overcome with new imaging techniques and dyes, with the challenge being in maintaining the simplicity and low cost of the technology.

The second class of microfluidics methods, based on surface-confined immunoassays (Fig. 7.1c, d), has several advantages and therefore is more advanced, with two specific methods having been reported; these technologies include the single-cell barcode chip (SCBC) [30] and the microengraving technique [31]. Advantages of these techniques include a higher level of multiplexing, and the ability to assay for secreted cytoplasmic and membrane proteins as well as still being able to fluorescently stain the cell. These methods also enable the incorporation of cell interaction parameters into protein measurements.

The operation of microengraved single-cell proteomics chips is performed by trapping cells on a specially designed antibody-coated chip containing hundreds of thousands of nanowells, all of which are collectively sealed by an antibody-coated glass slide. The cells in nanowells can be cultured on-chip for hours to allow cytokine release, and these cells are also accessible to immunostaining. This technology can detect approximately three secreted proteins using sandwich-ELISA, as well as permit the culture and recovery of specific cells [31]. Thus, this technique possesses notable similarities with ELISpot. Also of note is the ability to perform single-cell kinetic studies [32] on the temporal secretion of

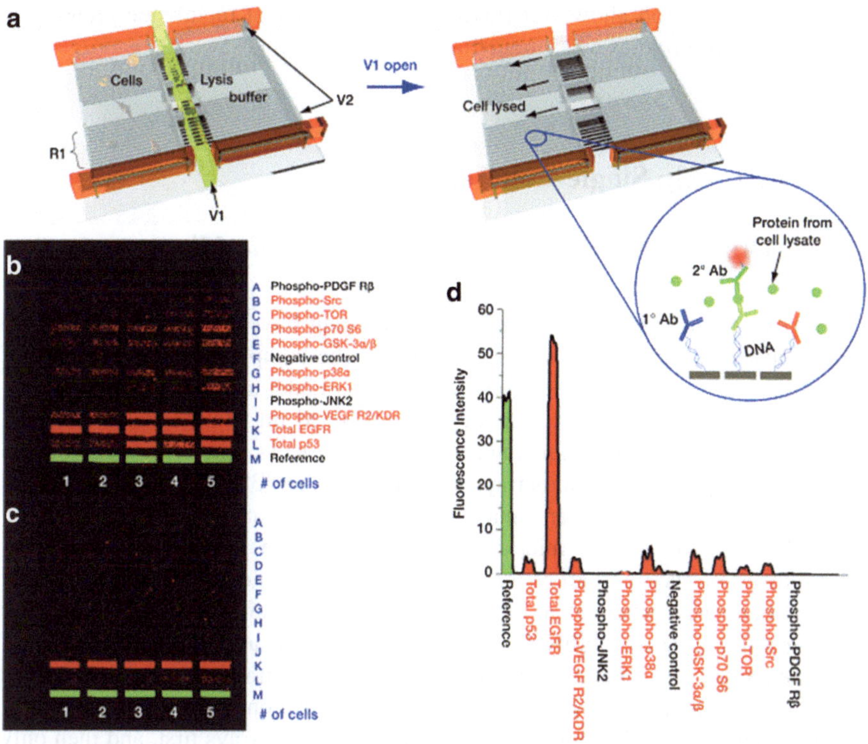

Fig. 7.2 An illustration of single-cell intracellular protein analysis (**a**) where intracellular proteins are analyzed by lysing the cells and allowing the released proteins to bind to a DNA-antibody conjugate in a barcode pattern within the chamber. V1 is the valve for lysis buffer release, V2 is the valve for chamber isolation control, and R1 is a DNA barcode array converted to an antibody array using DEAL. (**b**, **c**) are contrast-enhanced images of the barcode using a successful surface chemistry (**b**) and an unsuccessful surface chemistry (**c**) while (**d**) represents the respective fluorescence intensity profile. Reprinted with permission from "Chemistries for Patterning Robust DNA Microbarcodes Enable Multiplex Assays of Cytoplasm Proteins from Single Cancer Cells," by Shin YS et al. 2010, *Chemphyschem* 11. Copyright 2010 by John Wiley and Sons

cytokines; this is carried out by replacing the glass cover of a microchip after designated periods of time.

The single-cell barcode chip (SCBC) is a similar but more advanced technology. The operation of the chip, processing of SCBC data, and applications in cell interaction studies will be discussed in later sections of this chapter. In general, capture antibodies are patterned in an array fashion and such antibody arrays are supplied to single-cell microwells or microchambers, allowing for high-throughput immunoassays on the entrapped single cells (Fig. 7.1d). The platform may be extended to include analyses of pairs or larger groups of cells, thus providing information on cell–cell communication. A single microchip can contain as many as 10^4 microchambers for analysis with volumes ranging from 0.1 to 2 nL

[12, 33]. The method allows for sandwich-type ELISA of membrane proteins as well as secreted proteins and intracellular proteins [18].

7.4 Microchip Single-Cell Technologies

7.4.1 *High-Density, Multiplexed DNA Barcode Microarray*

Multiplexing is an essential component of emerging biotechnologies in both biological research and clinical studies because of the complex interconnectivity of biomolecules found within the biological systems under study [34–37]. The conventional, standard multiplex technologies are based on commercially available antibody microarray slides generated from array printers. There have been some successful integrations of microarrays with microfluidics [38], although the practice is not commonplace. However, a major concern that remains with the incorporation of such microarrays into microfluidic devices is the stability of antibodies immobilized on the microarray substrates, as it is well known that antibodies can be degraded easily when taken out of physiological conditions and can lose their binding affinity. This stability factor subsequently translates to the sensitivity of immunoassays. In contrast to antibodies, single-stranded oligonucleotides (ssDNAs) are more relatively stable under ambient conditions in a laboratory setting. Hence, a feasible approach is to pattern ssDNA arrays first, and then only right before the single-cell immunoassay is carried out will the ssDNA arrays be converted into antibody arrays. The conversion is achieved through the hybridization of complementary DNA-antibody conjugates with the ssDNA array—this technique was established with the generation of a DNA-encoded antibody library (DEAL). Bailey et al. demonstrated this approach by patterning arrays in a barcode-like fashion using microfluidics, resulting in antibody arrays at 10x higher density that what could typically be generated [34, 39]. The careful design and screening of ssDNAs in conjunction with the multiplexity of this method have given rise to reliable platforms for analyzing proteins with high sensitivity. For the single-cell measurements, however, the higher multiplexity requires proportional increase of barcode array dimension, which will adversely affect the sensitivity when assaying only hundreds of copy numbers of functional proteins. Thus, the current method limits the detection multiplexity to 42 [40].

There are two formats of barcode arrays developed: 1D array and 2D array. While the 2D array is relatively difficult to fabricate and the procedure is lengthy, the array size is ~10 times less than that of 1D array, leading to significantly enhanced detection limits when assaying single-cell proteins using a SCBC chip integrated with a 2D array. To fabricate the barcode array, ssDNA is patterned on a glass substrate modified with poly-L-lysine (PLL). The PLL coating renders a positively charged surface to the glass substrate. Microfluidic patterning is done by flowing parallel channels of ssDNA to form specific bands of immobilized DNA on the glass surface to generate a 1D barcode-like pattern. Flow is directed using specially fabricated molds to form polydimethylsiloxane (PDMS) chambers for guiding the fluid. After the 1D barcode is generated with the first PDMS mold, the substrate is baked to ensure strong adsorption of the ssDNA to the PLL layer. To proceed from 1D array to 2D array, a second mold is then used to flow ssDNAs perpendicularly to the previously patterned barcode, resulting in a 2D array ($n \times m$) patterned at high density [41]. The array results from the points where these two patterns intersect with each element having a "unique" molecular identity.

Surface chemistries for use in small scale but highly multiplexed technologies can present a challenge. Uniformity and reproducibility are the key factors for quantification of low copy numbers of proteins from single cells. To address this problem, Shin et al. explored several different chemistries, all done in microfluidic chips, to pattern the protein microarray [42]. Ultimately, it was discovered that by patterning the primary ssDNA in dimethyl sulfoxide by flowing the oligonucleotides on a poly-L-lysine-modified surface and then drying and baking the barcode chip, they could develop a uniform surface using the DEAL technique in more than 90 % of the trials with a process with the potential to be automated. They tested this technology by detecting eight proteins from a single-cell lysate, as well as nine proteins from five cells (Fig. 7.2). Their research resulted in the generation of high-density, high-uniformity DNA barcode patterns and served as the backbone for further microfluidic microchip protein detection.

The DNA array elements were initially 20 μm wide with 50 μm pitch [42]. These DNA barcodes were subsequently converted to antibody microarrays using DEAL [39]. Using these antibody microarrays, proteins are typically detected in a sandwich-type assay, where they are captured by the arrayed antibodies and then recognized by biotinylated detection antibodies. Subsequently, a fluorophore-conjugated streptavidin is used for visualization. Such microarrays can be easily customized due to the use of DNA-antibody conjugates made in-house.

7.4.2 The Single-Cell Barcode Chip

The single-cell barcode chip (SCBC) is a revolutionary microfluidic technology advancing the field in high-throughput multiplexing. In this technology, a microchip is integrated with a barcode antibody microarray that is formed using perpendicular flow patterning of a DNA library followed by conversion to an antibody microarray using the aforementioned DEAL technique [34, 39, 42]. Each chamber encapsulates a complete set of the microarray, so that all the microchambers of a chip are subjected to exactly the same sandwich-ELISA procedures. A single or specific number of cells can be entrapped within microchambers with a total volume of <2 nL for detection of analytes of interest. To assay secreted proteins, cells are incubated in the SCBC for at least 6 h, and then the microarray slide is developed. To assay intracellular proteins, cell lysis buffer supplemented with phosphatase/protease inhibitors is introduced into the microchambers of SCBC after a sufficient incubation period, after which the microarray slide readout is generated. In combination with microscopy imaging, single cells fluorescently stained in microchambers could be counted, and their corresponding microchamber addresses can be recorded. Such information is needed in correlating cell number with the protein signal. The microchambers with 0 cell number are usually taken as the background or the base line against which protein signals will be measured.

The first version of SCBC had relatively low throughput—it was able to analyze about 100 cells per chip [18], with each specific protein being analyzed twice to generate statistically representative data. This generation of SCBC operated using a system of microfluidic pneumatic valves and a two-layer microfluidic assembly [43]. It featured a total of 120 isolated microchambers for the entrapment of single or multiple cells on one microchip. The downside to having two or more array repeats per microchamber (relatively larger than later versions, this microchamber had a volume greater than 2 nL) is that it usually gives rise to fluctuations in the fluorescence readout. One complete set of the array may not have the same fluorescence intensity as the next array repeat, because a single cell might be in closer proximity to one of these sets, and consequentially more proteins are captured on that specific set. Shi and Wei carried out a thorough study on data analysis approaches that take these fluctuations into account [19]. They applied Monte Carlo simulations in order to extract the real biological signals while excluding measurement noises.

7.4.2.1 Valveless Microchip Design

A more novel variation of the SCBC has a sophisticated microchip design that obviates the need for pneumatic valves [33]. The valveless method allows 30-fold more cell assays than the previous design. The access of cells, buffers, and antibody solutions, etc. to microchambers is controlled by altering the height of PDMS posts in the microchip. The main points of contact of the microchip with the slide are these PDMS posts, which are deformable and able to be compressed under mechanical stress. When there is no pressure applied, the posts are non-deformed, resulting in an open state where all of the microchambers are open to allow flow of solutions and cells through the chip. Cells are loaded onto the chip in this open state. When slight pressure is applied, the PDMS posts are compressed; this is referred to as the close-I state and it enables chemical communication between the microchannels and microchambers without allowing the cells to move between chambers. Upon application of sufficiently high pressure, the PDMS posts completely collapse, allowing the main body of the microchip to be in contact with the glass substrate, thereby sealing microchambers. This is referred to as the close-II state. Cells are incubated in the close-II state while the antibody array captures secreted proteins. The application of pressure is regulated using a setup with adjustable screws that control glass pistons or clamps that press against the microchip. Specific angles of rotation for the screws correspond to different states of the microchip.

The operation of the clamp setup must be optimized for the microchip device, and the switching between states must be normalized.

7.4.2.2 Multiplexed Single-Cell Detection

Using the valveless SCBC, protein detection limits are usually in the hundreds of copies per cell, and this is comparable to conventional ELISA. In this platform, the array readout is consistent across the entire microchip, which is important for generating large amounts of data for statistical analysis. In order to translate fluorescent signals into protein copy numbers, calibration curves must be generated with known concentrations of recombinant proteins. These calibration curves would also provide the linear detection range for each protein.

Considering that assays for multiple proteins are carried out simultaneously, it is crucial to evaluate the platform for cross-reactivity. Validation procedures should be performed at the DNA patterning level and at the sandwich-ELISA stage. Moreover, the ssDNA building blocks for the arrays should have orthogonal sequences to minimize cross talk between array elements.

7.4.3 *Quantifying Cell–Cell Interactions in Microchip Platforms*

Cell–cell interactions play a significant role in a range of physiological processes. Research in the past decade has shed light onto a number of cell–cell interactions within a tumor microenvironment that have been found to influence metastasis, angiogenesis, and regulation of immune response [44]. Some notable examples are chemokine signaling between tumor-associated macrophages and cancer cells, as well as interactions between nonmalignant cells and malignant tumor cells. Traditionally, these interactions are studied through the analysis of the transcriptome or proteome of different cell types cultured together [45, 46], as compared to cultures of pure cell types. Microchip-based technologies allow for an alternative to these bulk measurement procedures. Through microchip platforms, the cellular systems under study may be reduced to smaller sizes (e.g., two-cell systems; Fig. 7.3). Intercellular distances between interacting cells can be quantified with the use of simple microscopes and conventional cell-staining procedures. The panel of antibodies on the array supplied to the microchip can be chosen to consist of key players in signaling pathways. Typically, antibodies against phosphorylated proteins and secreted proteins are included. Quantitation of cell–cell interactions is performed by measuring copy numbers of such signaling proteins in every n-cell system as a function of intercellular distance [33].

A single high-throughput microchip experiment may yield data sets for varying numbers of interacting cells (e.g., 0–3 cells per microchamber). Although cells are

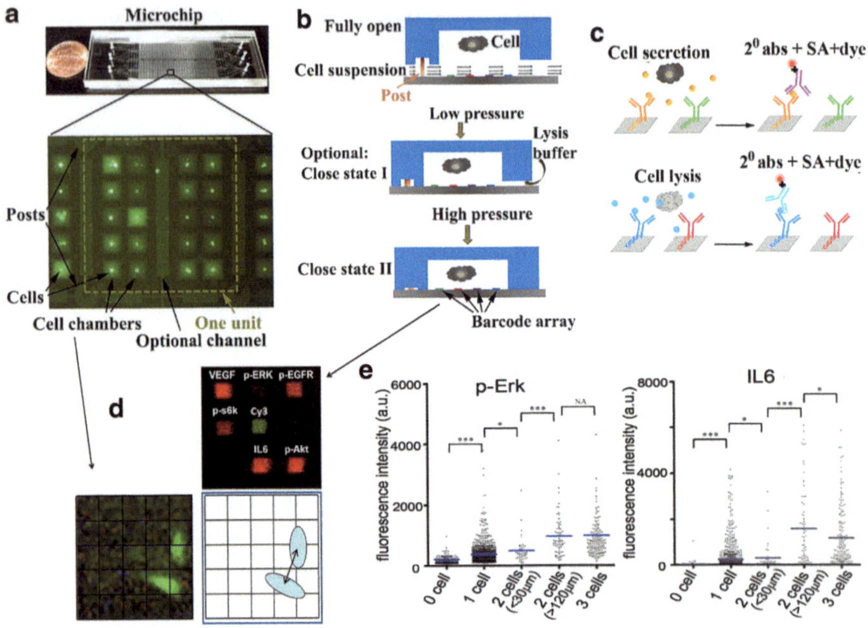

Fig. 7.3 An overview of the SCBC platform. (**a**) A digital photo of the PDMS microchip with a fluorescence micrograph of a 20-microchamber unit (there are 435 of these per individual chip) with the central channel housing the cell lysate. (**b**) Schematic of the chip operation from the fully open to closed state. Cells are loaded in the open state, lysed in the close state I, and completely isolated in the close state II. (**c**) The stepwise illustration of protein capture and detection using a sandwich-type immunoassay, detecting secreted, cytoplasmic, or membrane proteins. (**d**) Fluorescence data collected from each microchamber, including the position of the cells and the fluorescence intensity from the respective assays. (**e**) Fluorescence intensity data for analyzed proteins, specifically secreted interleukin-6 (IL-6), and cytoplasmic phosphorylated signal-related kinase (p-ERK). Reprinted with permission from "Quantitating Cell–Cell Interaction Functions with Applications to Glioblastoma Multiforme Cancer Cells," by Wang, J. et al. 2012, *Nano Lett* 12. Copyright 2012 by the American Chemical Society

randomly loaded into each microchamber, the number of cells enclosed per microchamber may be regulated by controlling the total cell count of the cell suspension introduced to the microchip. For a typical distribution, 2000 microchambers contain one cell, 1000 chambers contain two cells, 500 chambers contain three cells, etc. Microchambers that do not contain cells can be used for measuring background signals in fluorescence measurements. Protein detection is achieved through sandwich-ELISA performed on-chip. Each *n*-cell experiment is subjected to the immunoassay; thus each microchamber should be supplied with at least one complete antibody panel. Cells are incubated in the microchip for a fixed period of time, after which levels of secreted signaling proteins are quantitated. To measure copy numbers of intracellular phosphorylated proteins, cell lysis buffer supplemented with phosphatase and protease inhibitors may be added to the chip.

The protein copy numbers obtained in these n-cell experiments can provide insight into the intricacies of the cell–cell interactions. Interactions may be characterized by the inhibition or activation of signaling proteins. For two-cell interaction studies, inhibition is observed when the average copy number of a given protein for a two-cell system is lower than that corresponding to two one-cell (isolated) cultures. On the other hand, when the two-cell system yields a higher average copy number, the protein (i.e., its secretion by the cell or its phosphorylation) is regarded as "activated."

7.5 Data Collection, Analysis, and Model Development for Single-Cell Technologies

Microchip platforms call for extensive data collection and processing methods owing to the large body of data obtained from multiplexed on-chip assays. Some general approaches for gathering microarray fluorescence readouts and tracking cells in microchips are described here. Comparison and quantitation of fluorescence intensities are discussed. A guide for formulating algorithms that can translate fluorescence readouts into protein copy numbers is provided. Finally, the development of a particle interaction model applicable to cancer cell–cell interaction studies is discussed.

7.5.1 SCBC Data Collection and Statistical Analysis

On-chip single-cell experiments require imaging the microchip to determine the cell population corresponding to each microchamber and to measure intercellular distances in microchamber co-cultures. Conventional cell imaging methods such as phase-contrast microscopy or fluorescence microscopy may be easily employed to track cells in the optically transparent PDMS microchip. Cell suspensions may be incubated with tracker dyes such as calcein AM or 5-chloromethylfluorescein diacetate (CMFDA) before loading to the chip. Positions of cells may be recorded at various time points as required in distance-dependent interaction studies.

After performing sandwich-ELISA on-chip, the microarray slide is detached from the PDMS microchip and carefully washed in preparation for fluorescence scanning. A microarray scanner such as Axon GenePix is used to generate the fluorescence readout of the slide with two excitation wavelengths, one that corresponds to cyanine-3 dye (532 nm, for visualizing the spatial reference array element) and another wavelength that corresponds to the streptavidin-conjugated fluorescent label used for binding with biotinylated detection antibodies on the sandwich-ELISA platform. As an example, when streptavidin-Alexa-Fluor-647 is used in sandwich-ELISA, the microarray scanner may be set to an excitation wavelength of 635 nm. Microarray slide images are digitized (e.g., 16 bits with a scale of 0–65535) and the scale is selected depending on the requirements of subsequent data processing procedures that utilize custom-made programs for protein quantification.

Custom algorithms may be developed with computational tools such as Matlab. Automation of the data processing is achieved by using an algorithm that can associate microarray spots with microarray address labels and protein identification tags.

Statistical analysis is performed for comparing protein copy numbers under different conditions. Statistical tools should be used to establish statistical uniqueness of 0-, 1-, 2-, 3-cell, etc. data sets. For interaction studies that relate protein copy number with intercellular separation, there is a need to validate the uniqueness of different pairwise interaction data sets as a function of intercellular distance. Two-tailed student's t test assuming unequal variance is used to evaluate whether the difference in copy number is significant. A P value less than 0.05 % is taken as statistically significant. Other values such as $P < 0.01$ and $P < 0.001$ are also typically used. Mean protein levels are also reported with standard error of the mean, which is defined as the standard deviation of the mean copy number divided by the square root of cell number.

7.5.2 Cell–Cell Interaction Model Development

Wang et al. further extended cell–cell interaction quantitation by developing a model that is able to predict three-cell interactions from two-cell experiments [33]. In the model, each cell is treated as a particle that is able to exert some "force" or influence other particles as a function of distance. Protein levels are taken as a measure of such influence or strength of interaction, and are correlated with intercellular separations. Data from two-cell experiments are used to give rise to protein-dependent interaction functions. Normalized protein levels are plotted against separation distances and the relationship is expressed in the form of a quadratic equation (see Eq. 7.1) where a, b, and c are fitted constants and L_{AB} is cell–cell separation distance:

$$f(L_{\mathrm{AB}}) = a + b(L_{\mathrm{AB}}) + c(L_{\mathrm{AB}})^2 \qquad (7.1)$$

So far, this model has been applied to glioblastoma multiforme cancer cells and is limited to protein copy numbers from 6-h incubation periods.

7.5.3 *Signaling Networks Generated by Single-Cell Data Analysis*

The fluctuations of functional proteins across many cells imply the protein-protein connections and how a signal is transduced. For example, in a cell population, the copy number ratios between protein 1 and protein 2 remain approximately the same for all the single cells. Then with enough statistics, it is safe to conclude that protein 1 and protein 2 are in one signaling pathway. This is a simple approach to understand the network generated by single-cell functional protein assays. However, a signaling protein might be the joint point of multiple signaling pathways, and therefore the ratios between protein copies may not be exactly integers. This is confirmed by the single-cell network analysis results (Fig. 7.4). The thickness of connection lines and the color of edges reflect the strength of protein-protein correlations and the relative abundance of those proteins, respectively. The cutoff threshold is usually set at 0.5. Under various conditions, the correlation maps, or signaling networks, are significantly changed, indicating that two individual cells are strongly influencing each other by communications.

As mentioned above, each functional protein may take part in multiple signaling pathways, and the signal flow between any two proteins might be mediated by others. Therefore, the more meaningful derivation of signaling networks is via decomposition of the whole network into different tiers to approach the real picture of signaling flow. Shin et al. employed information theory, specifically Le Chatelier's principle, to profile the signaling networks from single-cell data, and provided an approach to quantitatively predict the role of pertubations and characterization of a protein-protein interaction network [47]. The theoretical prediction is made by seeking that distribution of copy numbers that is of maximal entropy, meaning that the distribution is as uniform as possible subject to a given mean number of copies. This means that at the very global maximum of the entropy, the probabilities of the different proteins are not equal. Rather, as in any multicomponent system at thermal equilibrium, each protein will be present in proportion to its partition function where the partition function is the effective thermodynamic weight of a species at thermal equilibrium (for supplementary reading on partition functions, the reader is referred to Statistical Mechanics (2011) by Pathria and Beale). The whole system, a single cell (or a small colony), is regarded as not being in an equilibrium state because it is under the action of constraints. When the constraints are present the system is in that state of equilibrium that is possible under the constraints. Here a single cell is a confined system, and the protein copies in this confined system are quantitated, so the modified Le Chatelier's principle combined with certain mathematic manipulations leads to newly constructed signaling networks with multiple tiers, with each tier representing the weight of the sub-network in the whole system. This theory has been successfully applied to phosphoprotein networks in response to hypoxia [19] as well as cancer cell interactions at various distances [48].

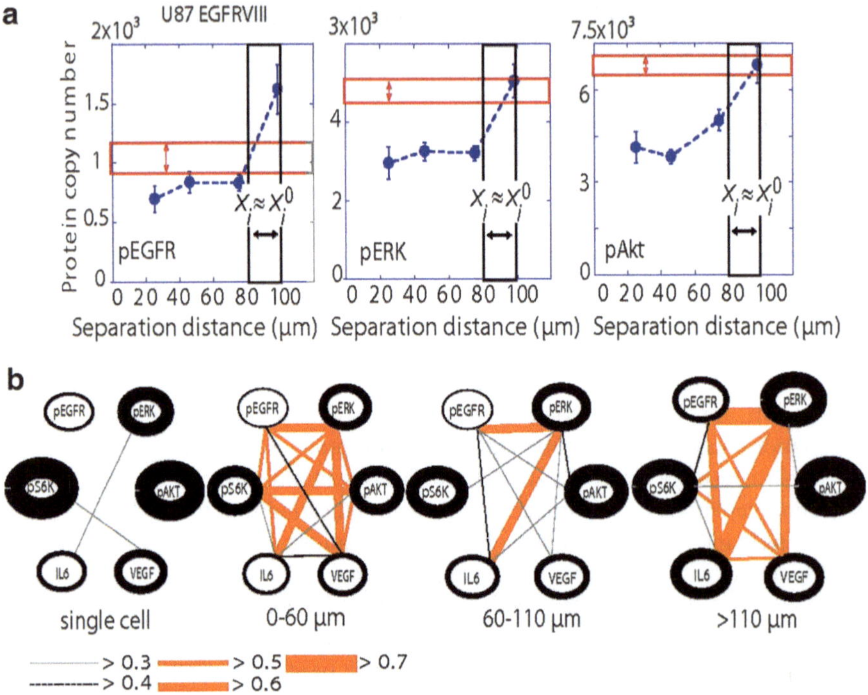

Fig. 7.4 Comparisons of noninteracting U87EGFRvIII single cells with interacting cell pairs. (**a**) Mean values of protein levels, as measured for U87EGFRvIII single cells, are compared against measurement of those same proteins for cell pairs, at cell separation distance ranges close to the steady state (80–100 μm for U87EGFRvIII cells) or deviating significantly (20–30 μm for U87EGFRvIII cells). (**b**) Protein–protein coordination maps were generated using U87EGFRvIII two-cell and single-cell data. The thickness of the lines encircling the protein names reflects the relative abundance of those proteins. The thickness and color of the edges reflect the extent of the protein–protein coordination ($P < 0.05$) as provided in the key below the networks. Reprinted with permission from "Glioblastoma cellular architectures are predicted through the characterization of two-cell interactions," by Kravchenko-Balasha, N. et al. 2014, *Proc Natl Acad Sci USA* 111. Copyright 2014 by the National Academy of Sciences

7.6 Conclusions and Perspectives

The advances in single-cell proteomics technologies have been rapid within recent years, with a few fresh perspectives offered on cellular heterogeneity and cell–cell interactions. These technologies are essential for advancing cell signaling studies, and have important potential in clinical and diagnostic applications. The clinical and biomedical drivers of this technology are propelling this field at an ever-faster rate, aiming at inexpensive and easy-to-use methods that will be available to a broad audience. So far, only a few microfluidic proteomics methods including the SCBC have matched the statistics and throughput of cytometry tools. Moreover, the robustness of these methods still have some room for improvement. The single-cell barcode technology affords the capability to enhance the multiplexity to 42 proteins detected simultaneously, higher than the most state-of-the-art multiparameter flow cytometry. It is likely that, in the future, the multiplexity of a single-cell proteomic technology can advance to over 100, as the newly introduced Cytof already demonstrates such a capacity. Beyond 100, all single-cell technologies including microchips and Cytof will be limited by the availability of antibodies and their cross talk, which is a common problem in the multiplexed assays [49]. Thus, the development of high-performance, inexpensive, and robust antibodies is equally imperative. Compared with genomic analyses, the cost of protein assays per amino acid is generally 100 times higher.

The future of single-cell functional proteomics technologies is expected to be more diversified. The reason is that many microfluidic tools are developed specifically to address certain technology bottlenecks and are driven by clinical needs, which are changing over time. For instance, the rapid growth of the immunotherapy field has pushed forward the two-cell functional proteomic assays [50]. The increasing use of targeted drugs in cancer therapies, coupled with modest performance of those drugs in the clinic, requires point-of-use single-cell tools to analyze phosphoprotein signaling networks that are targeted by those therapies. Furthermore, the major proteomics tool, mass spectrometry, has been becoming more accurate and quantitative for clinical diagnostics, and may continuously place a key role in single-cell analysis in the future.

References

1. Irish JM, Kotecha N, Nolan GP (2006) Innovation—mapping normal and cancer cell signalling networks: towards single-cell proteomics. Nat Rev Cancer 6(2):146–155
2. Altschuler SJ, Wu LF (2010) Cellular heterogeneity: do differences make a difference? Cell 141(4):559–563
3. Rawlings JS, Rosler KM, Harrison DA (2004) The JAK/STAT signaling pathway. J Cell Sci 117(8):1281–1283
4. Monti M, Orru S, Pagnozzi D, Pucci P (2005) Functional proteomics. Clin Chim Acta 357 (2):140–150
5. Wei W, Shin YS, Ma C, Wang J, Elitas M, Fan R et al (2013) Microchip platforms for multiplex single-cell functional proteomics with applications to immunology and cancer research. Genome Med 5(8):75
6. Yu J, Zhou J, Sutherland A, Wei W, Shin YS, Xue M et al (2014) Microfluidics-based single-cell functional proteomics for fundamental and applied biomedical applications. Annu Rev Anal Chem 7:275–295
7. Chattopadhyay PK, Gierahn TM, Roederer M, Love JC (2014) Single-cell technologies for monitoring immune systems. Nat Immunol 15(2):128–135
8. Visvader JE, Lindeman GJ (2008) Cancer stem cells in solid tumours: accumulating evidence and unresolved questions. Nat Rev Cancer 8(10):755–768
9. Morgan RA, Dudley ME, Wunderlich JR, Hughes MS, Yang JC, Sherry RM et al (2006) Cancer regression in patients after transfer of genetically engineered lymphocytes. Science 314 (5796):126–129

10. Kershaw MH, Westwood JA, Darcy PK (2013) Gene-engineered T cells for cancer therapy. Nat Rev Cancer 13(8):525–541
11. Ma C, Fan R, Ahmad H, Shi QH, Comin-Anduix B, Chodon T et al (2011) A clinical microchip for evaluation of single immune cells reveals high functional heterogeneity in phenotypically similar T cells. Nat Med 17(6):738–743
12. Ma C, Cheung AF, Chodon T, Koya RC, Wu ZQ, Ng C et al (2013) Multifunctional T-cell analyses to study response and progression in adoptive cell transfer immunotherapy. Cancer Discov 3(4):418–429
13. Zhao JL, Ma C, O'Connell RM, Mehta A, DiLoreto R, Heath JR et al (2014) Conversion of danger signals into cytokine signals by hematopoietic stem and progenitor cells for regulation of stress-induced hematopoiesis. Cell Stem Cell 14(4):445–459
14. Hanahan D, Weinberg RA (2011) Hallmarks of cancer: the next generation. Cell 144 (5):646–674
15. Logue JS, Morrison DK (2012) Complexity in the signaling network: insights from the use of targeted inhibitors in cancer therapy. Gene Dev 26(7):641–650
16. Gottesman MM (2002) Mechanisms of cancer drug resistance. Annu Rev Med 53:615–627
17. Holohan C, Van Schaeybroeck S, Longley DB, Johnston PG (2013) Cancer drug resistance: an evolving paradigm. Nat Rev Cancer 13(10):714–726
18. Shi QH, Qin LD, Wei W, Geng F, Fan R, Shin YS et al (2012) Single-cell proteomic chip for profiling intracellular signaling pathways in single tumor cells. Proc Natl Acad Sci U S A 109 (2):419–424
19. Wei W, Shi QH, Remacle F, Qin LD, Shackelford DB, Shin YS et al (2013) Hypoxia induces a phase transition within a kinase signaling network in cancer cells. Proc Natl Acad Sci U S A 110(15):E1352–E1360
20. Perfetto SP, Chattopadhyay PK, Roederer M (2004) Innovation—seventeen-colour flow cytometry: unravelling the immune system. Nat Rev Immunol 4(8):648–655
21. De Rosa SC, Herzenberg LA, Herzenberg LA, Roederer M (2001) 11-Color, 13-parameter flow cytometry: identification of human naive T cells by phenotype, function, and T-cell receptor diversity. Nat Med 7(2):245–248
22. Bendall SC, Simonds EF, Qiu P, Amir EAD, Krutzik PO, Finck R et al (2011) Single-cell mass cytometry of differential immune and drug responses across a human hematopoietic continuum. Science 332(6030):687–696
23. Newell EW, Sigel N, Bendall SC, Nolan GP, Davis MM (2013) Cytometry by time-of-flight shows combinatorial cytokine expression and virus-specific cell niches within a continuum of CD8(+) T cell phenotypes (vol 36, pg 142, 2012). Immunity 38(1):198–199
24. Newell EW, Sigal N, Bendall SC, Nolan GP, Davis MM (2012) Cytometry by time-of-flight shows combinatorial cytokine expression and virus-specific cell niches within a continuum of CD8(+) T cell phenotypes. Immunity 36(1):142–152
25. Czerkinsky CC, Nilsson LA, Nygren H, Ouchterlony O, Tarkowski A (1983) A solid-phase enzyme-linked immunospot (ELISPOT) assay for enumeration of specific antibody-secreting cells. J Immunol Methods 65(1–2):109–121
26. Bao N, Wang J, Lu C (2008) Recent advances in electric analysis of cells in microfluidic systems. Anal Bioanal Chem 391(3):933–942
27. Fu AY, Spence C, Scherer A, Arnold FH, Quake SR (1999) A microfabricated fluorescence-activated cell sorter. Nat Biotechnol 17(11):1109–1111
28. Faley SL, Copland M, Wlodkowic D, Kolch W, Seale KT, Wikswo JP et al (2009) Microfluidic single cell arrays to interrogate signalling dynamics of individual, patient-derived hematopoietic stem cells. Lab Chip 9(18):2659–2664
29. Brouzes E, Medkova M, Savenelli N, Marran D, Twardowski M, Hutchison JB et al (2009) Droplet microfluidic technology for single-cell high-throughput screening. Proc Natl Acad Sci U S A 106(34):14195–14200
30. Ahmad H, Sutherland A, Shin YS, Hwang K, Qin LD, Krom RJ et al (2011) A robotics platform for automated batch fabrication of high density, microfluidics-based DNA

microarrays, with applications to single cell, multiplex assays of secreted proteins. Rev Sci Instrum 82(9):094301
31. Nguyen CQ, Ogunniyi AO, Karabiyik A, Love JC (2013) Single-cell analysis reveals isotype-specific autoreactive B cell repertoires in Sjogren's syndrome. Plos One 8(3):e58127
32. Varadarajan N, Julg B, Yamanaka YJ, Chen HB, Ogunniyi AO, McAndrew E et al (2011) A high-throughput single-cell analysis of human CD8(+) T cell functions reveals discordance for cytokine secretion and cytolysis. J Clin Invest 121(11):4322–4331
33. Wang J, Tham D, Wei W, Shin YS, Ma C, Ahmad H et al (2012) Quantitating cell-cell interaction functions with applications to glioblastoma multiforme cancer cells. Nano Lett 12 (12):6101–6106
34. Fan R, Vermesh O, Srivastava A, Yen BKH, Qin LD, Ahmad H et al (2008) Integrated barcode chips for rapid, multiplexed analysis of proteins in microliter quantities of blood. Nat Biotechnol 26(12):1373–1378
35. Heath JR (2005) NanoSystems biology and cancer. Abstr Pap Am Chem S 230:U187–U
36. Heath JR, Davis ME (2008) Nanotechnology and cancer. Annu Rev Med 59:251–265
37. Khalil IG, Hill C (2005) Systems biology for cancer. Curr Opin Oncol 17(1):44–48
38. Zheng GF, Patolsky F, Cui Y, Wang WU, Lieber CM (2005) Multiplexed electrical detection of cancer markers with nanowire sensor arrays. Nat Biotechnol 23(10):1294–1301
39. Bailey RC, Kwong GA, Radu CG, Witte ON, Heath JR (2007) DNA-encoded antibody libraries: a unified platform for multiplexed cell sorting and detection of genes and proteins. J Am Chem Soc 129(7):1959–1967
40. Lu Y, Xue Q, Eisele MR, Sulistijo ES, Brower K, Han L et al (2015) Highly multiplexed profiling of single-cell effector functions reveals deep functional heterogeneity in response to pathogenic ligands. Proc Natl Acad Sci U S A 112(7):E607–E615
41. Vermesh U, Vermesh O, Wang J, Kwong GA, Ma C, Hwang K et al (2011) High-density, multiplexed patterning of cells at single-cell resolution for tissue engineering and other applications. Angew Chem Int Edit 50(32):7378–7380
42. Shin YS, Ahmad H, Shi QH, Kim H, Pascal TA, Fan R et al (2010) Chemistries for patterning robust DNA microbarcodes enable multiplex assays of cytoplasm proteins from single cancer cells. Chemphyschem 11(14):3063–3069
43. Thorsen T, Maerkl SJ, Quake SR (2002) Microfluidic large-scale integration. Science 298 (5593):580–584
44. Tlsty TD, Coussens LM (2006) Tumor stroma and regulation of cancer development. Annu Rev Pathol-Mech 1(1):119–150
45. Gagliano N, Costa F, Cossetti C, Pettinari L, Bassi R, Chiriva-Internati M et al (2009) Glioma-astrocyte interaction modifies the astrocyte phenotype in a co-culture experimental model. Oncol Rep 22(6):1349–1356
46. Bonavia R, Inda MD, Cavenee WK, Furnari FB (2011) Heterogeneity maintenance in glioblastoma: a social network. Cancer Res 71(12):4055–4060
47. Shin YS, Remacle F, Fan R, Hwang K, Wei W, Ahmad H et al (2011) Protein signaling networks from single cell fluctuations and information theory profiling. Biophys J 100 (10):2378–2386
48. Kravchenko-Balasha N, Wang J, Remacle F, Levine RD, Heath JR (2014) Glioblastoma cellular architectures are predicted through the characterization of two-cell interactions. Proc Natl Acad Sci U S A 111(17):6521–6526
49. Li HY, Bergeron S, Juncker D (2012) Microarray-to-microarray transfer of reagents by snapping of two chips for cross-reactivity-free multiplex immunoassays. Anal Chem 84 (11):4776–4783
50. Elitas M, Brower K, Lu Y, Chen JJ, Fan R (2014) A microchip platform for interrogating tumor-macrophage paracrine signaling at the single-cell level. Lab Chip 14(18):3582–3588

Chapter 8
Microfluidics for DNA and Protein Analysis with Multiplex Microbead-Based Assays

Wanqing Yue and Mengsu Yang

Abstract Early screening and diagnosis of diseases require the development of sensitive, reliable, and inexpensive high-throughput assays. There is a growing need for innovative diagnostic technologies that provide accurate detection of a broad range of diseases and enhance laboratory productivity. Developments in micro- and nanotechnology have advanced the "lab-on-a-chip" concept towards a new generation of point-of-care diagnostic devices that enable parallel detection of multiple analytes in small-volume samples with high sensitivity in a short time. These features fulfill some of the important criteria of bioanalysis used for biochemical studies, environmental analyses, and clinical diagnostics. However, the multiplexing capability of microfluidic-based assay is limited compared to conventional flow cytometric assays, particularly for encoded microbead-based assays, which are capable of simultaneously analyzing multiple analyte. It is possible to incorporate the microbead-based assays into microfluidic devices in order to retain all the advantages of the microdevice and significantly improve multiplexing capability. In this chapter, we summarize the latest development in microbead-based assays and discuss the integration of the microbead technology with microfluidic devices. Applications of the integrated approach in multiplexed protein and DNA analysis are growing rapidly in the areas of biomarker research, cancer screening, and disease diagnosis. The prospects for future development and commercialization of these microbead-based microfluidic devices are also discussed.

W. Yue, Ph.D.
Department of Biomedical Sciences, City University of Hong Kong, 83 Tat Chee Avenue, Kowloon, Hong Kong, People's Republic of China

School of Sciences, China Pharmaceutical University, 639 Longmian Avenue, Jiangning District, Nanjing, People's Republic of China
e-mail: helen.ywq@gmail.com

M. Yang, Ph.D. (✉)
Department of Biomedical Sciences, City University of Hong Kong, 83 Tat Chee Avenue, Kowloon, Hong Kong, People's Republic of China

Key Laboratory of Biochip Technology, Biotech and Health Centre, Shenzhen Research Institutes of City University of Hong Kong, Shenzhen, People's Republic of China
e-mail: bhmyang@cityu.edu.hk

© Springer International Publishing Switzerland 2016

C. Lu, S.S. Verbridge (eds.), *Microfluidic Methods for Molecular Biology*,
DOI 10.1007/978-3-319-30019-1_8

Keywords Microfluidics • Microbead-based assay • Multiplex analysis • Biomarker research • Clinical diagnosis

8.1 Introduction

Micro total analysis systems (μTAS) [1] have been rapidly developed during the last 20 years, which integrate sequential analytical processes including sample pretreatment, separation, reaction, detection, and data analysis in a single miniaturized chip [2, 3]. The highly integrated microfluidic devices featured significant advantages including low reagent consumption, high-throughput analysis, and portability for in situ use, which have played an important role in combinatorial screening for catalyst generation [4], cell-based assay [5], as well as molecular level analysis of DNA [6, 7] and proteins [8]. Confined in the microscale structures, the behaviors of flow deviate from those in macro-scale [9], which provides an opportunity to control fluid and transport soluble factors to specific locations for precisely creating artificial environments for analytical requirement. Microfluidics technology enables rapid and reproducible measurements of small volumes of sample while eliminating the labor-intensive procedure and potentially error-prone manipulation. Moreover, miniaturized devices with low power consumption have potential applications for point-of-care diagnostics. Commercially available products from companies such as Agilent and Fluidigm are equipped with measuring systems and custom-design microfluidic chips feasible for clinic applications.

Early screening and diagnosis of diseases require the development of sensitive, reliable, and inexpensive high-throughput assays [10–15]. High-throughput analysis can be achieved by parallel detection of one analyte for multiple samples, or by simultaneous detection of various analytes from one sample [16]. Bead-based assays, by which different specific probes are tethered on encoded particles via chemical or physical means [13] and detected with flow cytometric methods, have demonstrated feasibility and versatility in multiplex analysis of biomolecules such as genotyping [17], gene expression profiling [18], enzymatic assays [19], and immunoassays [20].

An alternative approach to implement microbead-based assays is to incorporate the assays in microfluidic devices [21, 22], which is accomplished by manipulating the microbeads in a microfluidic device for decoding and quantification of multiple analytes [13]. The main advantages of integrating microbead-based assays into microfluidics are short analysis time, small volume, and high sensitivity. Due to the large surface-to-volume ratio, fast reaction kinetics and high binding capacity enable the microbead-based assays in microfluidic format to be suitable for rapid and sensitive detection of various biomolecules [13]. Moreover, compared with surface-treated microfluidic-based analysis, analytical targets are easily expanded by addition of extra microbeads conjugated with relevant probes, making the multiplexing easily achievable in the microfluidic-based platform. Furthermore, the low cost and better quality control of the beads offer a flexible selection of probe sets for analytes [14]. These advantages are vital for rapid and accurate detection of disease biomarkers to provide point-of-care testing for clinical diagnoses. Sensitive detection of pathogenic viruses and bacteria also demands these qualities for early warning and preparation against biological warfare or epidemic outbreak.

In this chapter, we discussed the working principle and the latest progress in the integration of the microbead-based assay into microfluidic devices. While the potential applications of such an integrated platform are numerous, we focused on applications in multiplexed protein and DNA analysis which are important for biomarker research, cancer screening, and disease diagnosis. We also presented an outlook for future development and commercialization of these microbead-based microfluidic devices.

8.2 Microbead-Based Technology

8.2.1 Barcoding of Microbeads for Multiplex Assays

Multiplex assays refer to the capability of measuring multiple analytes from the same biological sample in a single test. Microbeads with different labels and surface probes have been demonstrated as a promising approach to accommodate simultaneous detection of multiple analytes. Microbeads could be optically barcoded that are identified by specific spectra, or graphically barcoded that are identified by deciphering the specific geometric patterns. Chromophores, fluorophores, or quantum dots are tethered on the surface of microbeads or incorporated inside the microbeads to optically encode microbeads, which could be identified by the spectrum and/or the intensity of the colors [23, 24]. Organic fluorescence dyes [25], quantum dots (Fig. 8.1a) [24], styrene monomers with different Raman spectra (Fig. 8.1b) [26, 27], and DNA-based monomer fluorescent nanostructures [28] have been reported to optically barcode the microbeads. By loading microbeads with precise proportions of two or more types of dyes, a group of microbeads could be created and mapped in a bead mixture by measuring the intensities of individual dye (Fig. 8.1c) [29, 30]. Instead of fluorescent dyes, quantum dots have also been used to identify microbeads. Quantum dots have narrow and symmetric emission spectra with high stability against photo bleaching, which have more than 20 times brighter emission than organic dyes. The emission wavelength of quantum dots can be controlled by varying the diameter and only a single light source is required for simultaneous excitation of different quantum dots. The use of 1.2 μm polystyrene beads encoded with quantum dots was demonstrated for multiplex assays, in which quantum dots with six colors and six intensity levels were used as barcode to generate more than 10,000 different codes [24]. Subsequently, quantum dots encoded mesoporous polystyrene beads and surfactant-coated quantum dots were developed, which could be analyzed by flow cytometer with the detection capability of 1000 bead/s [31].

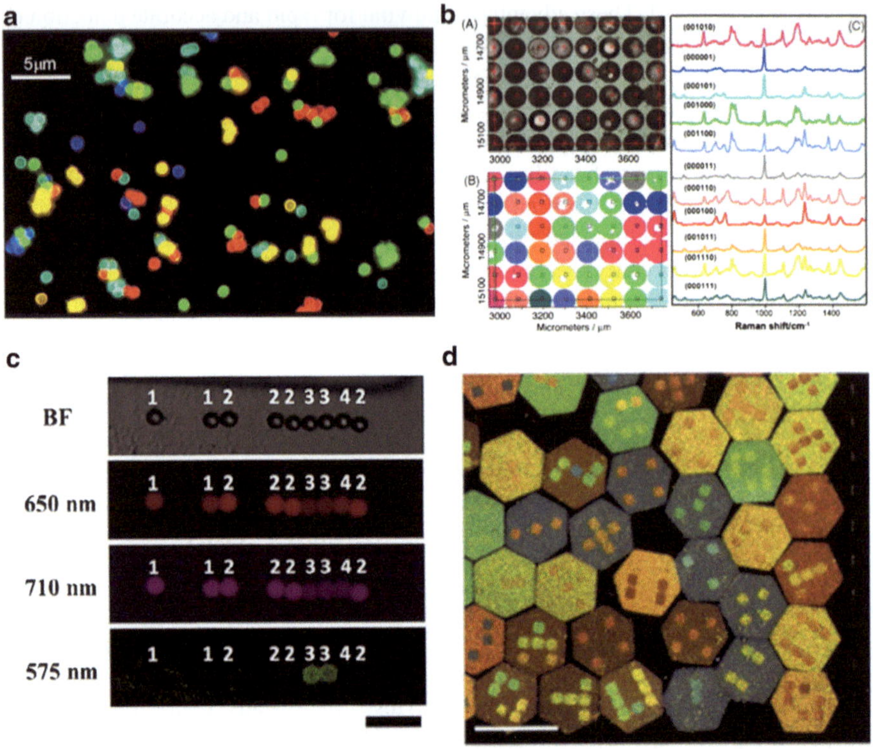

Fig. 8.1 Different barcoded microbeads for multiplex assays. (**a**) Quantum dot-encoded polystyrene microbeads. Reprinted by permission from Macmillan Publishers Ltd: Nature Biotechnology [24], copyright (2001). (**b**) Raman mapping of the barcoded microarray. Reprinted from [27], Copyright (2007), with permission from Elsevier. (**c**) Microbead encoded by two fluorescent dyes with the emission wavelengths of 650 nm and 710 nm, respectively. The detection signal was recorded by the labeled fluorescent molecules with the emission wavelength of 575 nm on the surface of microbeads. Reprinted from [30], Copyright (2014), with permission from Elsevier. (**d**) Graphically and optically patterned microbeads made by magnetic photonic crystals. Reprinted by permission from Macmillan Publishers Ltd: Nature Materials [32], copyright (2010)

Graphically encoded microbeads are often tens to hundreds of microns in diameter or length with a well-defined orientation in order to be accurately decoded. Recently, the combination of optically and graphically barcoded microparticles has been reported and the barcoded microparticles were read out by an optical imaging platform (Fig. 8.1d) [32, 33]. Theoretically a larger pool of beads could be constructed by quantum dots and/or graphically encoding technology, with several unique codes generated by increasing the number of dyes, controlling the ratio, and creating various patterns. However, many limitations in the compatibility of dyes, manufacture feasibility, and detection sensitivity reduced the number to around 100 unique groups available for coding and decoding. Currently, commercially available microbead-based assay kits, such as xMAP (Luminex, TX, USA) and Cytometric Bead Array CBA (BD Bioscience, CA, USA), are using this barcoding technique, which enable simultaneous measurement of cell signaling molecules, cytokines, and antibodies in various types of body fluids.

8.2.2 Surface Functionalization of Microbeads

A variety of biocompatible materials have been used for fabrication of bead-based solid support for capture of target biomolecules, such as silica [34], polystyrene (PS) [35], agarose [36], and paramagnetic materials [37]. Methods for increasing the binding capacity in beads have also been intensively studied, which focused on modifying the chemical functionalization of the bead surface, tuning the bead size, and physically compressing the bead. Using different surface chemistry such as UV/ozone treatment [38], silanization [39], covalent coupling [40], and adsorption [41], a vast number of molecules including antibodies, antigens, DNA, RNA, and other small molecules can be easily attached to bead surface. Covalent coupling is the most popular approach for immobilization of biomolecules. The solution-based binding kinetics has been extensively studied, making the coupling approach available for different applications [42, 43]. Covalently coupled, probes are favorably presented on the surface of the bead, leaving the binding moieties available for interaction with target molecules. Specific coating of probes on designated microbeads eliminates the "cross talk" between microbeads, allowing multiplexed analysis in one sample. As the probe molecules are permanently bound, they will not desorb or leach over time [44]. Recently, physical compression of beads has been reported to enhance binding capability in beads. A twofold increase in mass transfer rates for binding of biotin-functionalized quantum dots to streptavidin-coated agarose beads has been achieved by periodically compressing and expanding the beads [45]. Similarly, the adsorption rates could be increased when compressing chitosan beads in solutions containing dye molecules or nanoparticles [46]. Diameter of the beads is also critical for surface modification. Smaller size of beads presents larger surface area per unit weight while larger beads present more surface area per bead, such that the amount of reagent required for coating is different. In addition, solid beads could be porous and nonporous. The porous beads can be densely packed with immobilized probes due to their large surface-to-volume ratios than nonporous counterparts [47, 48]. The microporous geometry allows for greater capture efficiency and has potential to offer lower limits of detection compared to nonporous beads [49].

The principles of biomolecular recognition are applied to design the probes on the microbead surface for detection of analytes. Recognitions between antigen and antibody, hybridization of DNA strands with complementary sequences, DNA-protein interaction, as well as sugar-lectin binding have been widely used in the design of microbead-based assays. Moreover, the combination of DNA amplification techniques and microbeads has been reported. For example, Li et al. developed a platform for quantification of single-nucleotide polymorphisms (SNPs) in clinical samples by microbead-based rolling circle amplification assay

(RCA) [50]. Similarly, the presence of a protein and DNA could also be simultaneously detected on microbead surfaces [51]. In this approach, a two-layer sandwich assay was performed on the antibody-modified microbead for detection of protein and the presence of DNA in the same sample could be captured and detected by RCA strategy (Fig. 8.2a). Capture antibody-conjugated microbeads are used to detect target protein followed by binding of biotin-labeled secondary antibodies. Biotinylated capture DNA probes, which act as recognition element and amplification primers, are then attached to the secondary antibodies via a biotin-avidin bridge. A padlock probe is then hybridized onto the primer and terminated in the presence of a target DNA sequence, followed by ligation as a loop by a DNA ligase. During RCA, fluorescently labeled signal probes are hybridized to each amplified sequence and the microbeads could be imaged by a fluorescence microscope. Simultaneous detection of IL-8 and target DNA were evaluated and the detection limits were down to 10 fM and 1 pM for protein and DNA, respectively. This tool offers the capability of measuring various combinations of analytes to provide both proteomic and genomic information with a single test. Moreover, as the microbeads are individually identified, multiplexed assay is readily performed by using an array of microbeads with different capture antibodies and fluorescent-labeled probes.

Single-molecule assay could be accomplished in which the readout is based on counting microbeads tethered to DNA molecules. Tethered bead technology was initially developed to study biophysical properties of DNA molecules and enzymes that act on DNA, at the single molecule level [52–54]. DNA molecules of known length are attached via one end to a surface and the other end attached to a micron-sized bead that can be pulled to stretch the DNA. Pulling can be effected via liquid flow (drag force on the bead), laser trapping, or magnetic force if the bead is paramagnetic. Tethered beads exhibit damped and confined Brownian motion in the absence of external force, which are easily identified in light microscopy. The motion of tethered beads in flow is recognizable with a low-power optical microscope and distinguishable from that of beads that stick nonspecifically to the surface. Recently, a proof-of-principle immune sandwich assay was described for detection of single immune complexes based on the above principle [55].

To further decrease limit of detection and increase sensitivity, signal amplification strategies incorporated with nanoparticles [56, 57] have also been intensively used in the microbead-based assays. We have reported a method based on multienzyme-nanoparticle amplification for sensitive bead-based virus genotyping (Fig. 8.2b) [17]. The microbeads were functionalized with the capture probes and electron-rich proteins, and employed the HRP-functionalized gold nanoparticles as label on a secondary DNA probe. By using "sandwich" hybridization, the enzyme-functionalized gold nanoparticle (AuNPs) labels were brought close to the surface of microbeads. The oxidation of biotin-tyramine by hydrogen peroxide resulted in the deposition of multiple biotin moieties onto the surface of beads. The detection signals were amplified where the large surface area of AuNP carriers increased the amount of HRP bound per sandwiched hybridization. The approach could discriminate and genotype 10 copies/μL HPV genomic DNA. This assay was further integrated into a microfluidic device and the enzyme-nanoparticle-conjugated

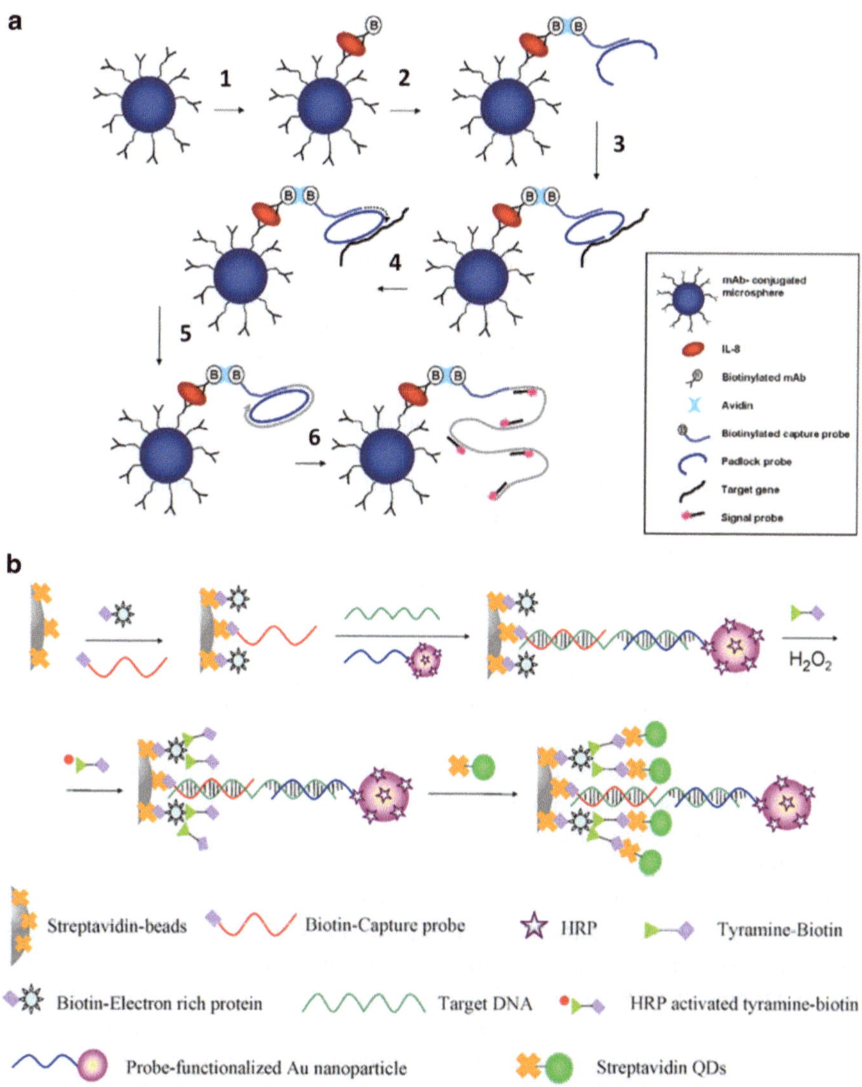

Fig. 8.2 (**a**) Microbead-based rolling circle amplification (RCA) assay for simultaneous detection of target DNA sequence and protein. (1) Monoclonal antibody-conjugated microbeads captured the target protein followed by secondary biotin-labeled antibodies. (2) The biotinylated DNA probes are captured via a biotin-avidin bridge and the padlock probes are then hybridized onto the DNA probes. (3) The padlock probes are terminated in proximity by the target gene. (4) The padlock probes are ended by DNA ligase, followed by capture probe extension by polymerase. (5) Rolling circle amplification of the ligated probes proceeded on the bead surface. (6) The Cy3-tagged detection probes are hybridized on the RCA product. Reprinted with permission from [51], Copyright (2009) American Chemical Society. (**b**) Multienzyme nanoparticle amplification for sensitive bead-based genotyping. Reprinted from [17], Copyright (2011), with permission from Elsevier

assay could discriminate as low as 1 fM HPV DNA, which was 1000 times more sensitive than that of the off-chip assay.

8.2.3 Recognition of Microbeads

Flow cytometry presents a powerful platform for microbead-based assays. Luminex has developed flow cytometry-based instrument for multiplexed microbead-based assays (Fig. 8.3a). The instrument consists of one flowing stream system and two laser beams. One beam decodes the microbeads and the other provides the quantification information by measuring the reporter fluorescent intensities [58]. Microdroplet technology has also attracted great interests for single-cell analysis, protein crystallization, enzyme kinetics, and cellular bioassays [8]. With the combination of microdroplet technology, BEAMing (beads, emulsion, amplification, and magnetics) related strategies (Fig. 8.3b) [59–63] have been extensively explored to analyze human genetic variation [59]. Streptavidin-labeled magnetic beads coated with oligonucleotide primers were mixed with template DNA and PCR components, and trapped into the emulsion droplets, each of which containing less than one template and one bead on average. PCR reactions were carried out by thermal cycling of the emulsions, followed by subsequently breaking the droplets and collecting the magnetic beads, which contained multiple copies of the DNA fragment identical in sequence to the original template. Recovery of the magnetic beads allowed hybridization of fluorescent-labeled oligonucleotide probes capable of specific detection and quantification of genetic sequence variants [59, 64], direct determination of polymerase error rates, and identification of transcription factor targets [60].

Other optical reading platforms such as microscopic imaging system and fiber-optic system have been introduced for microbead analysis and decoding. Walt's group developed an optic fiber platform and demonstrated its feasibility for genomic and proteomic analysis of clinical samples [51, 65–68]. The core of individual fibers was selectively etched to generate wells of predefined depth, which were used to assemble microbeads in array on the distal end of the fiber. Thousands of individual fibers embedded with one type of microbeads were further assembled as fiber bundles, which enables massive detection in parallel. The feasibility of this approach has been demonstrated by multiplexed analysis of DNA and protein in human saliva sample. For DNA detection, hybridization of fluorescent-labeled target to the probe with complementary strand was performed in the high-density microbead array. Each specific hybridization event was detected by emission of the fluorescent signal localized to the probe during the hybridization process. This platform has also been demonstrated for immunoassays, in which target-specific monoclonal antibodies were immobilized on the surface of the different microbeads in the fiber bundles for simultaneous measurement of multiple proteins [66, 69].

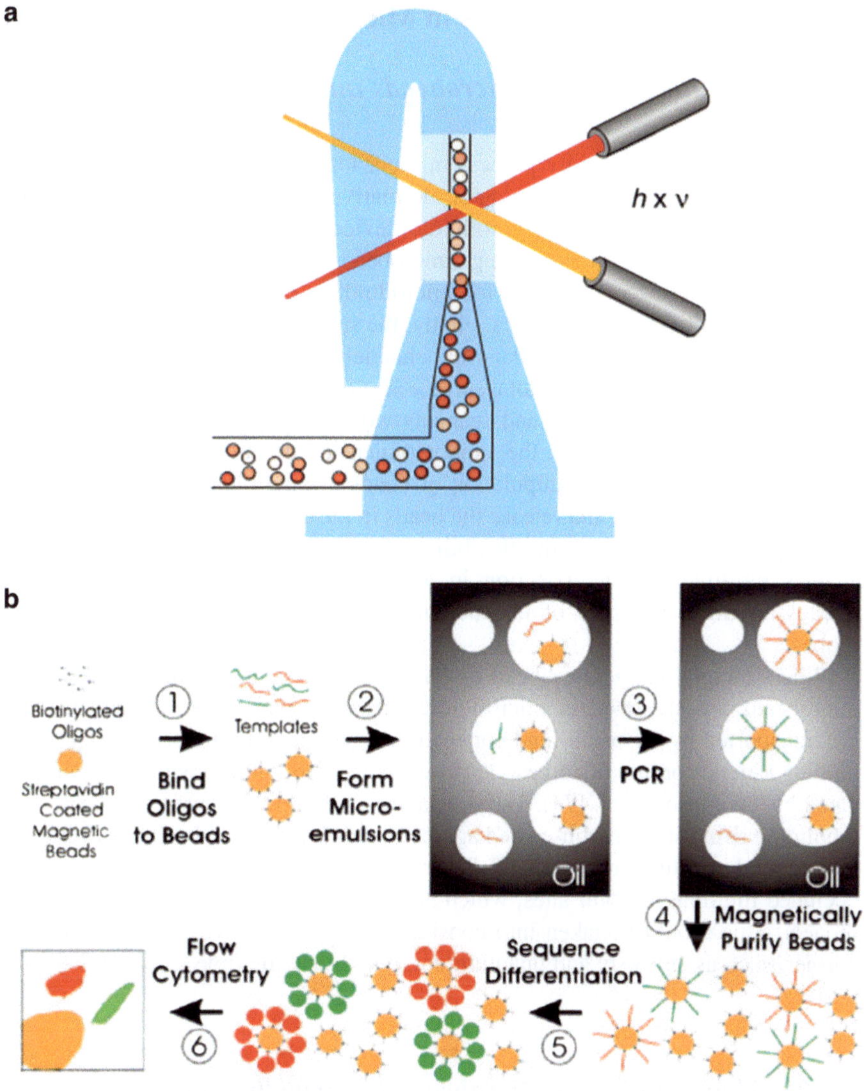

Fig. 8.3 (**a**) Flow cytometry of the multiplexed bead analysis. Two laser beams were equipped in the system. One laser beam decodes the microbeads and the other provides the quantification information by quantifying the reporter fluorescent intensities. Reprinted from [58], Copyright (2001), with permission from Elsevier. (**b**) Schematics of BEAMing (beads, emulsion, amplification, and magnetics) technique. (1) Biotinylated oligonucleotides (oligos) are bound to magnetic beads through biotin-streptavidin bridge. (2) Microemulsions are generated by stirring oligo-bound beads, PCR components, and template DNA with an oil-detergent mixture. On average, each aqueous compartment contains less than one template and one bead. (3) PCR is performed in the microemulsions in compartments with both bead and DNA template. (4) The resulting emulsions are broken and the beads are purified with a magnet. (5) After denaturation, the beads are incubated with oligos that can distinguish between the sequences of templates and fluorescent labeled. (6) Fluorescent beads are counted with flow cytometry [59], Copyright (2003) National Academy of Sciences, USA

8.3 Microbead-Based Assays in Microfluidics

8.3.1 The Integration of Microbeads into Microfluidics

A flow cytometer can rapidly process microbeads, but the high cost and lack of portability hinder its applications. An alternative approach towards microbead-based arrays is integrating in microfluidic devices [21, 22, 70]. There are three major advantages in incorporating microbead technique into microfluidics: increased surface-to-volume ratio for microfluidics, easy control of microbeads, and flexible multiple functionalities. Firstly, the surface-to-volume ratio of beads is much greater than the surface of flat microchannels. For example, 1 g of microbeads with a diameter of 0.1 μm has a total surface area of 60 m^2 [71]. Higher efficiency of interactions between samples and reagents would increase the sensitivity of the detection platform. Secondly, the similar dimension of microchannels and beads enables easy and precise manipulation of beads in microfluidics. Many different strategies to immobilize and release the beads in the channels have been developed [72, 73]. Last but not least, the flexibility of surface modification of microbeads enables multiple analyte detection in a single microfluidic design by simply employing different types of encoded microbeads with various target-specific probes.

When integrating a microbead-based assay into a microfluidic device, the microbeads are stationed in microfluidic structures while free-flowing analytes are transported by diffusion- and convection-driven flow to the bead surface. The binding kinetics on modified bead surfaces in a microfluidics context was investigated by Thompson et al. [74]. Target analyte binding to the surface of beads was monitored using streptavidin-coated agarose beads immobilized on microwells and biotinylated quantum dots as a label. The results indicated that a single quantum dot occluded multiple receptor sites, which is commonly overlooked in most kinetics models. This should be taken into consideration in assays where a large binding partner, such as antibody and quantum dot, may block multiple receptor sites.

8.3.2 Manipulation of Microbeads in Microfluidics

Microfluidics allows the transport, manipulation, and analysis of microbeads in the flow channels with the dimension of microns. Flow in microfluidics can be generated by mechanical forces, electrokinetic forces, capillary forces, or centrifugal forces. Manipulation of microbeads by microfluidic devices enables multi-functionalities of mixing, sorting, isolation, immobilization, and detection of microbeads.

The most common method of trapping microbeads is the use of a mechanical barrier based on the size of the particles, where microbeads could be isolated in special constriction structures, such as microfilters [75, 76], microwells [77, 78], double grooves [79], dam structure [80, 81], sandbag [82] and sieving microstructure [83], and single mechanical trap array [84, 85]. Microfilters featured with a rain drop bypass architecture configuration were designed [76] to have a substantially lower pressure drop to trap protozoan cells with high efficiency. Poly(ethylene glycol) (PEG) microwells within microfluidic channels were fabricated by capillary force lithography [77] to dock microbeads within predefined locations. Double microgrooves [79] with various depth ratios were developed by two-step process to achieve controlled double-level cell trapping with shear protection. Particle retention in the double grooves depended linearly on inlet flow speed, with the slope inversely proportional to the sheltering due to the groove geometry. A grid of pillars fabricated by deep reactive etching was used to confine the beads for reaction and analysis [86]. We have designed a dam structure [80] in parallel to the fluid flow as shown in Fig. 8.4a for docking and alignment of microbeads and biological cells. Two parallel channels were separated by a dam, which allows particle docking in desired locations while exposing to minimal stress. Hydrodynamic analysis [81] was carried out to compare the effects of dam dimensions on microbeads and cell

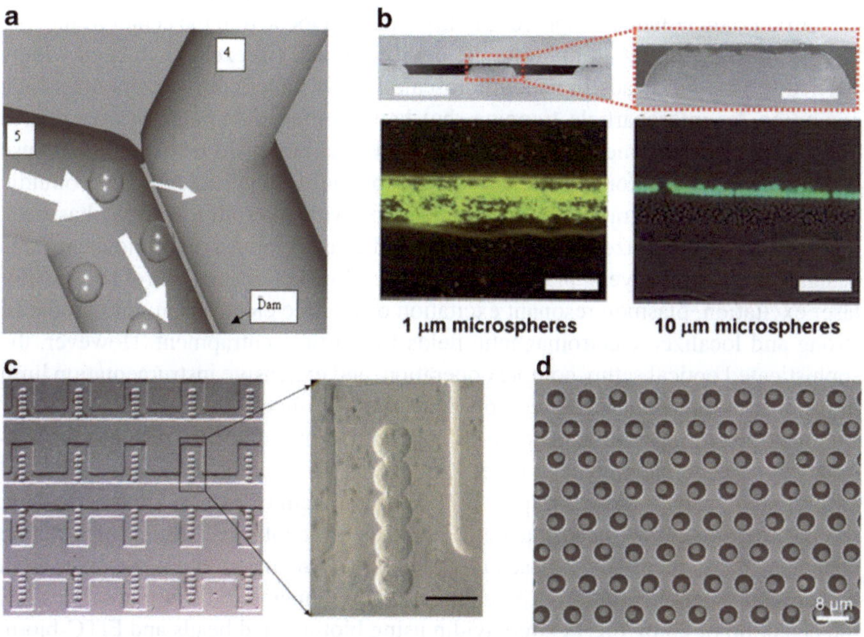

Fig. 8.4 Manipulation of microbead in microfluidics. (**a**) Simulated dam structure for particle entrapment. Reprinted with permission from [80], Copyright (2002) American Chemical Society. (**b**) Sieving microstructure for microbead with different sizes. Reproduced from [83] with permission of The Royal Society of Chemistry. (**c**) Five "sandbag" elements for individual microparticles and cell entrapment. The scale bar is 100 μm. Reprinted from [100], Copyright (2010), with permission from Elsevier. (**d**) Digital microfluidic printing of single superparamagnetic bead array. The scale bar is 8 μm. Reproduced from [102] with permission of The Royal Society of Chemistry

docking and behaviors under different flow velocities and shear stress. While most of the physical components require microstructures with fine resolution, we have further developed a single-step microfabrication method to prepare constriction microstructures on a printed circuit board (PCB) master (Fig. 8.4b). The constriction microstructures were generated by controlling the etching time of two microchannels separated by a finite distance on the PCB and the PDMS replica featured microsieves that could be used for size-dependent trapping of microspheres, biological cells, and formation of water-in-oil droplets. Such constriction microstructure enabled trapping of microbead entities with diameters of 1–10 μm, which is suitable for most microbead-based bioanalysis [83].

Recently, optical manipulations have been combined with microfluidics for biophysical and biomechanical characterizations of particles and cells [87–89]. Optical tweezers use a highly focused laser beam to trap small dielectric spherical particles, which experience both scattering force produced by the photons striking the particles along the propagation direction and gradient force induced by a gradient of field intensity. Scattering and gradient forces employed on the microbeads depend on the wavelength of the laser beam and the particle size. When the size of particles is much larger than the wavelength of the irradiation, the particles are tethered to spheres or ellipsoids and located in the center of the trap field [90]. Integrated with microfluidic device, optical tweezers have been widely applied to position beads in nano/microscale [91–95]. Optical tweezers offer high resolution for microparticle trapping, but have limited manipulation area due to tight focusing requirements [93]. The scaling limitation could be overcome by use of a diode laser bar for rapid manipulation of multiple particles in microfluidic device [94]. An integration of plasmonic traps with microfluidics [95] has been developed to immobilize dielectric spheres and yeast cells. An optofluidic chip was designed with three layers: plasmonic substrate, flow layer, and control layer. After laser excitation, plasmon resonant excitation of the free electrons in gold produced strong and localized electromagnetic fields for particle entrapment. However, the sophisticated optical setup, complex operation, and expensive instrumentation limit the application of optical trapping of microbeads in microfluidics.

An external magnetic field could be applied to a microfluidics system to immobilize and release paramagnetic beads [96], which was demonstrated to capture and enrich rare events in blood samples [97]. With the help of external magnetic field, the positions of biotinylated magnetic particles are controlled in the microfluidic channels, and sequentially exposed to various flow streams including sample, wash buffer, and labeling solution. The continuous-flow magnetic bead reactor enables the detection of 20 ng/mL of streptavidin using biotinylated beads and FITC-biotin to capture streptavidin in solution in a sandwich ELISA assay within 60 s [98]. This approach gives an added advantage of capturing and releasing the beads with target analytes and performing the wash, concentration, and other procedural steps in the microfluidics by simply controlling the position of the external magnetic fields.

Single-bead manipulation has become very important in biological research particularly in single-cell analysis. Many microfluidic structures have the capability for individually manipulating single microbeads or cells. For example, a high-density cell isolation array [84] was developed where individual particles were trapped and occluded the open region, decreasing the fraction of streamlines through the two-layer cup-shaped PDMS trapping sites, which resulted in the self-sealing quality for high quantity of single-cell isolation. The high-density single-cell array was applied to determine single-cell enzyme kinetics with fluorescence labeling and washing technique. We have designed sandbag structures [82] during the isotropic etching process by undercutting effect in a single-step photolithography. The sandbag structures (Fig. 8.4c) were utilized to immobilize microbeads and cells for high-throughput assays [99, 100] and cell communication studies [101].

8.3.3 Droplet Microfluidics for Microbead Analysis

Droplet microfluidic systems enable the formation of high-density microreactors with the sample volume of pico- or nano-liter, thus allowing fast mixing, enhancing thermal transfer and chemical reaction rates, and preventing cross-contamination and sample loss [8]. When trapped in droplets, microbeads can be transported back and forth over an array of hydrophilic entrapment units to obtain high loading efficiencies with single bead. Reagents can be transported over the beads several times while capillary and/or magnetic forces hold the beads into place. Witters et al. introduced droplet microfluidics for manipulating single superparamagnetic beads in femtoliter-sized wells for single-molecule detection [102]. They developed an electrowetting-on-dielectric-based digital microfluidic chip, in which femtoliter droplets can be generated in a high-throughput fashion by transporting microliter-sized droplet containing superparamagnetic beads over a Teflon-AF surface that contains hydrophilic-in-hydrophobic micropatterns. By exploiting the selective wettability of the hydrophilic patches, thousands of picoliter- to femtoliter-size droplets were passively generated per second, each of which containing a single superparamagnetic bead. Loading efficiencies of single beads in microwells were obtained as much as 98 % with a CV of 0.9 % in less than 1 min (Fig. 8.4d), which is much higher than previously reported methods (40–60 %). The superparamagnetic beads were used for capturing single-protein molecules by fluorescent detection of biotinylated enzyme β-galactosidase on streptavidin-coated beads with a linear dynamic range of 4 orders of magnitude ranging from 10 aM to 90 fM.

Another high-throughput immunoagglutination assay that relies on a combination of magnetic beads, droplet microfluidics, and magnetic tweezers has been developed in a fully automated manner (Fig. 8.5a) [103]. Water-in-oil droplets containing antibody-coated magnetic beads (MBs) and analytes were generated and transported in Teflon tubing. When passing in between magnetic tweezers, the MBs in the droplets were magnetically mixed with analyte such that the agglutination rate and kinetics were enhanced. When released from the magnetic field, the internal recirculation flows in the droplet-induced shear forces that allowed MBs redispersing in the droplet. In the presence of a specific analyte, MBs stayed in the

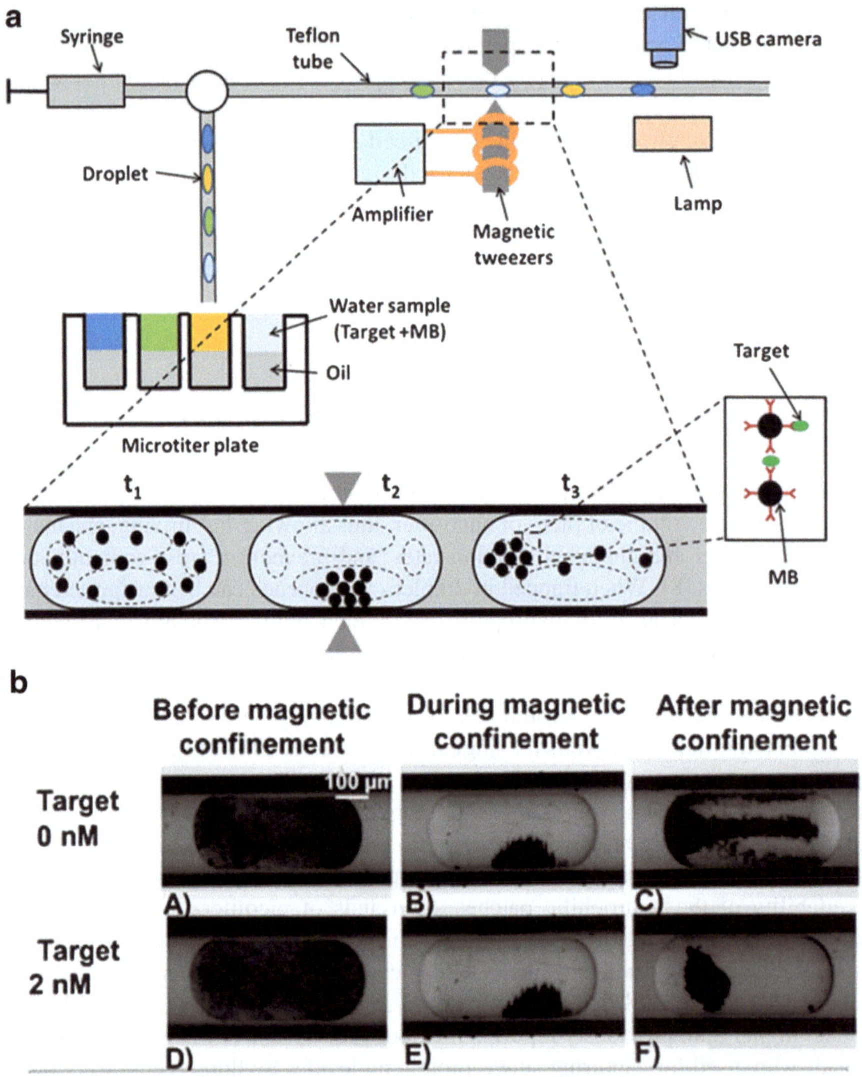

Fig. 8.5 Magnetic bead-based immunoagglutination in droplets. (**a**) Droplets are generated in Teflon tube sequentially as aspirating defined volumes of oil and water sample containing target analyte and magnetic beads. The droplets are transported towards the magnetic tweezers, which generate magnetic field in the confinement to enhance the aggregation. After the magnetic confinement, the magnetic beads are redispersed due to the internal recirculation flows (*dotted lines*). (**b**) Bright-field micrographs of magnetic bead agglutination with analyte in droplet with the concentration of 0 and 2 nM before, during, and after magnetic confinement. Reproduced from [103] with permission of The Royal Society of Chemistry

aggregated state while in the case of a nonspecific analyte, redispersion of particles occurred (Fig. 8.5b). The feasibility of the approach was demonstrated by the detection of biotinylated phosphatase alkaline in 100 nL sample volumes with the sampling size of 300 assays per hour and the limit of detection down to 100 pM. This immunoagglutination assay offers major advantages regarding the reagent consumption and the platform cost.

8.4 Applications of Microbead-Based Assays in Microfluidics for Bioanalysis

8.4.1 Immunoassays

Immunoassays, such as the enzyme-linked immunosorbent assay (ELISA), are the most commonly used biochemical test in medical diagnosis. Traditionally, these assays are carried out without sample replenishment in microliter-volume wells. Recently, microfluidic-based immunoassays have been developed with fluorometric and colorimetric measurement, electrochemical detection, and surface plasmon resonance. Direct functionalization of microchannel surface with antibodies to capture biomolecules in the sample is one of the common approaches of performing an immunoassay in microfluidics. For multiplexing, multiple microchannels conjugated with distinct antibodies are required to enable the detection of different targets in the same sample. Although many surface patterning methods, including photolithography liftoff techniques [104, 105], micro-contact printing [106, 107], and fluidic patterning [108], are available to pattern proteins on the substrates, sophisticated patterning procedure and flow control are required for multiplexed analysis. Barcoded microbead-based microfluidics immunoassay enables multiplexed analysis in microchannels without the need for surface patterning. In addition, the sensitivity could be enhanced by at least two orders of magnitude with the capability of minimal consumption of reagents in microfluidics [109].

A microfluidic fluorescence bead assay has been developed to accurately measure antibody-antigen binding kinetics [110]. The association and dissociation rate constants from antibody-antigen interactions spanning nearly 4 orders of magnitude in equilibrium binding affinity (30 pM–100 nM) were measured. Complete antibody-antigen binding kinetics were measured as little as 8×10^4 antibodies (~132 zeptomoles) immobilized on a single bead and less than 2×10^6 antibodies (~3 attomoles) loaded into the microfluidic device. A reduction in detection limit and sample consumption of 4 orders of magnitude could be achieved when compared to surface plasmon resonance (SPR) spectroscopy and alternative measurement techniques. By use of bead-based assays in microfluidics, multiplexed protein detection of TNF-α, CXCL2, IL-6, and IL-1b has been reported [111] with the sensitivity of 1000 pg/mL with the sample volume of 4.7 nL. A capillary-based system (~100 μm diameter), in which the antibody-coupled microbeads were captured on a 0.5 μm filter, was developed to detect protein expression of receptor tyrosine kinase in breast cancer tissues [112]. Our group has developed a microfluidic platform for immobilization of single-layer microbead array for multiplexed high-throughput analysis of biomolecules [30]. The microfluidic device comprised eight microbead-trapping units, where microbeads were immobilized in a linear array format by the exertion of a negative pressure in the control channel. Multiplexed assays were achieved by using a mixture of different spectrally encoded microbeads functionalized with specific probes, followed by on-chip reaction and detection. Six cancer biomarkers including leptin, carcinoembryonic antigen (CEA), carbohydrate antigen 125 (CA125), interleukin 8 (IL-8), hepatocyte growth factor (HGF), and human epididymis protein (HE4) have been simultaneously discriminated with good correlation with flow cytometry system. Compared with the off-chip protocols, the on-chip analysis exhibited better reacti on efficiency, higher sensitivity, and wider linear detection range.

With the advances in both microfluidic synthesis and analysis, sensitive and high-throughput multiplexed detection of clinically relevant protein panels in complex media has been demonstrated [37]. Stop-flow lithography method was leveraged to simultaneously synthesize and functionalize particles with a fluorescent barcoded region and a spatially segregated probe region embedded with different capture antibodies (Fig. 8.6a). Covalent binding of capture antibodies using a heterobifunctional poly(ethylene glycol) linker enabled one-step synthesis using only small amounts of reagents. The functionalized hydrogel particles were incubated with the sample to capture the analyte and measured using fluorescent-labeled detection antibodies for quantification. The feasibility of the device was demonstrated by detection of 1–8 pg/mL amounts of three cytokines (interleukin-2, interleukin-4, and tumor necrosis factor alpha) in multiplexed assays without the need for filtration or blocking agents [113].

Ultralow concentration of proteins could also be discriminated using aforementioned multienzyme-nanoparticle amplification and quantum dot labels [114]. Based on a dual-signal amplification strategy, a microfluidic bead-based immune sensor was developed to discriminate as low as 0.2 pg/mL AFP in 10 mL of undiluted calf serum. A 500-fold decrease in detection limit compared to the off-chip test and 50-fold decrease in limit of detection were demonstrated compared to microfluidic bead-based immunoassay using single-labeled HRP-conjugated capture antibody. Otieno et al. incorporated an online chamber to capture cancer biomarker on magnetic beads with 300,000 enzyme labels and 40,000 antibodies into a microfluidic immunoarray [115]. Protein analytes are captured from serum or other biological samples in the stirred magnetic bead in the capture chamber. In the detection chamber, the beads were captured on gold nanoparticle film sensors pre-coated with antibody and detected by amperometry. Detection of 5 fg/mL for interleukin-6 (IL-6) and 7 fg/mL for IL-8 in serum was demonstrated in the microfluidic system, with good accuracy and correlations with standard ELISAs. In the microscale, fluids exhibit laminar flow, i.e., fluidic streams flow parallel to each other, and mixing occurs only by diffusion. Although diffusion distance in microchannels is significantly reduced compared to conventional microliter well

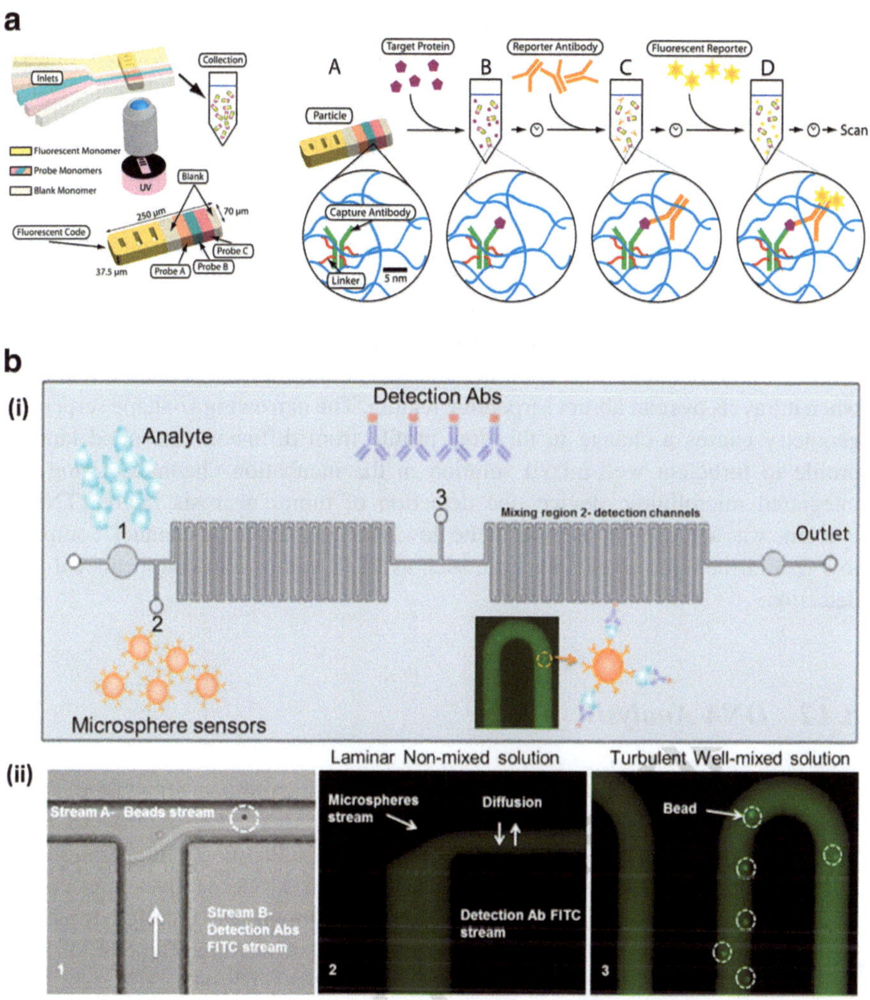

Fig. 8.6 Schematic of synthesis of barcoded hydrogel particles in a microfluidic device for protein analysis. (**a**) Six pressure-driven inlet streams were combined into a single channel where the three probe hydrogel particles were formed by polymerization under the UV exposure defined by a transparency mask. The hydrogel particles were collected in the tube and washed before storage or use (*left*). Premade particles containing probes are incubated with sample containing the target protein (*purple pentagon*), washed, and reacted with biotinylated reporter antibody (*orange*) to form a sandwich with the protein target. After removal of the unbound reporter antibody, fluorescent report of streptavidin phycoerythrin (SAPE, *yellow stars*) is bound to the complexes after incubation (*right*). Reprinted with permission [113] from Copyright (2011) American Chemical Society. (**b**) Microfluidic microbead-based system for real-time monitoring of the biomolecules. (i) Schematics of the device. (ii) (*1*) Microscopic and (*2*) fluorescent images of the flow in inlet 3 and (*3*) fluorescent image of detection channels. In inlet 3, laminar non-mixing solution was observed while in the detection channel turbulent well-mixed solution was obtained and the image of microbeads with sandwich complex formation of target analytes was captured and analyzed. Reprinted from [118], Copyright (2015), with permission from Elsevier

plate, diffusion is still the determent step in terms of detection time at low sample concentration [116], such that incubation step is still required for capture analyte and fluorescent labels. One of the strategies to improve the analyte capture and detection of antibody binding is to integrate mixing elements in the microfluidic device [117]. Cohen et al. integrated microbead-based assay in a well-mixed microfluidic device, which enabled near-real-time diagnostics of clinically relevant analytes [118]. The diffusion-controlled immunoassays could be overcome by introducing the microbead-based assay into the mixing microchannels with turbulent flow profiles. The flow profile in the developed microfluidic device consists of laminar flow and turbulent profiles in distinct regions of the device (Fig. 8.6b). Laminar flow occurs when beads stream and detection antibody streams in parallel layers with no disruption between the layers. The fluid flow was altered markedly when it travels over an abrupt serpentine feature. The narrowing U-shape serpentine geometry causes a change in the flow profile from diffusion-controlled laminar profile to turbulent well-mixed solution in the incubation channels. Using this integrated microfluidic device, the detection of tumor necrosis factor (TNF)-α cytokine was achieved in seconds in the flow through incubation channel, compared to 1–2 h in the non-mixed solutions, such as ELISA, thus allowing near-real-time detection.

8.4.2 DNA Analysis

Nucleic acid analyses have been widely used in clinical diagnostics, food safety, and forensics. Integrated with microbead-based assay, the microbead-based microfluidics has been widely applied to nucleic acid sample preparation and detection. Solid-phase extraction (SPE), in which DNA samples are usually absorbed onto solid insoluble materials to remove impurities, and eluted out later with buffer, is the most common method for DNA purification. The SPE of DNA samples could be implemented in microfluidic devices. Christel et al. fabricated arrays of high-aspect-ratio oxidized silica pillars as solid phase for DNA binding [119]. The DNA binding capacity was improved in silica-coated pillars where the surface areas within the channel were increased by 300–600 % [120]. Although the micropillars increased the purification efficiency, the process of deep reactive ion etching for fabrication of the oxidized silica pillar is complex with very high fabrication costs. To reduce the complexity and cost for SPE of DNA samples in microfluidics, a capillary-based microfluidics device packed with silica microbeads was developed, which is capable of adsorption and desorption of pictogram to nanogram amounts of DNA samples [121]. The feasibility of the device was demonstrated by extraction DNA from white blood cells with an efficiency of 70 % while removing 80 % of proteins. A method named laser-irradiated magnetic bead system (LIMBS) was reported that combined laser irradiation and magnetic beads to rapidly and efficiently extract DNA from biological samples and perform rt-PCR in one device [122]. By integrating it on a centrifugal microfluidic device, pathogen DNA isolation from whole blood sample was achieved [123]. A microfluidic magnetic particle-based device was developed with the multifunctional units of genomic DNA isolation, asymmetric PCR, and electrochemical sequence-specific detection, which was capable of multiplexed pathogen identification [124].

Single-nucleotide polymorphisms (SNPs) could also be identified in microbead-based microfluidic analysis systems. An automated magnetic bead-based microfluidic platform was reported with the integration of leukocyte purification, genomic DNA extraction, and on-chip PCR [125]. Two types of magnetic beads were utilized in the system, one for leukocyte purification and the other for DNA purification. Based on the apyrase-mediated allele-specific primer extension with quantum dots as labels [126], single-nucleotide discrimination was demonstrated with high discrimination specificity and sensitivity of 0.5 pM in a microbead array microfluidic platform. Another genotyping approach has been developed by incorporating solid-phase-based extraction reactions and PCR in a single microfluidic chamber, which is further coupled to MALDI-TOF MS, where the nucleotides at SNP site on both a mutated and an unmutated HBB gene were recognized [127].

Advances in modern genomic research depend heavily on applications of various devices for automated high- or ultra-throughput assays. A bead-based microfluidic platform for fully automated and programmable DNA microarrays has been developed, where the use of beads as solid support for DNA immobilization and hybridization offers a high surface-to-volume ratio and makes the microfluidic chips completely reusable [128]. Uniform microfluidic architecture for large-scale integration and automation of DNA microarrays was demonstrated and applied for single-nucleotide polymorphism analysis and DNA sequencing by synthesis without the need for fluorescent removal step.

8.4.3 Other Biomolecules

While proteins and DNAs are two major classes of important biomarkers in blood or serum, other small molecules such as metabolites, lipids, and hormones regulate physiological and behavioral activities in human body. Microfluidic devices in combination with microbeads can be applied to detect small molecules using enzyme-based or fluorescent readouts with sensitive detection ranges. For example, a microfluidic device integrated with glass beads with enzymes was developed for the detection of glucose in the range of 1–10 mM. The hydrogen peroxide generated from the enzyme reaction between glucose in the sample and glucose oxidase immobilized on the surface of microbeads could be quantified by Amplex Red with fluorescent readout [129].

For small molecule analysis, it is necessary to utilize signal amplification strategies in order to detect trace amount of target analyte. Zhang et al. developed an aptamer-mediated microfluidic bead-based sensor for simultaneous detection of adenosine and cocaine using multienzyme-linked nanoparticle amplification and quantum dot labels [130]. Microbeads functionalized with the aptamers and modified electron-rich proteins were arrayed within a microfluidic channel and were connected with the horseradish peroxidases (HRP) and capture DNA probe derivative gold nanoparticles (AuNPs) via hybridization. The conformational transition of aptamer induced by target–aptamer complex contributes to the displacement of functionalized AuNPs and decreases the fluorescence signal of microbeads. In this approach, increased binding events of HRP on each nanosphere and enhanced mass transport capability inherent from microfluidics are integrated for enhancing the detection sensitivity of analytes. The developed aptamer-based microfluidic bead array sensor could discriminate as low as 0.1 pM of adenosine and 0.5 pM cocaine, and showed a 500-fold increase in detection limit of adenosine compared to the off-chip assay.

8.5 Conclusion and Future Perspectives

In this chapter, we have summarized recent breakthroughs in integration of microbead-based techniques with microfluidic devices for multiplexed detection of various clinically relevant agents, from barcoded technology for multiplexed microbead-based assay to bead functionalization approaches, and from manipulation of microbeads in microfluidics to practical applications of these devices in bioanalysis. We believe that the combination of these two technologies can create powerful devices that may overcome the limitations in current diagnostic strategies in terms of portability, cost, sensitivity, and reliability. While many proof-of-concept prototypes have been demonstrated using low-cost portable microbead-based device for multiplexing analysis of biomolecules, there are many issues and challenges for successful application of multiplexed microbead analysis in microfluidic devices. The extent of multiplexing is limited by the number of unique codes that can be generated by both encoding and decoding strategies, and the sensitivity of detection equipment. Many barcoding technologies have shown promise for multiplexing, but their performances in practical applications are seldom evaluated, as most microfluidic prototypes only employed a few types of encoded microbeads, which is far beyond the coding capability of the multiplexing technologies.

Currently, highly integrated microdevices with optical, electrical, hydrodynamic, magnetic, mechanical techniques have been widely applied in microbead manipulation research. Optical and/or magnetic forces manipulate microbeads without contact and contamination, but complicated setup and expensive instrumentation limit their application in portable devices. Physical entrapment is preferred to manipulating microbeads in a simple and effective way, but in most microfluidic devices, microbeads are captured for collective measurement. These methods created a mass of entrapped microbeads in cluster that cannot be easily analyzed individually, which would pose a problem in multiplexed analysis when encoded microbeads are incorporated, as the fluorescent interference from other beads might affect the decoding of the microbeads, thus causing inaccuracy of the multiplexed analysis. Single-layer bead array or other highly ordered array patterned with individual microbeads are preferred. In these approaches, the optimal size of the encoded beads is with a diameter of a few microns to submicrons for easy manipulation in microchannels as well as optical identification and quantification. While many approaches have been developed for immobilization and patterning of individual beads under flow conditions in bench scale, the next obstacle in the commercialization of microbead-based microfluidic device would lie in the manufacturing process of such device and automatic bead immobilizations for massive production.

In addition, lab-based optical or fluorescent microscopes are used in most prototypes of microbead-based microfluidic devices for bioanalysis. It is essential to develop portable fluorescent detectors with bead visualization function integrated in microfluidic devices for the purpose of point-of-care diagnostics. Alternatively, novel strategies for bead recognition and quantification with reliable validation which are compatible for integration in portable device are preferred. Lastly, in order to create the "sample-in-result-out" platform for sensitive detection of biomolecules, multifunctional modules including sample preparation, accumulation, separation, and analyte capture should be integrated into one microfluidic device. As the feasibilities of many prototype microfluidic devices have been demonstrated to achieve one or more functions, it is promising to develop novel microfluidic-based platforms integrating microbead-based assay for practical point-of-care testing. There is no doubt that additional efforts should be placed on fundamental research and practical application of multiplexed microbead-based microfluidic device.

References

1. Manz A, Graber N, Widmer HM (1990) Miniaturized total chemical-analysis systems—a novel concept for chemical sensing. Sens Actuators B 1:244–248
2. Auroux PA, Iossifidis D, Reyes DR, Manz A (2002) Micro total analysis systems. 2. Analytical standard operations and applications. Anal Chem 74:2637–2652
3. Vilkner T, Janasek D, Manz A (2004) Micro total analysis systems. Recent developments. Anal Chem 76:3373–3385
4. Holden MA, Cremer PS (2005) Microfluidic tools for studying the specific binding, adsorption, and displacement of proteins at interfaces. Annu Rev Phys Chem 56:369–387
5. El-Ali J, Sorger PK, Jensen KF (2006) Cells on chips. Nature 442:403–411
6. Chen L, Manz A, Day PJ (2007) Total nucleic acid analysis integrated on microfluidic devices. Lab Chip 7:1413–1423
7. McDonald JC, Duffy DC, Anderson JR, Chiu DT, Wu H, Schueller OJ et al (2000) Fabrication of microfluidic systems in poly(dimethylsiloxane). Electrophoresis 21:27–40
8. Song H, Chen DL, Ismagilov RF (2006) Reactions in droplets in microfluidic channels. Angew Chem Int Ed Engl 45:7336–7356
9. Atencia J, Beebe DJ (2005) Controlled microfluidic interfaces. Nature 437:648–655

10. Mani V, Chikkaveeraiah BV, Patel V, Gutkind JS, Rusling JF (2009) Ultrasensitive immunosensor for cancer biomarker proteins using gold nanoparticle film electrodes and multienzyme-particle amplification. ACS Nano 3:585–594
11. Jain KK (2005) Nanotechnology-based lab-on-a-chip devices. In: Encyclopedia of diagnostic genomics and proteomics, vol. 2. Marcel Dekkar Inc., New York, pp. 891–895
12. Hsu HY, Joos TO, Koga H (2009) Multiplex microsphere-based flow cytometric platforms for protein analysis and their application in clinical proteomics—from assays to results. Electrophoresis 30:4008–4019
13. Derveaux S, Stubbe BG, Braeckmans K, Roelant C, Sato K, Demeester J et al (2008) Synergism between particle-based multiplexing and microfluidics technologies may bring diagnostics closer to the patient. Anal Bioanal Chem 391:2453–2467
14. Braeckmans K, De Smedt SC, Leblans M, Pauwels R, Demeester J (2002) Encoding microcarriers: present and future technologies. Nat Rev Drug Discov 1:447–456
15. Wilson R, Cossins AR, Spiller DG (2006) Encoded microcarriers for high-throughput multiplexed detection. Angew Chem Int Ed Engl 45:6104–6117
16. Situma C, Hashimoto M, Soper SA (2006) Merging microfluidics with microarray-based bioassays. Biomol Eng 23:213–231
17. Zhang H, Liu L, Li CW, Fu HY, Chen Y, Yang MS (2011) Multienzyme-nanoparticles amplification for sensitive virus genotyping in microfluidic microbeads array using Au nanoparticle probes and quantum dots as labels. Biosens Bioelectron 29:89–96
18. Lawrie GA, Robinson J, Corrie S, Ford K, Battersby BJ, Trau M (2006) Multiplexed microsphere diagnostic tools in gene expression applications: factors and futures. Int J Nanomedicine 1:195–201
19. Holmes D, She JK, Roach PL, Morgan H (2007) Bead-based immunoassays using a microchip flow cytometer. Lab Chip 7:1048–1056
20. Puig O, Caspary F, Rigaut G, Rutz B, Bouveret E, Bragado-Nilsson E et al (2001) The tandem affinity purification (TAP) method: a general procedure of protein complex purification. Methods 24:218–229
21. Lion N, Reymond F, Girault HH, Rossier JS (2004) Why the move to microfluidics for protein analysis? Curr Opin Biotechnol 15:31–37
22. Tudos AJ, Besselink GAJ, Schasfoort RBM (2001) Trends in miniaturized total analysis systems for point-of-care testing in clinical chemistry. Lab Chip 1:83–95
23. Cook EB, Stahl JL, Lowe L, Chen R, Morgan E, Wilson J et al (2001) Simultaneous measurement of six cytokines in a single sample of human tears using microparticle-based flow cytometry, allergies vs. non-allergies. J Immunol Methods 254:109–118
24. Han MY, Gao XH, Su JZ, Nie S (2001) Quantum-dot-tagged microbeads for multiplexed optical coding of biomolecules. Nat Biotechnol 19:631–635
25. Dunbar SA, Jacobson JW (2000) Application of the Luminex LabMAP in rapid screening for mutations in the cystic fibrosis transmembrane conductance regulator gene: a pilot study. Clin Chem 46:1498–1500
26. Fenniri H, Chun S, Ding LH, Zyrianov Y, Hallenga K (2003) Preparation, physical properties, on-bead binding assay and spectroscopic reliability of 25 barcoded polystyrene-poly (ethylene glycol) graft copolymers. J Am Chem Soc 125:10546–10560
27. Bravo-Vasquez JP, Alvarez-Puebla RA, Hicham F (2007) Self-encoded polymer beads for microarray technologies. Sensor Actuator B Chem 125:357–359
28. Lee JB, Roh YH, Um SH, Funabashi H, Cheng WL, Cha JJ et al (2009) Multifunctional nanoarchitectures from DNA-based ABC monomers. Nat Nanotechnol 4:430–436
29. Prabhakar U, Eirikis E, Davis HM (2002) Simultaneous quantification of proinflammatory cytokines in human plasma using the LabMAP (TM) assay. J Immunol Methods 260:207–218
30. Yue WQ, Zou H, Jin QH, Li CW, Xu T, Fu HY et al (2014) Single layer linear array of microbeads for multiplexed analysis of DNA and proteins. Biosens Bioelectron 54:297–305

31. Gao X, Nie S (2004) Quantum dot-encoded mesoporous beads with high brightness and uniformity: rapid readout using flow cytometry. Anal Chem 76:2406–2410
32. Lee H, Kim J, Kim H, Kim J, Kwon S (2010) Colour-barcoded magnetic microparticles for multiplexed bioassays. Nat Mater 9:745–749
33. Dejneka MJ, Streltsov A, Pal S, Frutos AG, Powell CL, Yost K et al (2003) Rare earth-doped glass microbarcodes. Proc Natl Acad Sci U S A 100:389–393
34. Zhao Y, Zhao X, Sun C, Li J, Zhu R, Gu Z (2008) Encoded silica colloidal crystal beads as supports for potential multiplex immunoassay. Anal Chem 80:1598–1605
35. Sochol RD, Casavant BP, Dueck ME, Lee LP, Lin L (2011) A dynamic bead-based microarray for parallel DNA detection. J Micromech Microeng 21:54019–54026
36. Chou J, Lennart A, Wong J, Ali MF, Floriano PN, Christodoulides N et al (2012) Modeling analyte transport and capture in porous bead sensors. Anal Chem 84:2569–2575
37. Tekin HC, Gijs MA (2013) Ultrasensitive protein detection: a case for microfluidic magnetic bead-based assays. Lab Chip 13:4711–4739
38. Yusilawati AN, Maizirwan M, Sopyan I, Hamzah MS, Ng KH, Wong CS (2011) Surface modification of polystyrene beads by UV/ozone treatment. Adv Mater Res 264:1532–1537
39. Badley RD, Ford WT, McEnroe FJ, Assink RA (1990) Surface modification of colloidal silica. Langmuir 6:792–801
40. Lin J, Siddiqui JA, Ottenbrite RM (2001) Surface modification of inorganic oxide particles with silane coupling agent and organic dyes. Polym Adv Technol 12:285–292
41. Liu XD, Tokurab S, Harukia M, Nishia N, Sakairia N (2002) Surface modification of nonporous glass beads with chitosan and their adsorption property for transition metal ions. Carbohydr Polym 49:103–108
42. Lund V, Schmid R, Rickwood D, Hornes E (1988) Assessment of methods for covalent binding of nucleic acids to magnetic beads, Dynabeads, and the characteristics of the bound nucleic acids in hybridization reactions. Nucleic Acids Res 16:10861–10880
43. Wolf SF, Haines L, Fisch J, Kremsky JN, Dougherty JP, Jacobs K (1987) Rapid hybridization kinetics of DNA attached to submicron latex particles. Nucleic Acids Res 15:2911–2926
44. Siiman O, Burshteyn A, Insausti ME (2001) Covalently bound antibody on polystyrene latex beads: formation, stability, and use in analyses of white blood cell populations. J Colloid Interface Sci 234:44–58
45. Thompson JA, Bau HH (2011) Pulsating bead-based assay. Anal Chem 83:2858–2861
46. Ouyang A, Liang J (2014) Tailoring the adsorption rate of porous chitosan and chitosan carbon nanotube core–shell beads. RSC Adv 4:25835–25842
47. Song T, Zhang Q, Lu C, Gong X, Yang Q, Li Y et al (2011) Structural design and preparation of high-performance QD-encoded polymer beads for suspension arrays. J Mater Chem 21:2169–2177
48. Yang Z, Fu Z, Yan F, Liu H, Ju H (2008) A chemiluminescent immunosensor based on antibody immobilized carboxylic resin beads coupled with micro-bubble accelerated immunoreaction for fast flow-injection immunoassay. Biosens Bioelectron 24:35–40
49. Sung D, Yang S, Park JW, Jon S (2013) High-density immobilization of antibodies onto nanobead-coated cyclic olefin copolymer plastic surfaces for application as a sensitive immunoassay chip. Biomed Microdevices 15:691–698
50. Li J, Zhong W (2007) Typing of multiple single-nucleotide polymorphisms by a microsphere-based rolling circle amplification assay. Anal Chem 79:9030–9038
51. Konry T, Hayman RB, Walt DR (2009) Microsphere-based rolling circle amplification microarray for the detection of DNA and proteins in a single assay. Anal Chem 81:5777–5782
52. Schafer DA, Gelles J, Sheetz MP, Landick R (1991) Transcription by single molecules of RNA polymerase observed by light microscopy. Nature 352:444–448
53. Cluzel P, Lebrun A, Heller C, Lavery R, Viovy JL, Chatenay D et al (1996) DNA: an extensible molecule. Science 271:792–794
54. Abbondanzieri EA, Greenleaf WJ, Shaevitz JW, Landick R, Block SM (2005) Direct observation of base-pair stepping by RNA polymerase. Nature 438:460–465

55. Silver J, Li Z, Neuman K (2015) Tethered-bead, immune sandwich assay. Biosens Bioelectron 63:117–123
56. Hu C, Yue W, Yang M (2013) Nanoparticle-based signal generation and amplification in microfluidic devices for bioanalysis. Analyst 138:6709–6720
57. Zhang Y, Guo Y, Xianyu Y, Chen W, Zhao Y, Jiang X (2013) Nanomaterials for ultrasensitive protein detection. Adv Mater 25:3802–3819
58. Joos TO, Stoll D, Templin MF (2002) Miniaturised multiplexed immunoassays. Curr Opin Chem Biol 6:76–80
59. Dressman D, Yan H, Traverso G, Kinzler KW, Vogelstein B (2003) Transforming single DNA molecules into fluorescent magnetic particles for detection and enumeration of genetic variations. Proc Natl Acad Sci U S A 100:8817–8822
60. Kojima T, Takei Y, Ohtsuka M, Kawarasaki Y, Yamane T, Nakano H (2005) PCR amplification from single DNA molecules on magnetic beads in emulsion: application for high-throughput screening of transcription factor targets. Nucleic Acids Res 33(17):e150
61. Diehl F, Li M, He YP, Kinzler KW, Vogelstein B, Dressman D (2006) BEAMing: single-molecule PCR on microparticles in water-in-oil emulsions. Nat Methods 3:551–559
62. Li M, Chen WD, Papadopoulos N, Goodman SN, Bjerregaard NC, Laurberg S et al (2009) Sensitive digital quantification of DNA methylation in clinical samples. Nat Biotechnol 27:858–863
63. Li M, Diehl F, Dressman D, Vogelstein B, Kinzler KW (2006) BEAMing up for detection and quantification of rare sequence variants. Nat Methods 3:95–97
64. Diehl F, Li M, Dressman D, He YP, Shen D, Szabo S et al (2005) Detection and quantification of mutations in the plasma of patients with colorectal tumors. Proc Natl Acad Sci U S A 102:16368–16373
65. LaFratta CN, Walt DR (2008) Very high density sensing arrays. Chem Rev 108:614–637
66. Blicharz TM, Siqueira WL, Helmerhorst EJ, Oppenheim FG, Wexler PJ, Little FF et al (2009) Fiber-optic microsphere-based antibody array for the analysis of inflammatory cytokines in saliva. Anal Chem 81:2106–2114
67. Walt DR (2010) Fibre optic microarrays. Chem Soc Rev 39:38–50
68. Konry T, Walt DR (2009) Intelligent medical diagnostics via molecular logic. J Am Chem Soc 131:13232–13233
69. Walt DR, Blicharz TM, Hayman RB, Rissin DM, Bowden M, Siqueira WL et al (2007) Microsensor arrays for saliva diagnostics. Ann N Y Acad Sci 1098:389–400
70. Vo-Dinh T, Cullum B (2000) Biosensors and biochips: advances in biological and medical diagnostics. Fresenius J Anal Chem 366:540–551
71. Verpoorte E (2003) Beads and chips: new recipes for analysis. Lab Chip 3:60n–68n
72. Yi CQ, Li CW, Ji SL, Yang MS (2006) Microfluidics technology for manipulation and analysis of biological cells. Anal Chim Acta 560:1–23
73. Peterson DS (2005) Solid supports for micro analytical systems. Lab Chip 5:132–139
74. Thompson JA, Bau HH (2010) Microfluidic, bead-based assay: theory and experiments. J Chromatogr B Analyt Technol Biomed Life Sci 878:228–236
75. Zhu L, Zhang Q, Feng HH, Ang S, Chauc FS, Liu WT (2004) Filter-based microfluidic device as a platform for immunofluorescent assay of microbial cells. Lab Chip 4:337–341
76. Lay C, Teo CY, Zhu L, Peh XL, Ji HM, Chew BR et al (2008) Enhanced microfiltration devices configured with hydrodynamic trapping and a rain drop bypass filtering architecture for microbial cells detection. Lab Chip 8:830–833
77. Khademhosseini A, Yeh J, Jon S, Eng G, Suh KY, Burdick JA et al (2004) Molded polyethylene glycol microstructures for capturing cells within microfluidic channels. Lab Chip 4:425–430
78. Yeon JH, Park JK (2009) Drug permeability assay using microhole-trapped cells in a microfluidic device. Anal Chem 81:1944–1951
79. Khabiry M, Chung BG, Hancock MJ, Soundararajan HC, Du YN, Cropek D et al (2009) Cell docking in double grooves in a microfluidic channel. Small 5:1186–1194

80. Yang MS, Li CW, Yang J (2002) Cell docking and on-chip monitoring of cellular reactions with a controlled concentration gradient on a microfluidic device. Anal Chem 74:3991–4001
81. Yang J, Li CW, Yang MS (2004) Hydrodynamic simulation of cell docking in microfluidic channels with different dam structures. Lab Chip 4:53–59
82. Li CW, Cheung CN, Yang J, Tzang CH, Yang MS (2003) PDMS-based microfluidic device with multi-height structures fabricated by single-step photolithography using printed circuit board as masters. Analyst 128:1137–1142
83. Yue WQ, Li CW, Xu T, Yang MS (2011) Integrated sieving microstructures on microchannels for biological cell trapping and droplet formation. Lab Chip 11:3352–3355
84. Di Carlo D, Aghdam N, Lee LP (2006) Single-cell enzyme concentrations, kinetics, and inhibition analysis using high-density hydrodynamic cell isolation arrays. Anal Chem 78:4925–4930
85. Wlodkowic D, Faley S, Zagnoni M, Wikswo JP, Cooper JM (2009) Microfluidic single-cell array cytometry for the analysis of tumor apoptosis. Anal Chem 81:5517–5523
86. Andersson H, Wijingaart WVD, Enoksson P, Stemme G (2000) Micromachined flow-through filter-chamber for chemical reactions on beads. Stemme Sensor Actuator B Chem 67:203–208
87. Birkbeck AL, Flynn RA, Ozkan M, Song DQ, Gross M, Esener SC (2003) VCSEL arrays as micromanipulators in chip-based biosystems. Biomed Microdevices 5:47–54
88. Chiou PY, Ohta AT, Wu MC (2005) Massively parallel manipulation of single cells and microparticles using optical images. Nature 436:370–372
89. MacDonald MP, Spalding GC, Dholakia K (2003) Microfluidic sorting in an optical lattice. Nature 426:421–424
90. Ashkin A, Dziedzic JM (1987) Optical trapping and manipulation of viruses and bacteria. Science 235:1517–1520
91. Enger J, Goksor M, Ramser K, Hagberg P, Hanstorp D (2004) Optical tweezers applied to a microfluidic system. Lab Chip 4:196–200
92. Umehara S, Wakamoto Y, Inoue I, Yasuda K (2003) On-chip single-cell microcultivation assay for monitoring environmental effects on isolated cells. Biochem Biophys Res Commun 305:534–540
93. Ozkan M, Wang M, Ozkan C, Flynn R, Birkbeck A, Esener S (2003) Optical manipulation of objects and biological cells in microfluidic devices. Biomed Microdevices 5:61–67
94. Vestad T, Marr DWM, Munakata T (2004) Flow resistance for microfluidic logic operations. Appl Phys Lett 84:5074–5075
95. Huang L, Maerkl SJ, Martin OJF (2009) Integration of plasmonic trapping in a microfluidic environment. Opt Express 17:6018–6024
96. Grover WH, Mathies RA (2005) An integrated microfluidic processor for single nucleotide polymorphism-based DNA computing. Lab Chip 5:1033–1040
97. Furdui VI, Harrison DJ (2004) Immunomagnetic T cell capture from blood for PCR analysis using microfluidic systems. Lab Chip 4:614–618
98. Peyman SA, Iles A, Pamme N (2009) Mobile magnetic particles as solid-supports for rapid surface-based bioanalysis in continuous flow. Lab Chip 9:3110–3117
99. Li CW, Yang J, Yang MS (2006) Dose-dependent cell-based assays in V-shaped microfluidic channels. Lab Chip 6:921–929
100. Xu T, Li CW, Yao XS, Cai GP, Yang MS (2010) Microfluidic formation of single cell array for parallel analysis of Ca2+ release-activated Ca2+ (CRAC) channel activation and inhibition. Anal Biochem 396:173–179
101. Xu T, Yue WQ, Li CW, Yao XS, Yang MS (2013) Microfluidics study of intracellular calcium response to mechanical stimulation on single suspension cells. Lab Chip 13:1060–1069
102. Witters D, Knez K, Ceyssens F, Puers R, Lammertyn J (2013) Digital microfluidics-enabled single-molecule detection by printing and sealing single magnetic beads in femtoliter droplets. Lab Chip 13:2047–2054

103. Teste B, Ali-Cherif A, Viovy JL, Malaquin L (2013) A low cost and high throughput magnetic bead-based immuno-agglutination assay in confined droplets. Lab Chip 13:2344–2349
104. Spargo BJ, Testoff MA, Nielsen TB, Stenger DA, Hickman JJ, Rudolph AS (1994) Spatially controlled adhesion, spreading, and differentiation of endothelial-cells on self-assembled molecular monolayers. Proc Natl Acad Sci U S A 91:11070–11074
105. Healy KE, Thomas CH, Rezania A, Kim JE, McKeown PJ, Lom B et al (1996) Kinetics of bone cell organization and mineralization on materials with patterned surface chemistry. Biomaterials 17:195–208
106. Chen CS, Mrksich M, Huang S, Whitesides GM, Ingber DE (1997) Geometric control of cell life and death. Science 276:1425–1428
107. Singhvi R, Kumar A, Lopez GP, Stephanopoulos GN, Wang DIC, Whitesides GM et al (1994) Engineering cell-shape and function. Science 264:696–698
108. Jeon NL, Baskaran H, Dertinger SKW, Whitesides GM, Van de Water L, Toner M (2002) Neutrophil chemotaxis in linear and complex gradients of interleukin-8 formed in a microfabricated device. Nat Biotechnol 20:826–830
109. Huang NT, Truxal SC, Tung YC, Hsiao AY, Luker GD, Takayama S et al (2010) Multiplexed spectral signature detection for microfluidic color-coded bioparticle flow. Anal Chem 82:9506–9512
110. Singhal A, Haynes CA, Hansen CL (2010) Microfluidic measurement of antibody-antigen binding kinetics from low-abundance samples and single cells. Anal Chem 82:8671–8679
111. Diercks AH, Ozinsky A, Hansen CL, Spotts JM, Rodriguez DJ, Aderem A (2009) A microfluidic device for multiplexed protein detection in nano-liter volumes. Anal Biochem 386:30–35
112. Yu XB, Hartmann M, Wang QA, Poetz O, Schneiderhan-Marra N, Stoll D et al (2010) μFBI: a microfluidic bead-based immunoassay for multiplexed detection of proteins from a μL sample volume. Plos One 5(10), pii:e13125
113. Appleyard DC, Chapin SC, Doyle PS (2011) Multiplexed protein quantification with barcoded hydrogel microparticles. Anal Chem 83:193–199
114. Zhang H, Liu L, Fu X, Zhu Z (2013) Microfluidic beads-based immunosensor for sensitive detection of cancer biomarker proteins using multienzyme-nanoparticle amplification and quantum dots labels. Biosens Bioelectron 42:23–30
115. Otieno BA, Krause CE, Latus A, Chikkaveeraiah BV, Faria RC, Rusling JF (2014) On-line protein capture on magnetic beads for ultrasensitive microfluidic immunoassays of cancer biomarkers. Biosens Bioelectron 53:268–274
116. Parsa H, Chin CD, Mongkolwisetwara P, Lee BW, Wang JJ, Sia SK (2008) Effect of volume- and time-based constraints on capture of analytes in microfluidic heterogeneous immunoassays. Lab Chip 8:2062–2070
117. Hu G, Gao Y, Li D (2007) Modeling micropatterned antigen-antibody binding kinetics in a microfluidic chip. Biosens Bioelectron 22:1403–1409
118. Cohen N, Sabhachandani P, Golberg A, Konry T (2015) Approaching near real-time biosensing: microfluidic microsphere based biosensor for real-time analyte detection. Biosens Bioelectron 66:454–460
119. Christel LA, Petersen K, McMillan W, Northrup MA (1999) Rapid, automated nucleic acid probe assays using silicon microstructures for nucleic acid concentration. J Biomech Eng 121:22–27
120. Cady NC, Stelick S, Batt CA (2003) Nucleic acid purification using microfabricated silicon structures. Biosens Bioelectron 19:59–66
121. Tian H, Huhmer AF, Landers JP (2000) Evaluation of silica resins for direct and efficient extraction of DNA from complex biological matrices in a miniaturized format. Anal Biochem 283:175–191

122. Lee JG, Cheong KH, Huh N, Kim S, Choi JW, Ko C (2006) Microchip-based one step DNA extraction and real-time PCR in one chamber for rapid pathogen identification. Lab Chip 6:886–895
123. Cho YK, Lee JG, Park JM, Lee BS, Lee Y, Ko C (2007) One-step pathogen specific DNA extraction from whole blood on a centrifugal microfluidic device. Lab Chip 7:565–573
124. Yeung SW, Lee TMH, Cai H, Hsing IM (2006) A DNA biochip for on-the-spot multiplexed pathogen identification. Nucleic Acids Res 34(18):e118
125. Lien KY, Liu CJ, Lin YC, Kuo PL, Lee GB (2009) Extraction of genomic DNA and detection of single nucleotide polymorphism genotyping utilizing an integrated magnetic bead-based microfluidic platform. Microfluid Nanofluid 6:539–555
126. Zhang H, Fu X, Liu L, Zhu ZJ, Yang K (2012) Microfluidic bead-based enzymatic primer extension for single-nucleotide discrimination using quantum dots as labels. Anal Biochem 426:30–39
127. Zhu J, Palla M, Ronca S, Warpner R, Ju J, Lin Q (2013) A MEMS-based approach to single nucleotide polymorphism genotyping. Sensor Actuator A Phys 195:175–182
128. Penchovsky R (2013) Programmable and automated bead-based microfluidics for versatile DNA microarrays under isothermal conditions. Lab Chip 13:2370–2380
129. Kim DN, Lee Y, Koh WG (2009) Fabrication of microfluidic devices incorporating bead-based reaction and microarray-based detection. Sens Actuator B Chem 137:305–312
130. Zhang H, Hu X, Fu X (2014) Aptamer-based microfluidic beads array sensor for simultaneous detection of multiple analytes employing multienzyme-linked nanoparticle amplification and quantum dots labels. Biosens Bioelectron 57:22–29

Chapter 9
Single-Cell Phenotypic Screening in Inverse Metabolic Engineering

A.E. Vasdekis and G. Stephanopoulos

Abstract Contrary to classical approaches centering on debottlenecking flux-limiting steps in a metabolic pathway, inverse metabolic engineering (IME) aims at identifying and modulating *all* gene factors that contribute to an optimal phenotype. Within IME, mutant libraries are generated and screened in order to select mutants with the desired phenotype. The screening process is traditionally performed using microtiter well plates, a laborious and expensive process of limited throughput. Here, we review emerging screening methods that address these throughput and cost-effectiveness shortcomings, but also operate at the single-cell level. We discuss the importance of single-cell analyses in *IME* and detail two specific single-cell screening approaches: the first is *fluorescence-activated cell sorting* for phenotypic discrimination based on cytosolic or cell-membrane-bound products. The second is *droplet microfluidics* for screening of cells capable of overproducing secreted products or overconsuming substrates, properties that require confinement to isolate mutants with specific secretory phenotypes.

Keywords Metabolic engineering • Single-cell analysis • Microfluidics • Flow cytometry

9.1 Introduction

Substantial advances in DNA technology have enabled the successful reconstruction of the genomes of many organisms in recent years [1–4]. As first demonstrated by Cohen et al., similar DNA technologies have also enabled the insertion of new genes into host organisms to perform novel functions, or incapacitate existing ones [5]. In addition to numerous biomedical applications (e.g., immunoengineering and phage therapy [6]), such genome engineering approaches have permitted the

A.E. Vasdekis (✉)
Department of Physics, University of Idaho, Moscow, ID 83844, USA
e-mail: andreasv@uidaho.edu

G. Stephanopoulos
Department of Chemical Engineering, Massachusetts Institute of Technology, Cambridge, MA 02139, USA

© Springer International Publishing Switzerland 2016
C. Lu, S.S. Verbridge (eds.), *Microfluidic Methods for Molecular Biology*,
DOI 10.1007/978-3-319-30019-1_9

transformation of living cells to biochemical factories for producing high-value organic compounds. In essence, the design of the intracellular reaction networks forms the field of *metabolic engineering*, typically implemented through protein knockout, modification, or overexpression [7–9].

9.1.1 Rational Metabolic Engineering

As a result of rapid advances in biotechnology and computational methods facilitating strain construction, physiology description, and function prediction, several metabolic engineering strategies have become possible [9]. Among them, the most traditional approach is to first identify and then modulate flux-limiting steps in a metabolic pathway. This—classical strategy—is termed "rational" or constructive metabolic engineering [10–12]. Rational metabolic engineering has met with success, but is limited to those cases with well-identified kinetics and regulation, such as amino acids, products of central carbon metabolism, antibiotics, and others [13, 14]. Furthermore, it is now known that the desired overproducing or overtolerant phenotype does not originate from a single gene, but is rather a multigenic trait [15]. Such high-order interactions between genes are not completely understood to date, thereby placing limitations on the overall potential of recombinant technologies and rational design.

9.1.2 Inverse Metabolic Engineering

Combinatorial metabolic engineering is another approach towards strain construction. This takes the aforementioned complex gene interactions into account [16] and accelerates strain evolution through genome shuffling, error-prone PCR, and other random mutagenesis until the desired phenotype is expressed [17]. Several successful strain improvements of this type have been reported, such as the tylosin production in *Streptomyces fradiae* [18], penicillin production in *Penicillium chrysogenum* [19], as well as acid and ethanol tolerance in *Lactobacillus* [20] and *Escherichia coli*, respectively [21]. To a similar end, *inverse metabolic engineering* (IME) emerged approximately a decade ago. In IME, mutational strategies can be very targeted to specific places of the chromosome and the generated libraries are screened in a high-throughput fashion to identify the optimal phenotype, followed by cloning to the target strain [12]. Typically, IME is implemented through the following four steps (Fig. 9.1):

1. Selection of a mutational strategy that induces a large enough phenotypic diversity, the size of which is proportional to the probability of generating the desired phenotype

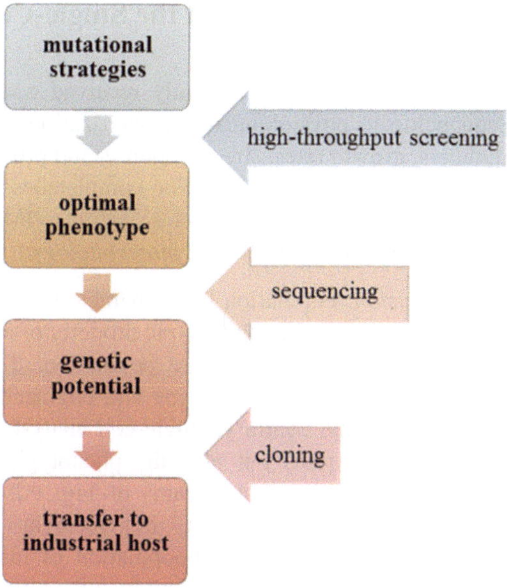

Fig. 9.1 A block diagram illustrating the steps involved in an inverse metabolic engineering strategy (*vertical blocks*) along with the biotechniques associated with each step (*horizontal arrows* on the right)

2. Identification of the desired phenotype through high-throughput screening of the generated mutant libraries
3. Determination of the genetic potential of the optimal phenotype by sequencing and quantitative polymerase chain reaction (qPCR)
4. Cloning to confer the genotype of the ideal phenotype to another strain

All four IME steps are of critical importance. Of special note is selecting the mutational strategy, which traditionally has been performed by *adaptation* and *selection* under specific growth conditions [21], the use of *mutagens* [22], or more modern approaches such as the *global transcription machinery engineering* (gTME) [23, 24]. Covering such aspects however is beyond the scope of this chapter, and several recent reviews are available on the topic [9, 25, 26]. Here, our objective instead is to review phenotypic screening technologies, with a focus on ultrahigh-throughput ones operating at the single-cell level. For completeness, we first discuss the importance of phenotypic screening at the single-cell level, as well as some of the foundational and most recent biophotonic screening methods. The chapter continues with a detailed overview of *fluorescence-activated cell sorting* (FACS) and *droplet microfluidics*, as well as their applications in IME.

9.2 Screening: Why at the Single-Cell Level?

Single-cell analysis has recently received considerable attention, ranging from sequencing to physiological characterizations, such as intracellular protein copy numbers, growth rates, stress response, and productivity. Several excellent reviews cover these topics [27–33]. For completeness, we summarize below the main advantages of single-cell analyses specific to IME. We term these as PUB, standing for *precision, uniqueness, and biotechnology*:

(a) *Precision*: Population-level characterizations (e.g., HPLC or NMR) typically report the median phenotypic property of a cell population due to the inherent averaging over the entire population. This is critical in determining the diversity between wild-type and engineered strains (dotted lines—Fig. 9.2) [24]. On the contrary, single-cell analysis determines not only the diversity between strains but also the phenotypic distribution within [34]. This is illustrated in the histograms of Fig. 9.2, where "*strain 1*" does exhibit a lower median productivity, in agreement with the population-level characterization. However, a subpopulation of insignificant size contains cells that clearly exceed the productivity of all cells in "*strain 2*." Identifying such subpopulations is critical in precision diversity characterization between strains.

(b) *Uniqueness*: During directed evolution, mutations take place in individual single cells at a specific rate, occasionally enabling a phenotype of improved performance. Similar to "a," such occurrences can be rather insignificant, shown to be as low as 1 in 10^4 cells [35]. Such rather infrequent individuals may or may not give rise to subpopulations that are detectable through conventional bulk cell screening techniques. In essence, this forms a "*needle in the haystack*" challenge that can be only addressed through single-cell measurements.

(c) *Biotechnology*: Single-cell analyses uniquely combine sensitive cell characterization and manipulation strategies. These range from advanced mass spectrometry [36] to sophisticated microfluidic architectures capable of dynamically handling one cell at a time [37–39]. It thus becomes clear that the need to detect and isolate low-abundance phenotypes necessitates significant advances in biotechnology, with implications in other life science fields. A representative example to this end is the emergence of single-cell sequencing [40, 41]: the development and sensitivity improvement of such detection methods eventually lead to the noninvasive genome analysis of fetal cells in the maternal blood for genetic disease screening [42].

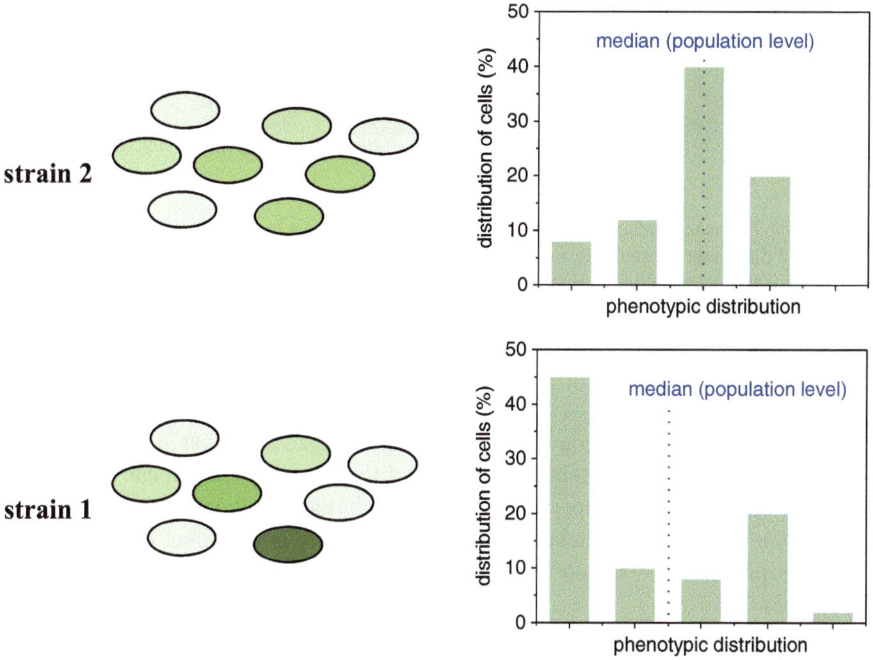

Fig. 9.2 A hypothetical model of cell-to-cell heterogeneity in two populations of "strain 1" (*lower*) and "strain 2" (*upper*); such phenotypic heterogeneity may manifest itself in traits such as growth rates and enzyme productivity. The histograms of the phenotypic distribution (*right*) illustrate the outcome of a single-cell characterization and contrast it to population-level characterizations as noted by the dotted lines in the same graphs

9.3 Biophotonics: High-Throughput Screening

Phenotypic screening is arguably one of the most important steps in IME, and possibly one of the most challenging [9]. This is due to the fact that conventional screening techniques, such as H^1- and C^{13}-NMR, liquid chromatography (e.g., HPLC), and mass spectrometry (e.g., GC-MS), are not yet compatible with high-throughput implementations. Secondly, detecting the production of small molecules does not comply with conventional fluorometric or colorimetric assays since such bioentities are typically non-light emitting. Finally, IME investigations generally cover a wide range of cell types and products. This hinders the development of a global screening approach, but rather enforcing the need for tailored ones for each specific case.

To address the aforementioned challenges, photonic biosensors have been developed and implemented for high-throughput IME applications. Such biosensors are chemical moieties that respond to both the copy number of the produced molecules and light, thereby indirectly reporting productivity through optical means. One of the foundational reports on high-throughput screening through

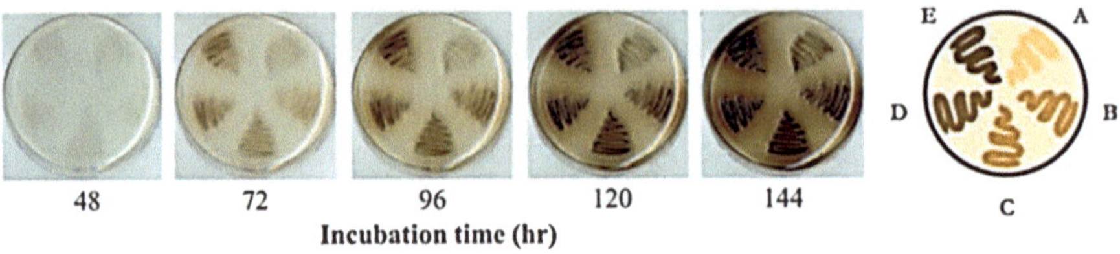

Fig. 9.3 L-Tyrosine production in five *Escherichia coli* strains (*A* to *E*). L-Tyrosine is linked to the black and diffusible pigment melanin through the introduction of a heterologous gene encoding a tyrosinase, thereby enabling visual distinction of the overproducing strains. Reproduced with permission from [43]

visual means was on the directed evolution of *Escherichia coli* for enhanced L-tyrosine production [43, 44]. To visualize tyrosine production, the authors implemented the heterologous expression of tyrosinase, a catalyst in the conversion of L-tyrosine to melanin. Melanin is a black, diffusible pigment that can be detected through both optical and spectroscopic means. In these embodiments, the product (L-tyrosine) to pigment (melanin) conversion enabled the indirect detection of L-tyrosine production—a colorless compound—through the visible melanin levels (Fig. 9.3).

More recently, whole-cell biosensors have been developed for detecting the intracellular levels of a produced compound. In these, the compound itself regulates the expression of a fluorescent protein vector linked to a promoter that is in turn transcriptionally regulated through the intracellular concentration of metabolites [45]. Consequently, transcription factors regulate the expression of a fluorescent protein in response to the intracellular metabolites (Fig. 9.4). Such promising approaches have been successfully employed in various applications, including the *Corynebacterium glutamicum* production of lysine [46], L-valine [47], and L-methionine [48]. While such sensors are extremely practical, care needs to be taken in their implementation with a focus on sensitivity, dynamic range, and protein maturation kinetics, which have been shown to vary even between closely related strains [49].

The melanin and whole-cell biosensor paradigms are merely two examples of how fusing biological engineering and photonics enabled screening platforms for identifying overproducing phenotypes. Several reviews and more complete articles are available on this topic [27, 50], including specific ones to flow cytometry and enzyme production quantification [51–53].

9.4 Fluorescence-Activated Cell Sorting

While flow cytometers (FC) emerged more than 50 years ago [54, 55], today they are still the cornerstone of most biotechnology laboratories. A typical flow cytometer is comprised of a "flow-cell," a microfluidic system that selects

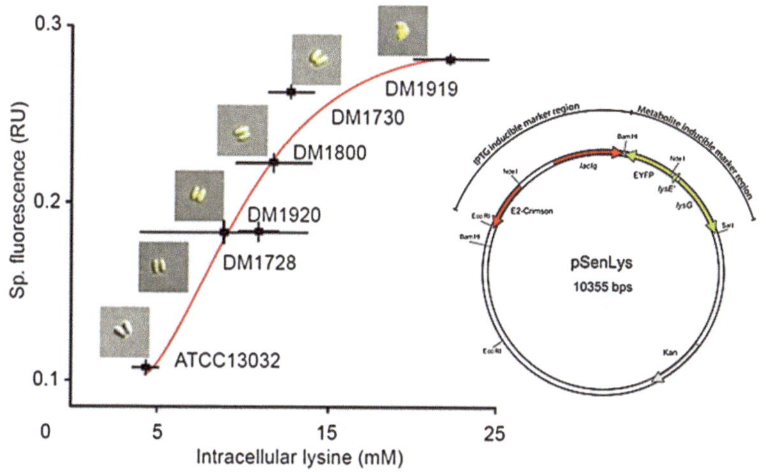

Fig. 9.4 Cytosolic L-lysine concentration in five different *C. glutamicum* strains. All strains carry the biosensor pSenLys (*inset*) with a specific fluorescence output from an eYFP protein (*y*-axis) coupled to the LysG transcriptional regulator that responds to different levels of intracellular lysine concentration (*x*-axis). Reproduced with permission from [46]

individual cells from a population and forces them to flow in a serial fashion in front of a source detector apparatus (Fig. 9.5a). In this way, the response from individual cells is measured over three different detectors, each collecting a specific optical signal, namely the fluorescence emanating from a fluorescent marker, side scattering ("SSC"—proportional to cell size) and forward scattering ("FWS"—proportional to internal cell complexity) [27, 56, 57]. In FC, individual cells are screened based on their morphological properties and intracellular or membrane-bound fluorophores at rates higher than 10^4 cells per second. Such high screening rates in turn enable an impressive ultrahigh-throughput analysis of up to 10^8 mutant variants per day.

In a minor modification to a typical FC apparatus, the cells may also be sorted depending on their fluorescent or light scattering properties (Fig. 9.5a). This modality, termed as FACS, has been particularly pertinent to directed evolution-mediated metabolic engineering [51, 59]. This is due to FACS' capability of both high-throughput screening and sorting, enabling thus consecutive rounds of diversification and selection. To this end, FACS has been critical in identifying genomic variants that are linked to metabolite overproduction, as recently demonstrated in L-lysine [46] and L-methionine [48] production using the whole-cell biosensor approach discussed in the previous section.

FACS has also been critical in directed enzyme evolution, whereas enzyme-producing cells are screened and selected based on their productivity of both membrane-bound and intracellular enzymes [52, 53, 60]. Both cases however necessitate intelligent fluorogenic substrate designs with enhanced selectivity, both spatially and chemically. Two representative examples include the

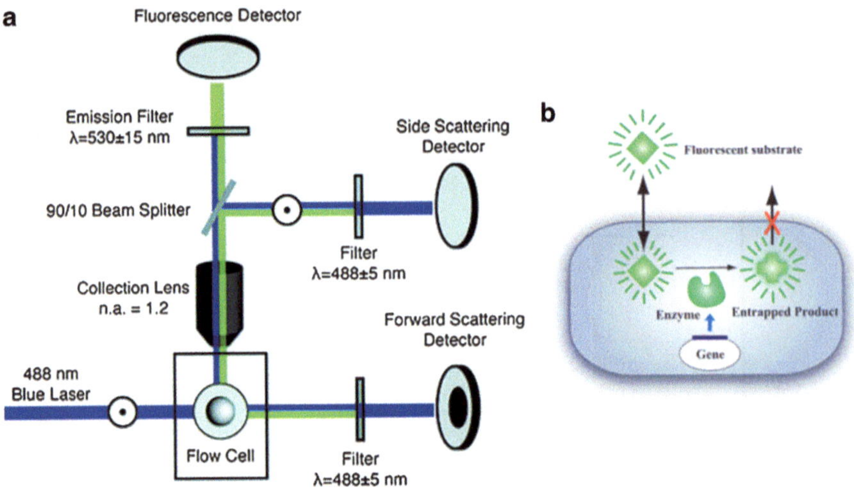

Fig. 9.5 (**a**) A typical flow cytometer apparatus, including a flow cell where individual cells flow in series, an excitation laser of a range of wavelengths, a collection lens, and three detectors measuring the fluorescence, side scattering and forward scattering. Reproduced with permission from [58]. (**b**) Product entrapment within cells through enzymatic modification of product membrane permeability, whereas enzymatic cleavage forbids the fluorescent product to escape the cell [53]; copyright (c) 2009 Wiley-VCH Verlag GmbH

membrane-bound protease production in *Escherichia coli* [61] and the intracellular glycosyltransferase production in *Campylobacter jejuni* [62]. In these, enzyme activity was, respectively, inferred through enzyme-catalyzed resonant color change of the substrate and through enzyme-modulated membrane permeability enforcing permanent cytosolic entrapment of the fluorescent product (Fig. 9.5b).

Finally, FACS has also been critical in ultrahigh-throughput screening of population response to extracellular stresses, such as solvents, acids, and high substrate concentrations [63–65]. It is worth noting that in such embodiments it is critical to include more than one physiology indicator, thus establishing substantial discrimination between physiological states. Such approaches are termed *multiparametric flow cytometry*, detailed in this recent review [51].

9.5 Droplet Microfluidics

While flow cytometry enables both high-content and high-throughput analysis of single cells, it is limited to probing biochemical events occurring within the intracellular environment and/or the cell membrane. This is due to the fact that prior to flow-through and detection, the cells are homogeneously mixed, thereby giving rise to a similarly homogenous extracellular environment. Therefore, the signature of individual cells is lost and thus flow cytometry is unable to interrogate extracellular events at the single-cell level, such as enzyme, metabolite, or product secretion and consumption [66].

On the contrary, droplet microfluidics do not suffer from such shortcomings. Therein, cells are individually compartmentalized within micro- to picoliter droplets of one fluid (typically buffer) within another immiscible one, such as oil [67]. This form of compartmentalization enforces no (or limited) exchange between the intra- with the extra-droplet environment, thereby ensuring that cells grow within a personalized environment, and that their extracellular phenotypic signature is preserved [68]. It is worth adding that droplet microfluidics not only enable directed evolution experiments of enzyme production that are not possible otherwise, but also are very effective in enhancing the throughput efficiency and cost-effectiveness by more than two orders of magnitude.

9.5.1 Droplet Formation

Making droplets (or drops) necessitates a junction between three (or four) microfluidic channels. One microchannel carries the formed droplets, another introduces the oil phase (or continuous phase) into the junction, and the remaining one (or two) introduce the aqueous phase [69, 70]. As illustrated in Fig. 9.6, a variety of junction geometries for droplet formation exist: the coaxial injection (Fig. 9.6a), the T-junction (Fig. 9.6b), and flow-focusing geometry (Fig. 9.6c) [71]. At this microfluidic junction, and strictly under passive means (i.e., no actuation necessary) as well as appropriate dimensional and flow field conditions, a continuous and evenly spaced series of droplets is formed. It is also possible to accurately control the dimensions of the individual droplets. In practice, this is achieved through the ratio of flow rates, viscosities, and dimensions of the dispersed and continuous-phase microchannels [72]. It is worth noting that a remarkable volume homogeneity between droplets is possible—typically within 1–3 % [73]—a significant parameter in the phenotypic comparison between single encapsulated cells as discussed in more detail below.

9.5.2 Droplet Microfluidics in IME

Droplet microfluidics have been critical in many aspects of life sciences, such as the directed in vitro enzyme evolution [74, 75] and single DNA molecule analysis [76]. In this section, we detail the use of single-cell droplet microfluidics for high-throughput analysis in IME. Generally, the first objective in single-cell analysis using droplet microfluidics is to accurately encapsulate one cell per droplet [67, 77]. However, this poses a challenge given that cell loading into droplets is random, governed by Poisson statistics. The latter dictates that under typical cell

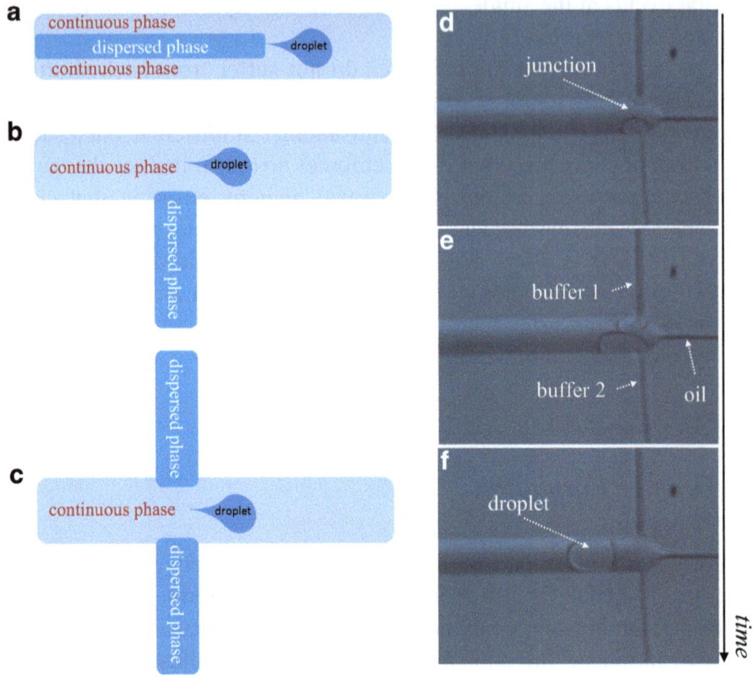

Fig. 9.6 (**a–c**) Schematics indicating the possible geometries for generating droplets in microfluidics, namely (**a**) the coaxial injection geometry, (**b**) the T-junction, and (**c**) the flow focusing geometry. Time-lapse microscopy images of a microfluidic channel in the flow-focusing geometry during droplet formation: (**d**); the junction comprises of two dispersed phase channels (buffer) and one continuous-phase channel (oil) (**e**). At an appropriate flow rate ratio between the three microchannels, a droplet is formed at the junction, and flows within the main channel (**f**)

density conditions, the probability of having one cell per droplet is significantly smaller than having zero cells, but higher than having two.

Two approaches have been proposed to address this shortcoming and ensure that single droplet screening coincides with single-cell screening. The first is to simply utilize low cell densities prior to encapsulation. This, in agreement with the loading Poisson distribution, ensures that the majority of observations stem either from empty droplets (87 %) or from droplets of single occupancy (16 %) [78]. This is particularly pertinent in experiments where it is possible to achieve very rapid droplet flow, thus addressing the low probability of one cell per droplet, whilst preserving high signal-to-background ratio. Another more elegant solution to ensuring single-cell occupancy per droplet is to organize the position of cells within the aqueous channel prior to droplet encapsulation. This is typically achieved by forcing the cells to travel at very high flow rates in microchannels with dimensions comparable to the cells. Such simple selection rules have been shown to enable single-cell loading efficiencies per droplet that exceed the 90 % levels [79].

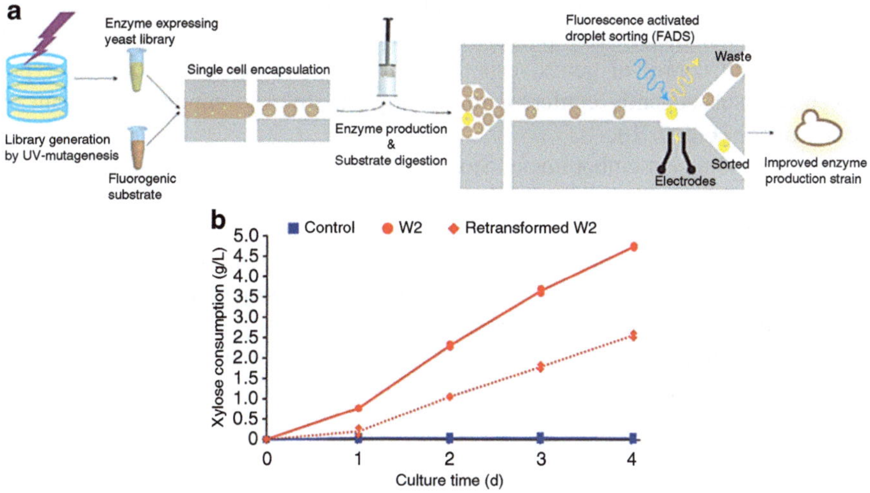

Fig. 9.7 (**a**) A typical assay in directed evolution using droplet microfluidics: following the generation of a mutant library, the droplets are loaded with cells and a fluorogenic enzyme substrate to quantify the levels of production through changes in the droplet's fluorescent properties. Following incubation, the droplets are introduced into a dielectrophoretic sorter, to isolate mutants of enhanced performance [80]—Reproduced by permission of The Royal Society of Chemistry. (**b**) The xylose consumption rate as a function of culture time for the yeast strain H131-A31. In comparison to the control strain, the plot confirms the xylose overconsumption in the selected mutant through droplet microfluidics (W2), as well as the strain harboring the W2 genotype (retransformed W2). Reprinted by permission from Macmillan Publishers Ltd: Nature Biotechnology [35], copyright (2014)

Specific to using droplet microfluidics in directed evolution, the droplets also need to contain a fluorescent sensor to quantify the levels of metabolite and enzyme secretion or consumption [50]. Subsequently, the cell-loaded droplets can be incubated for controlled durations, ranging from seconds to days, either on-chip or off-chip chambers, such as Eppendorf tubes or syringes. This conveniently separates cell loading from cell incubation and cell screening. Following a selected incubation time the cell-loaded droplets can be reintroduced in a microfluidic chip (or alternatively in a flow cytometer), rapidly flown through an optical detection apparatus and then sorted using dielectrophoretic means based on their optical properties (Fig. 9.7a).

Recently, droplet microfluidics has been successfully implemented in enriching enzyme-overproducing or substrate-overconsuming microbes. In the first embodiment, Wang et al. employed single-cell droplet microfluidics to screen mutant libraries for selecting as low as 1 cell per 10^4 of xylose-overconsuming *Saccharomyces cerevisiae* and L lactate-overproducing *Escherichia coli* [35]. Following cell loading and a 2-day-long microanaerobic incubation off-chip, the cell-loaded droplets were reintroduced in a microfluidic chip and fused with droplets containing a fluorogenic enzymatic assay. In this way, the compartmentalized cells were screened and sorted based on their metabolic profile as determined by the fluorogenic assay. Subsequently, the authors performed genetic screening by sequencing and quantitative polymerase chain reaction (qPCR). This identified the types and copies of genes responsible for the enhanced mutant phenotype, which was subsequently conferred to the original strain to confirm its enhanced performance (Fig. 9.7b).

In a second recent embodiment, Sjostrom et al. adapted droplet microfluidics to enrich yeast overproducing the α-amylase enzyme [80]. To this end, a yeast library was generated by UV mutagenesis, followed by co-encapsulating single cells with a starch fluorogenic substrate in individual droplets and off-chip incubation for fixed amounts of 1–2 h (Fig. 9.7a). The droplets were subsequently reintroduced in microfluidics, analyzed through optical means, and finally sorted based on their fluorescence properties. In this way, the authors demonstrated the evolution and selection of a stable strain with more than a twofold increase in α-amylase production.

9.6 Conclusions

With the advent of modern recombinant DNA technology, new ways have been identified to improve or modify cell function. However, the metabolism of a cell is highly complex, thereby placing strong limitations into the improvement possibilities using rational engineering approaches. Combinatorial and inverse approaches emerged to address these limitations, where directed evolution is employed to generate a large library of mutants. Such mutants are subsequently screened to isolate the ideal candidate, identify its genotype, followed by conferring this genetic potential to other strains.

Here, we reviewed ultrahigh-throughput screening methods that operate at the single-cell level. Our focus was primarily on flow cytometry, as well as droplet microfluidics that screen and select mutants based on their intracellular or extracellular phenotypic signature, respectively. Traditional screening approaches have been based on the compartmentalization of clonal populations using microtiter plates (96 or 384 wells). Therein, automation through colony pickers or liquid-handling robots enables a screening throughput of approximately 10^4 clones per day [81]. Flow cytometry and droplet microfluidics, however, have greatly surpassed that by achieving screening throughputs of approximately 10^8 variants per day, or 10^6 cells per hour [35, 53, 80, 82]. Such impressive metrics are not only applicable to fundamental research, but also greatly exceed current automation standards employed in industry.

References

1. Heller MJ (2002) DNA microarray technology: devices, systems, and applications. Annu Rev Biomed Eng 4:129–153
2. Lander ES et al (2001) Initial sequencing and analysis of the human genome. Nature 409 (6822):860–921
3. Ota T et al (2004) Complete sequencing and characterization of 21,243 full-length human cDNAs. Nat Genet 36(1):40–45
4. Blattner FR et al (1997) The complete genome sequence of Escherichia coli K-12. Science 277 (5331):1453–1462
5. Cohen SN et al (1973) Construction of biologically functional bacterial plasmids in-vitro. Proc Natl Acad Sci U S A 70(11):3240–3244
6. Levin BR, Bull JJ (2004) Population and evolutionary dynamics of phage therapy. Nat Rev Microbiol 2(2):166–173
7. Bailey JE (1991) Toward a science of metabolic engineering. Science 252(5013):1668–1675
8. Stephanopoulos GN, Aristidou AA, Nielsen J (1998) Metabolic engineering: principles and methodologies. Elsevier, San Diego
9. Woolston BM, Edgar S, Stephanopoulos G (2013) Metabolic engineering: past and future. Annu Rev Chem Biomol Eng 4:259–288
10. Brown PO, Botstein D (1999) Exploring the new world of the genome with DNA microarrays. Nat Genet 21:33–37
11. Datsenko KA, Wanner BL (2000) One-step inactivation of chromosomal genes in Escherichia coli K-12 using PCR products. Proc Natl Acad Sci U S A 97(12):6640–6645
12. Bailey JE et al (2002) Inverse metabolic engineering: a strategy for directed genetic engineering of useful phenotypes. Biotechnol Bioeng 79(5):568–579
13. Koffas MAG, Jung GY, Stephanopoulos G (2003) Engineering metabolism and product formation in Corynebacterium glutamicum by coordinated gene overexpression. Metab Eng 5(1):32–41
14. Nakamura CE, Whited GM (2003) Metabolic engineering for the microbial production of 1,3-propanediol. Curr Opin Biotechnol 14(5):454–459
15. Stephanopoulos G (2007) Challenges in engineering microbes for biofuels production. Science 315(5813):801–804
16. Stephanopoulos G (2002) Metabolic engineering by genome shuffling. Nat Biotechnol 20 (7):666–668
17. Biot-Pelletier D, Martin VJJ (2014) Evolutionary engineering by genome shuffling. Appl Microbiol Biotechnol 98(9):3877–3887
18. Zhang YX et al (2002) Genome shuffling leads to rapid phenotypic improvement in bacteria. Nature 415(6872):644–646
19. Barrios-Gonzalez J, Fernandez FJ, Tomasini A (2003) Microbial secondary metabolites production and strain improvement. Indian J Biotechnol 2(3):322–333
20. Patnaik R et al (2002) Genome shuffling of Lactobacillus for improved acid tolerance. Nat Biotechnol 20(7):707–712
21. Yomano LP, York SW, Ingram LO (1998) Isolation and characterization of ethanol-tolerant mutants of Escherichia coli KO11 for fuel ethanol production. J Ind Microbiol Biotechnol 20 (2):132–138
22. Smith KM, Liao JC (2011) An evolutionary strategy for isobutanol production strain development in Escherichia coli. Metab Eng 13(6):674–681
23. Alper H, Stephanopoulos G (2007) Global transcription machinery engineering: a new approach for improving cellular phenotype. Metab Eng 9(3):258–267
24. Klein-Marcuschamer D, Stephanopoulos G (2008) Assessing the potential of mutational strategies to elicit new phenotypes in industrial strains. Proc Natl Acad Sci U S A 105 (7):2319–2324

25. Liu WS, Jiang RR (2015) Combinatorial and high-throughput screening approaches for strain engineering. Appl Microbiol Biotechnol 99(5):2093–2104
26. Skretas G, Kolisis FN (2012) Combinatorial approaches for inverse metabolic engineering applications. Comput Struct Biotechnol J 3:e201210021
27. Vasdekis AE, Stephanopoulos G (2015) Review of methods to probe single cell metabolism and bioenergetics. Metab Eng 27:115–135
28. Lidstrom ME, Konopka MC (2010) The role of physiological heterogeneity in microbial population behavior. Nat Chem Biol 6(10):705–712
29. Kaern M et al (2005) Stochasticity in gene expression: from theories to phenotypes. Nat Rev Genet 6(6):451–464
30. Li G-W, Xie XS (2011) Central dogma at the single-molecule level in living cells. Nature 475 (7356):308–315
31. Zenobi R (2013) Single-cell metabolomics: analytical and biological perspectives. Science 342(6163):1201
32. Blake WJ et al (2003) Noise in eukaryotic gene expression. Nature 422(6932):633–637
33. Heinemann M, Zenobi R (2011) Single cell metabolomics. Curr Opin Biotechnol 22(1):26–31
34. Vasdekis E, Silverman AM, Stephanopoulos G (2015) Origins of cell-to-cell bioprocessing diversity and implications of the extracellular environment revealed at the single-cell level. Sci Rep 5:17689. doi:10.1038/srep17689
35. Wang BL et al (2014) Microfluidic high-throughput culturing of single cells for selection based on extracellular metabolite production or consumption. Nat Biotechnol 32(5):473–478
36. Doughty DM et al (2014) Probing the subcellular localization of hopanoid lipids in bacteria using NanoSIMS. PLos One 9(1):e84455
37. Di Carlo D, Wu LY, Lee LP (2006) Dynamic single cell culture array. Lab Chip 6 (11):1445–1449
38. Tan W-H, Takeuchi S (2007) A trap-and-release integrated microfluidic system for dynamic microarray applications. Proc Natl Acad Sci U S A 104(4):1146–1151
39. Vasdekis AE (2013) Single microbe trap and release in sub-microfluidics. RSC Adv 3 (18):6343–6346
40. Wu AR et al (2014) Quantitative assessment of single-cell RNRNA-sequencing methods. Nat Methods 11(1):41
41. Pushkarev D, Neff NF, Quake SR (2009) Single-molecule sequencing of an individual human genome. Nat Biotechnol 27(9):847–850
42. Fan HC et al (2012) Non-invasive prenatal measurement of the fetal genome. Nature 487 (7407):320
43. Santos CNS, Stephanopoulos G (2008) Melanin-based high-throughput screen for L-tyrosine production in Escherichia coli. Appl Environ Microbiol 74(4):1190–1197
44. Lutke-Eversloh T, Stephanopoulos G (2007) A semi-quantitative high-throughput screening method for microbial L-tyrosine production in microtiter plates. J Ind Microbiol Biotechnol 34 (12):807–811
45. Gu MB, Mitchell RJ, Kim BC (2004) Whole-cell-based biosensors for environmental biomonitoring and application. Biomanufacturing 87:269–305
46. Binder S et al (2012) A high-throughput approach to identify genomic variants of bacterial metabolite producers at the single-cell level. Genome Biol 13(5):R40
47. Blombach B et al (2008) Corynebacterium glutamicum tailored for high-yield L-valine production. Appl Microbiol Biotechnol 79(3):471–479
48. Mustafi N et al (2012) The development and application of a single-cell biosensor for the detection of L-methionine and branched-chain amino acids. Metab Eng 14(4):449–457
49. Hebisch E et al (2013) High variation of fluorescence protein maturation times in closely related Escherichia coli strains. PLos One 8(10):e75991
50. Dietrich JA, McKee AE, Keasling JD (2010) High-throughput metabolic engineering: advances in small-molecule screening and selection. In: Kornberg RD (ed) Annual review of biochemistry, vol 79. pp 563–590. doi:10.1146/annurev-biochem-062608-095938. http://www.annualreviews.org/doi/abs/10.1146/annurev-biochem-062608-095938

51. Tracy BP, Gaida SM, Papoutsakis ET (2010) Flow cytometry for bacteria: enabling metabolic engineering, synthetic biology and the elucidation of complex phenotypes. Curr Opin Biotechnol 21(1):85–99
52. Farinas ET (2006) Fluorescence activated cell sorting for enzymatic activity. Comb Chem High Throughput Screen 9(4):321–328
53. Yang GY, Withers SG (2009) Ultrahigh-throughput FACS-based screening for directed enzyme evolution. Chembiochem 10(17):2704–2715
54. Croslandtaylor PJ (1953) A device for counting small particles suspended in a fluid through a tube. Nature 171(4340):37–38
55. Kamentsky LA, Melamed MR, Derman H (1965) Spectrophotometer—new instrument for ultrarapid cell analysis. Science 150(3696):630
56. Shapiro HM (2003) Practical flow cytometry. John Wiley & Sons Inc., Hoboken, NJ
57. Taylor LD (2007) High content screening. In: Taylor LD, Haskins JR, Giuliano KA (eds) Methods in molecular biology. Humana Press Inc., Totowa, NJ
58. Fattaccioli J et al (2009) Size and fluorescence measurements of individual droplets by flow cytometry. Soft Matter 5(11):2232–2238
59. Diaz M et al (2010) Application of flow cytometry to industrial microbial bioprocesses. Biochem Eng J 48(3):385–407
60. Becker S et al (2004) Ultra-high-throughput screening based on cell-surface display and fluorescence-activated cell sorting for the identification of novel biocatalysts. Curr Opin Biotechnol 15(4):323–329
61. Olsen MJ et al (2000) Function-based isolation of novel enzymes from a large library. Nat Biotechnol 18(10):1071–1074
62. Aharoni A et al (2006) High-throughput screening methodology for the directed evolution of glycosyltransferases. Nat Methods 3(8):609–614
63. Papadimitriou K et al (2007) Acid tolerance of Streptococcus macedonicus as assessed by flow cytometry and single-cell sorting. Appl Environ Microbiol 73(2):465–476
64. Amanullah A et al (2003) Measurement of strain-dependent toxicity in the indene bioconversion using multiparameter flow cytometry. Biotechnol Bioeng 81(4):405–420
65. Da Silveira MG, Abee T (2009) Activity of ethanol-stressed Oenococcus oeni cells: a flow cytometric approach. J Appl Microbiol 106(5):1690–1696
66. Xiao H, Bao Z, Zhao H (2015) High throughput screening and selection methods for directed enzyme evolution. Ind Eng Chem Res 54(16):4011–4020
67. Guo MT et al (2012) Droplet microfluidics for high-throughput biological assays. Lab Chip 12(12):2146–2155
68. Shim J-U et al (2009) Simultaneous determination of gene expression and enzymatic activity in individual bacterial cells in microdroplet compartments. J Am Chem Soc 131 (42):15251–15256
69. Anna SL, Bontoux N, Stone HA (2003) Formation of dispersions using "flow focusing" in microchannels. Appl Phys Lett 82(3):364–366
70. Thorsen T et al (2001) Dynamic pattern formation in a vesicle-generating microfluidic device. Phys Rev Lett 86(18):4163–4166
71. Baroud CN, Gallaire F, Dangla R (2010) Dynamics of microfluidic droplets. Lab Chip 10(16):2032–2045
72. Garstecki P et al (2006) Formation of droplets and bubbles in a microfluidic T-junction—scaling and mechanism of break-up. Lab Chip 6(3):437–446
73. Nisisako T, Torii T, Higuchi T (2002) Droplet formation in a microchannel network. Lab Chip 2(1):24–26
74. Miller OJ et al (2006) Directed evolution by in vitro compartmentalization. Nat Methods 3(7):561–570
75. Turner NJ (2003) Directed evolution of enzymes for applied biocatalysis. Trends Biotechnol 21(11):474–478

76. Schaerli Y et al (2009) Continuous-flow polymerase chain reaction of single-copy DNA in microfluidic microdroplets. Anal Chem 81(1):302–306
77. Leung K et al (2012) A programmable droplet-based microfluidic device applied to multiparameter analysis of single microbes and microbial communities. Proc Natl Acad Sci U S A 109(20):7665–7670
78. Huebner A et al (2007) Quantitative detection of protein expression in single cells using droplet microfluidics. Chem Commun 12:1218–1220
79. Edd JF et al (2008) Controlled encapsulation of single-cells into monodisperse picolitre drops. Lab Chip 8(8):1262–1264
80. Sjostrom SL et al (2014) High-throughput screening for industrial enzyme production hosts by droplet microfluidics. Lab Chip 14(4):806–813
81. Sonderegger M, Schumperli M, Sauer U (2005) Selection of quiescent Escherichia coli with high metabolic activity. Metab Eng 7(1):4–9
82. Daugherty PS et al (1998) Antibody affinity maturation using bacterial surface display. Protein Eng 11(9):825–832

Chapter 10
Microfluidic Paper-Based Multiplexing Biosensors for Electrochemical Detection of Metabolic Biomarkers

Chen Zhao, Martin M. Thuo, and Xinyu Liu

Abstract Paper-based microfluidic devices, which have emerged as a low-cost analytical platform technology, promise simple yet powerful solutions for point-of-care (POC) disease diagnostics. In this chapter, we present a microfluidic paper-based electrochemical biosensor array capable of multiplexed detection of physiologically related metabolic markers. Compared with existing paper-based electrochemical devices, our design consists of a multiplexing array of eight electrochemical biosensors and works together with a portable, custom-made potentiostat for signal readout. The integrated platform, including the paper-based biosensor array and the portable potentiostat, is able to detect multiple target markers from one sample and carry out multiple measurements for each marker. Based on this platform, we conduct the detection of three metabolic biomarkers (i.e., glucose, lactate, and uric acid) in artificial urine, and achieve analytical performance comparable to existing commercial and paper-based analytical devices. As an integrated electrochemical system, the proposed multiplexing biosensor array will permit rapid and inexpensive disease diagnosis at the POC and/or in resource-limited regions.

Keywords Paper-based microfluidics • Biosensor • Electrochemical detection • Multiplexing • Metabolic markers • Point-of-care diagnostics

---

C. Zhao • X. Liu, Ph.D. (✉)
Department of Mechanical Engineering, McGill University, 817 Sherbrooke Street West, MD270, Montreal, QC H3A 0C3, Canada
e-mail: chen.zhao@mail.mcgill.ca; xinyu.liu@mcgill.ca

M.M. Thuo, Ph.D.
Department of Materials Science and Engineering, Iowa State University, 2220A Hoover, Ames, IA 50011, USA
e-mail: mthuo@iastate.edu

© Springer International Publishing Switzerland 2016

C. Lu, S.S. Verbridge (eds.), *Microfluidic Methods for Molecular Biology*,
DOI 10.1007/978-3-319-30019-1_10

10.1 Introduction

The control and treatment of infectious and noninfectious diseases require accurate quantification of disease-relevant biomarkers in human fluids such as blood, urine, and saliva. The current state-of-the-art diagnostic technologies depend on sophisticated clinical equipment and laboratory resources (e.g., electricity and clean water), and require long turnaround time and relatively high assay costs; thus, they cannot meet the requirements of medical care in rural areas and in the developing world, where low-cost, point-of-care (POC) diagnostic tests are urgently needed. For disease diagnosis in resource-limited settings, microfluidic paper-based analytical devices (μPADs), initially emerged as a low-cost diagnostic alternative for developing countries, have attracted significant interests worldwide due to their unique characteristics such as low cost, high portability, equipment independency, and ease of operation [1–5]. MicroPADs can be fabricated using simple and cost-effective facilities and operated by less-skilled users, which lowers the requirements for device manufacturing and medical personnel. These devices compose of single and multiple layers of paper substrates, can transport fluids in predefined two-dimensional (2D) and three-dimensional (3D) paper channels autonomously (by capillarity), and permit quantitative biomarker detections using different biosensing (e.g., colorimetric, electrochemical, and fluorescent) mechanisms [6–12].

Among the biosensing techniques explored in μPAD designs, colorimetric detection, a common method employed in the optical identification of chemical reactions, has been very popular in both qualitative and (semi-)quantitative paper-based tests in virtue of its simplicity and compatibility with camera-phone-based telemedicine [6, 13–15]. The electrochemical detection, which relies on electrically triggered chemical reactions for quantitative biomarker measurements, is attracting more and more research interests in μPAD designs because of its superior analytical performance such as high accuracy (reflected by limit of detection—LOD) and sensitivity [11, 16–18]. Through various electrochemical techniques, existing μPADs are able to provide more detailed information about the reaction process and the electron transfers among the analytes, including the current-time relationship, the electrochemical cell impedance, and the redox potentials of the reactive species [11, 18–21]. Electrochemical μPADs have been reported with much lower LODs and higher sensitivity than μPADs employing colorimetric detection. Moreover, electrochemical detection is not sensitive to ambient illumination conditions and impurities in samples (e.g., dust and insoluble particulate), making it particularly suitable for use in the field and/or dirty environments [22, 23].

Traditional electrochemical μPADs usually rely on a benchtop precision potentiostat for signal measurement, which is too bulky and expensive to use in resource-limited environments and at the POC. Handheld potentiostats are commercially available, but their costs are still relatively high (>$1000), making it less affordable in the developing world. Commercial products and custom-made solutions have been explored to further lower the equipment cost of electrochemical biosensors. The integration of electrochemical μPADs with a commercial glucose meter, a highly mature and inexpensive electrochemical reader, was proposed to detect different types of analytes other than glucose [24–27]. The "CheapStat," an inexpensive and open-source potentiostat, was also developed from a microcontroller-based electronic circuit [28]. It can be programmed to generate electrical waveforms for multiple electrochemical mechanisms, such as cyclic voltammetry, square-wave voltammetry, and anodic stripping voltammetry, and provides satisfactory analytical performance in electrochemical biosensing.

Even though the experimental strategies of using glucose meters and the CheapStat platform show great promises for low-cost and accurate bioanalysis on electrochemical μPADs, these instruments only accept one electrochemical cell at a time. Hence, repeated human operations (for changing devices, pipetting reagent solutions, and reading output signals) are needed while performing a large number of measurements. This hampers the throughput of disease diagnoses that require large-volume tests of multiple biomarkers, such as the clinical identification of some unknown symptoms that can be related to different diseases. For instance, a patient with long-lasting fever may need to undergo molecular diagnostic tests for several possible diseases such as tuberculosis, human immunodeficiency virus (HIV), and mononucleosis. To meet the potential requirement of using μPADs for high-volume diagnostic tests, we develop a paper-based electrochemical biosensor array, working with a custom-made, inexpensive electrochemical reader, for performing multiplexed biomarker detection in eight human fluid samples within a single run. Using the biosensor array, we demonstrate the simultaneous concentration measurements of three metabolic markers (glucose, lactate, and uric acid) in artificial urine (AU). A single biosensor array device can generate multiple data points for each of the three analytes, with comparable analytical performance to commercial meters and existing μPADs.

10.2 Materials and Methods

10.2.1 Materials

All chemicals were used as received without further purification. Potassium ferricyanide, D-(+)-glucose, sodium L-lactate, uric acid, glucose oxidase (from Aspergillus niger, 147.9 U/mg), lactate oxidase (from Pediococcus sp., ~39 U/mg), and uricase (from Candida sp., >2 U/mg) were purchased from Sigma-Aldrich (Oakville, ON, Canada). The AU solution was made according to a previously reported recipe [11], which contained 1.1 mM lactic acid, 2.0 mM citric acid, 25 mM sodium bicarbonate, 170 mM urea, 2.5 mM calcium chloride, 90 mM sodium chloride, 2.0 mM magnesium sulfate, 10 mM sodium sulfate, 7.0 mM potassium dihydrogen phosphate, 7.0 mM dipotassium hydrogen phosphate, and 25 mM ammonium chloride. All the chemicals for making AU were mixed in deionized (DI) water and the pH of this solution was adjusted to 6.0 with hydrochloric acid and sodium hydroxide. Chromatography paper (Whatman 1 Chr) was obtained from Sigma-Aldrich (Oakville, ON, Canada). Carbon ink (E3456) and silver ink (E1660) were purchased from Ercon (Wareham, MA). All electronic components for constructing the electrochemical reader were purchased from Mouser Electronics (Mansfield, TX), and printed circuit boards (PCBs) were fabricated through RushPCB Inc. (San Jose, CA).

10.2.2 Device Design and Fabrication

Figure 10.1a and b illustrates the photographs of a microfluidic paper-based biosensor array and a portable, custom-made potentiostat with a biosensor array inserted, respectively. Each biosensor array includes eight electrochemical cells patterned on a piece of cellulose paper, which are used to quantify three metabolic biomarkers including glucose (three measurements), lactate (three measurements),

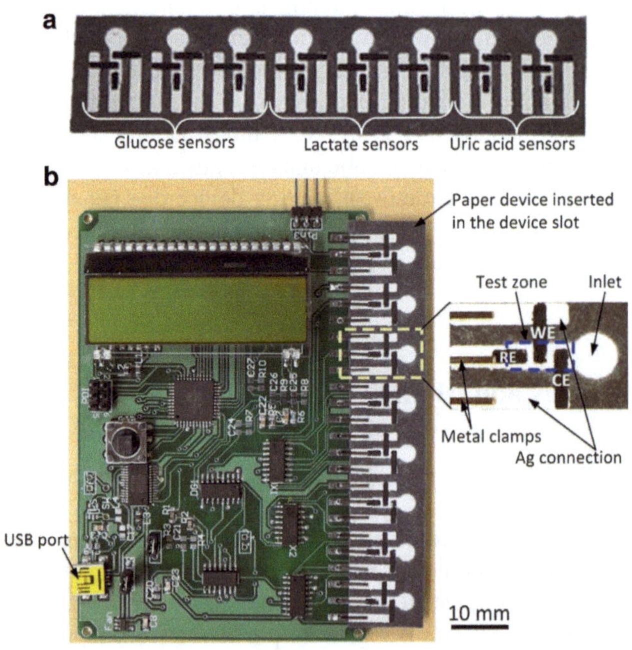

Fig. 10.1 Photographs of the paper-based electrochemical biosensor array and a custom-made handheld potentiostat for multiplexed detection of metabolic biomarkers. (**a**) A paper-based biosensor array including eight electrochemical cells. (**b**) A microcontroller-based multiplexing potentiostat with a paper-based biosensor array inserted in its device connection slot. The potentiostat has eight measurement channels of electrochemical signals, and each channel is connected to a group of three sensing electrodes of an electrochemical cell through metal clamps. *CE* counter electrode, *WE* working electrode, *RE* reference electrode

and uric acid (two measurements). As shown in the inset of Fig. 10.1b, each electrochemical cell consists of a patterned hydrophilic paper channel (as the test zone) and three screen-printed carbon electrodes (for electrochemical biosensing). The working electrode (WE), counter electrode (CE), and reference electrode (RE) were patterned on top of the paper test zone via stencil printing, and the paper channel was created by wax printing [29]. Silver strips were also screen-printed underneath the carbon electrodes (inset of Fig. 10.1b) to minimize the contact resistance between the electrodes and the metal clips on the circuit board of the potentiostat. To detect glucose, lactate, and uric acid in AU, we employed enzymatic reactions, which involve enzyme-catalyzed redox reactions of the three analytes and produce free electrons for electrochemical measurement. We pre-stored the corresponding enzymes and electron-transfer mediators in the test zones to react with the target analytes.

To fabricate a biosensor array device, the patterns of microfluidic channels and electrodes were firstly laid out in AutoCAD. We then printed wax patterns defining barriers of the microfluidic channels (unprinted areas correspond to the channels) on a piece of chromatography paper (Whatman 1 Chr) using a commercial solid wax printer (ColorQube 8570, Xerox). With the printed paper, we baked it on a hot plate at 150 °C for 2 min, during which the wax was melt and vertically penetrated the paper substrate to form complete hydrophobic barriers of the channels. After that, we sequentially patterned the silver strips and carbon electrodes on the surface of the chromatography paper through stencil printing [24]. According to the designed patterns of electrodes and silver strips, we cut out printing windows on single-sided adhesive stencil films (Grafix Frisket® film) with a laser cutter (VersaLASER VLS2.30, Universal Laser Systems). Then, we taped a stencil film on top of the wax paper (with the printing windows aligned with the test zones), filled the printing windows of the stencil with carbon or silver ink, removed the extra ink from the windows using a printing knife, peeled the stencil film off the paper, and finally baked the paper device on a hot plate at 65 °C for 20 min until the ink was dried. Both the printing and baking processes were conducted in a ventilated environment. The whole fabrication process can be completed in an hour, and five 1×8 array devices are typically fabricated from a single batch. Figure 10.1a shows a photograph of the paper-based biosensor array device.

10.2.3 Potentiostat Design

To expand the capabilities of the electrochemical detection on μPADs, we custom-made a low-cost, handheld potentiostat capable of conducting a variety of electrochemical detection protocols. The potentiostat has eight measurement channels and interfaces with the multiplexing biosensor array through a plug-and-play device slot. The potentiostat circuit was designed based on a microcontroller unit (MCU), with a similar architecture to the CheapStat [28]. Figures 10.1b and 10.2a show

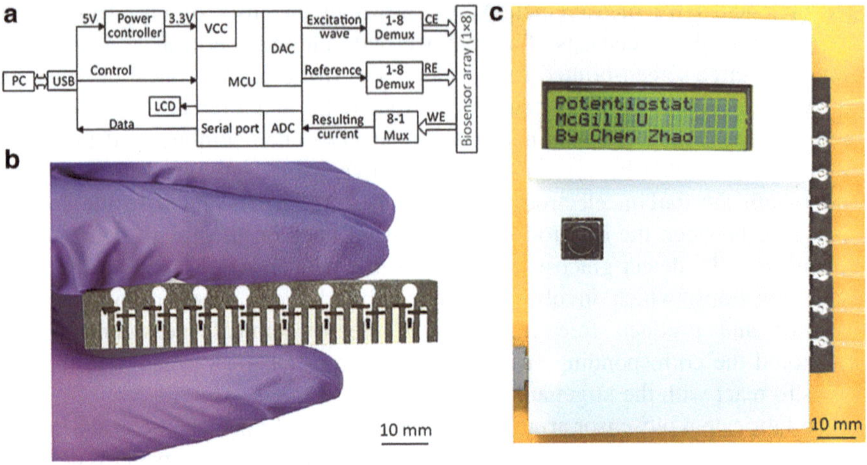

Fig. 10.2 (**a**) Schematic diagram of the potentiostat circuit architecture with eight measurement channels. (**b**) A multiplexing paper device held by fingers. (**c**) A photograph of the potentiostat inserted with a paper-based biosensor array

photographs of the potentiostat and its design schematic diagram. On the potentiostat, a low-power MCU (ATMEGA32, Atmel; with 12-bit A/D converters) is programmed to generate voltammetric waveforms for electrochemical excitations, and acquire resulting currents in response to the electrons transferred during redox reaction of the analytes. In contrast to the single-channel CheapStat [28], we add multiplexing capability to the potentiostat by integrating a multiplexer chip into the circuit, which enables automated electrochemical measurements from the eight electrochemical cells of the biosensor array. We designed a plug-and-play device slot to electrically interface the 24 electrodes of a 1×8 biosensor array device upon insertion, making the device connection fairly easy.

The potentiostat was set with six measurement ranges of the electrochemical currents (±10 μA, ±60 μA, ±100 μA, ±400 μA, ±600 μA, and ±1 mA) to accommodate different redox reactions that generate different levels of electrochemical current. These ranges can be easily adjusted via onboard jumpers. The smallest current measurement resolution is 39 nA with the measurement range of 0–10 μA. In the current design, we powered the handheld potentiostat through a universal serial bus (USB) cable (Fig. 10.1b), which also transmits acquired data to a host computer for analysis. To improve its portability, one can also program the MCU to directly display the results on the LCD screen, integrate a battery as the power, and add wireless communication module to eliminate the usage of USB. Figure 10.2c shows a photograph of the handhold potentiostat with a paper-based device inserted on the right side. The total cost of the electronic components used to construct the potentiostat is ~CAD $90 (based on commercial prices in small quantities).

10.2.4 *Marker Detection Principles*

As a proof-of-concept demonstration, we performed concentration measurements of D-glucose, L-lactate, and uric acid in AU samples, using a time-current mechanism (chronoamperometry). Corresponding enzymes such as glucose oxidase, lactate oxidase, and uricase and an electron-transfer mediator (potassium ferricyanide—$K_3[Fe(CN)_6]$) were pre-stored in the test zones (Fig. 10.1b) of the eight electrochemical cells in a dry form. To perform a test, we added a 4 μL drop of spiked AU sample to the inlet of each electrochemical cell, to allow the sample to wick to the test zone and react with the stored reagents. During the reaction, the enzyme catalyzed the oxidation of corresponding analyte into their oxidized products (glucose to gluconic acid, lactate to pyruvate, and uric acid to allantoin), with a concomitant reduction of $Fe(CN)_6{}^{3-}$ into $Fe(CN)_6{}^{4-}$ (Eqs. 10.1–10.3). The generated electrons were transferred between the WE and CE and quantified using chronoamperometry:

$$\text{Glucose} + 2K_3\left[Fe(CN)_6\right] + H_2O \overset{\text{GOx}}{\rightarrow} \text{Gluconic Acid} + 2K_4\left[Fe(CN)_6\right] \quad (10.1)$$

$$\text{Lactate} + 2K_3\left[Fe(CN)_6\right] + H_2O \overset{\text{LOx}}{\rightarrow} \text{Pyruvate} + 2K_4\left[Fe(CN)_6\right] \quad (10.2)$$

$$\text{Uric Acid} + 2K_3\left[Fe(CN)_6\right] + H_2O \overset{\text{UOx}}{\rightarrow} \text{Allantoin} + CO_2 + 2K_4\left[Fe(CN)_6\right] \quad (10.3)$$

10.2.5 *Device Preparation*

After device fabrication, test zones of the device were washed with 10 μL of alcohol to remove potential contaminants introduced during fabrication, and dried at room temperature. The alcohol was added carefully to prevent the etching of wax barrier. To prepare the device for experiments, reagent mixtures of enzymes and the electron-transfer mediator ($K_3[Fe(CN)_6]$) were freshly prepared at the following concentrations: (1) 250 U/mL glucose oxidase in 600 mM $K_3[Fe(CN)_6]$ and 1 M KCl; (2) 50 U/mL lactate oxidase in 200 mM $K_3[Fe(CN)_6]$ and 1 M KCl; and (3) 27 U/mL uricase in 100 mM $K_3[Fe(CN)_6]$ and 1 M KCl. Four microliter drops of particular reagent mixtures were added to the test zones of the device. After dried in a light-resistant incubator at 4 °C for 20 min, the device was ready for electrochemical detections.

10.3 Results and Discussion

10.3.1 Characterization of the Device Electrochemical Behavior

Before applying the paper-based biosensor array to detection of the three analytes in AU, we firstly characterized the electrochemical behavior of individual cells with the handheld potentiostat. Using 10 mM $K_3[Fe(CN)_6]$ in 1 M KCl as a model electroactive compound, each electrochemical cell on the array device was experimentally characterized through cyclic voltammetry (CV). In the characterization, a 6 μL drop of 10 mM $K_3[Fe(CN)_6]$ in 1 M KCl solution was firstly added to the inlet of an electrochemical cell and wicked to the test zone autonomously. Upon the application of a CV waveform, the $Fe(CN)_6^{3-}$ was reduced at the cathodic electrode and the resultant $Fe(CN)_6^{4-}$ was oxidized at the anodic electrode. The potentiostat was programmed to provide CV waves with six scan rates (50, 100, 200, 300, 400, and 500 mV/s) to the WE (vs. the graphite pseudo-RE), and measured the generated electrochemical currents in the current range of ±10 μA.

The current-potential data were acquired and analyzed on a computer connected with the potentiostat through the USB interface. A group of typical cyclic voltammograms are shown in Fig. 10.3a. The ratio of cathodic peak current (i_{pc}) and anodic peak current (i_{pa}) was calculated to be close to the unity ($i_{pa}/i_{pc} = 1.0$) for all the scan rates between 50 and 500 mV/s. The peak shapes of the cyclic voltammograms reveal typical reversible (Nernstian) electrochemical reaction [30]. For an n-electron-transfer process, the polarization overpotential (the difference between potentials at i_{pc} and i_{pa}) is $|E_{pc} - E_{pa}| = 57$ mV/n theoretically. The number of transferred electrons in the redox-active couple of $Fe(CN)_6^{3-}/Fe(CN)_6^{4-}$ is $n = 1$, giving an overpotential value of 57 mV. However, the polarization overpotential values measured from our devices ranged from 80 to 250 mV due to the wave shifts at the reduction and oxidation peaks. The shifting was attributed to the uncompensated resistance between the WE and RE, and also the relative instability of the graphite pseudo-RE in multiple CV scans. In this regard, we observed that different electrochemical cells in the same device and from different devices revealed fairly small variations in the overpotential value, when used at the first time. Since these paper-based devices were designed for one-time uses, the peak-shifting issue did not induce significant measurement errors in our experiments. Using fresh devices, we performed the experimental data of measuring glucose, lactate, and uric acid in AU (as presented in Sect. 10.3.2), and demonstrated reproducible analytical performance. Therefore, it is concluded that our paper-based biosensor array devices possess reliable biosensing capabilities for single-use diagnostic applications.

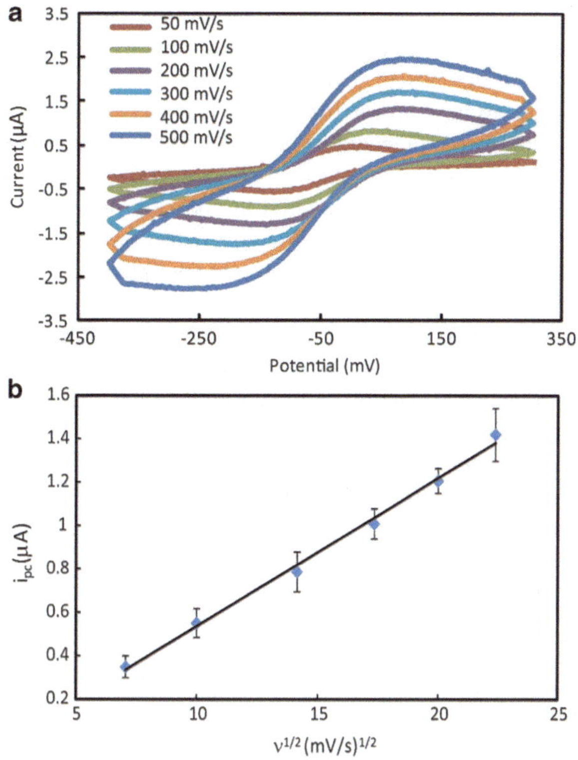

Fig. 10.3 Electrochemical characterization of the paper-based biosensor array. (**a**) CV curves of 10 mM potassium ferricyanide in 1 M potassium chloride solution in an electrochemical cell of the array device at six different scan rates. (**b**) A plot of the cathodic peak current i_{pc} versus the square root of the scan rate $\nu^{1/2}$, measured from the cyclic voltammograms ($n = 5$) (reprinted with permission from Ref. [31]. © 2013 National Institute for Materials Science)

10.3.2 Multiplexed Detection of Glucose, Lactate, and Uric Acid

Based on the biosensor array device and the handheld potentiostat, we quantified the concentrations of three clinically relevant metabolic markers (glucose, lactate, and uric acid) in AU using chronoamperometry. From a 1×8 biosensor array (Fig. 10.1a), three electrochemical cells were utilized for glucose detection, three cells for lactate detection, and two cells for uric acid detection. Chronoamperometry was chosen for analyte detection because it offers better accuracy and sensitivity than CV, and is easier to implement than other electrochemical techniques [11, 32]. In a chronoamperometry measurement, a 500 mV step potential was applied to the WE (vs. the graphite pseudo-RE), and generated the resulting current as a function of time. For the current device design, we observed that the current decayed from an initial high value to a low steady state within 90 s after applying the step potential. By taking the average of current values measured in 90–95 s post-step potential, we measured the steady-state Faradaic currents in detection of all the three markers at different concentrations.

The measured Faradaic current forms a linear relationship with the concentration of target analyte, which can be described by the Cottrell equation [30]. To accommodate the current levels in the detection, we adjusted the current measurement ranges to be ±60 μA for glucose and lactate, and ±10 μA for uric acid. The AU samples were prepared by adding the three markers to AU at concentrations in the ranges of 0–20 mM (glucose), 0–25 mM (lactate), and 0–10 mM (uric acid). These concentration ranges covered the possible levels of the three markers in clinical urine samples. To start a marker detection test, we inserted the device into the device slot of the potentiostat, and then pipetted 4 μL of spiked AU sample to the eight inlets on the biosensor array. We waited for 2 min to allow the sample to wick to the reaction area and thoroughly react with the pre-loaded reagents. Finally, a 500 mV step potential of chronoamperometry was applied to each electrochemical cell, and the experimental data of the electrochemical current vs. the time were collected to the host computer.

The typical chronoamperometric curves (at different concentrations) and the calibration plots for multiplexed detection of glucose, lactate, and uric acid in AU were shown in Figs. 10.4, 10.5, and 10.6. The experimental data reveal a linear relationship between the measured current and the analyte concentration for all the

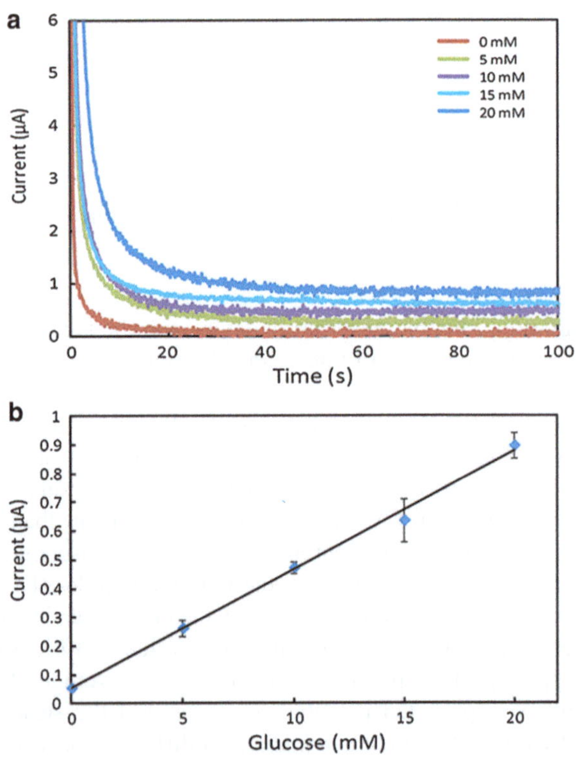

Fig. 10.4 (**a**) Chronoamperometric curves and (**b**) the calibration plot for glucose detections in AU. The current in (**b**) forms a linear relationship to the glucose concentration with regression equation: $y = 0.041x + 0.054$ ($R^2 = 0.996$, $n = 5$) (reprinted with permission from Ref. [31]. © 2015 National Institute for Materials Science)

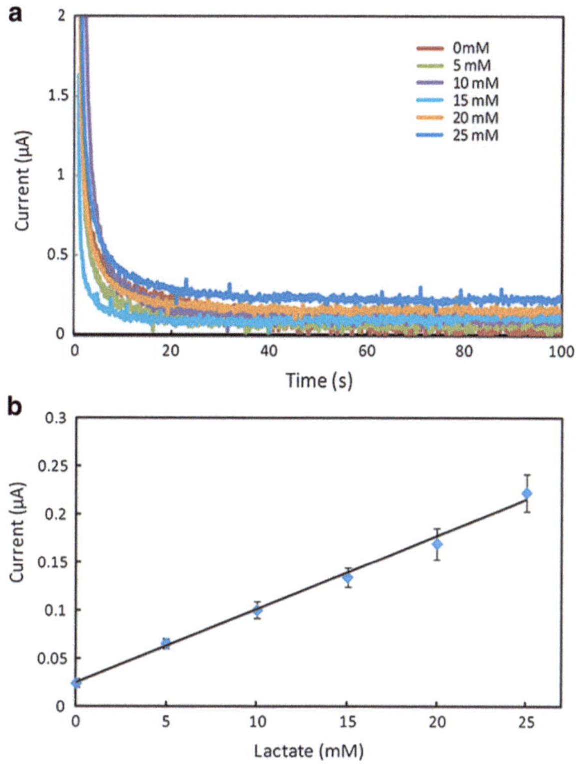

Fig. 10.5 (**a**) Chronoamperometric curves and (**b**) the calibration plot for lactate detections in AU. The current in (**b**) forms a linear relationship to lactate concentration with regression equation: $y = 0.0076x + 0.025$ ($R^2 = 0.995$, $n = 5$) (reprinted with permission from Ref. [31]. © 2015 National Institute for Materials Science)

three markers. The LOD was calculated to be the concentration that generated a current three times the standard deviation of currents measured in zero-concentration AU samples. In Table 10.1, we summarize the analytical performance of the biosensor array for detecting the three markers. The linear measurement ranges of our device cover the clinical relevant ranges of all three metabolic markers in clinical urine samples. The LOD values of our device for glucose (0.35 mM), lactate (1.76 mM), and uric acid (0.52 mM) were found to be comparable to those of commercial meters (glucose: 0.83 mM and lactate: 1.1 mM) and an existing μPAD (glucose: 0.21 mM, lactate: 0.36 mM, and uric acid: 1.38 mM). These results prove the sufficient biosensing capability of our device for quantitative detection of the three metabolic biomarkers. We believe that this multiplexing paper-based biosensor array and its handheld potentiostat will also be useful for performing electrochemical detection of other biomarkers in human fluids (e.g., antigens/antibodies, virus, and DNA segments) and chemicals/pollutants in environmental samples (e.g., heavy metal ions, bacteria, and toxins).

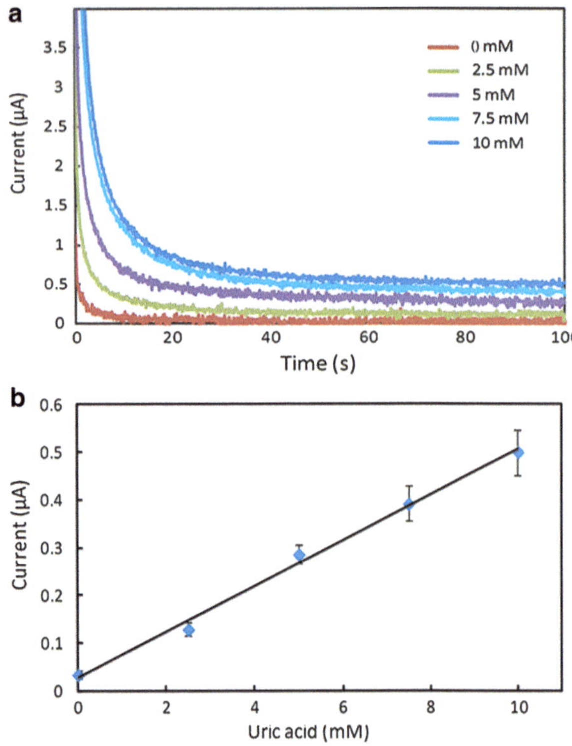

Fig. 10.6 (**a**) Chronoamperometric curves and (**b**) the calibration plot for uric acid detections in AU. The current in (**b**) forms a linear relationship to uric acid concentration with regression equation: $y = 0.048x + 0.029$ ($R^2 = 0.994$, $n = 5$) (reprinted with permission from Ref. [31]. © 2015 National Institute for Materials Science)

Table 10.1 Summary of the analytical performance of the biosensor array device and its comparison with that of commercial meters and an existing μPAD

	Analytical performance (mM)				
Analyte	Dynamic linear range (this work)	Clinically relevant ranges [18]	LOD (this work)	LOD (commercial meters [24])	LOD (Ref. [18])
Glucose	0–20	0.1–0.8	0.35	0.83	0.21
Lactate	0–25	5.5–22	1.76	1.1	0.36
Uric acid	0–10	1.5–4.4	0.52	N/A	1.38

10.4 Conclusion

In this work, a microfluidic paper-based electrochemical biosensor array (1×8) was developed to perform diagnostic tests of multiple biomarkers in a multiplexing and high-throughput fashion. To use the paper-based biosensor array for electrochemical detection at the POC, an inexpensive, handheld potentiostat was constructed. The electrochemical μPAD biosensor array and the potentiostat form a portable, self-contained, easy-to-operate electrochemical biosensing platform, which is particularly useful for low-cost diagnostic applications in resource-limited settings and/or at the POC. As a proof of concept, we used the platform for multiplexed detection of three metabolic markers including glucose, lactate, and uric acid. The experimental results showed that our device provides sufficient analytical performance (LOD and dynamic linear range) for clinical assays of all the three markers. This paper-based biosensing platform will have promising applications in POC diagnosis and on-site environmental tests.

In the current design, we used a USB cable to power the potentiostat, and transmit measured data to a host computer for analysis. We will soon integrate a battery into the potentiostat for self-powering and a Bluetooth module for wireless data transmission to a cell phone or a computer, and implement data analysis codes in the MCU to display the concentration data on the LCD.

References

1. Martinez AW, Phillips ST, Whitesides GM, Carrilho E (2010) Diagnostics for the developing world: microfluidic paper-based analytical devices. Anal Chem 82:3–10. doi:10.1021/ac9013989
2. Parolo C, Merkoci A (2013) Paper-based nanobiosensors for diagnostics. Chem Soc Rev 42:450–457. doi:10.1039/c2cs35255a
3. Li X, Ballerini DR, Shen W (2012) A perspective on paper-based microfluidics: current status and future trends. Biomicrofluidics 6:011301
4. Abe K, Suzuki K, Citterio D (2008) Inkjet-printed microfluidic multianalyte chemical sensing paper. Anal Chem 80:6928–6934. doi:10.1021/ac800604v
5. Zuo P, Li X, Dominguez DC, Ye B-C (2013) A PDMS/paper/glass hybrid microfluidic biochip integrated with aptamer-functionalized graphene oxide nano-biosensors for one-step multiplexed pathogen detection. Lab Chip 13:3921–3928
6. Martinez AW, Phillips ST, Butte MJ, Whitesides GM (2007) Patterned paper as a platform for inexpensive, low-volume, portable bioassays. Angew Chem Int Ed 46:1318–1320. doi:10.1002/anie.200603817
7. Martinez AW, Phillips ST, Whitesides GM (2008) Three-dimensional microfluidic devices fabricated in layered paper and tape. Proc Natl Acad Sci U S A 105:19606–19611. doi:10.1073/pnas.0810903105
8. Li X, Tian J, Nguyen T, Shen W (2008) Paper-based microfluidic devices by plasma treatment. Anal Chem 80:9131–9134
9. Yan J et al (2013) A microfluidic origami electrochemiluminescence aptamer-device based on a porous Au-paper electrode and a phenyleneethynylene derivative. Chem Commun 49:1383–1385
10. Liu X et al (2011) A portable paper-based microfluidic device for ELISA. In: I.E. 24th international conference on micro electro mechanical systems (MEMS). IEEE, p 75–78
11. Nie Z et al (2010) Electrochemical sensing in paper-based microfluidic devices. Lab Chip 10:477–483
12. Thom NK, Lewis GG, Yeung K, Phillips ST (2014) Quantitative fluorescence assays using a self-powered paper-based microfluidic device and a camera-equipped cellular phone. RSC Adv 4:1334–1340

13. Martinez AW et al (2008) Simple telemedicine for developing regions: camera phones and paper-based microfluidic devices for real-time, off-site diagnosis. Anal Chem 80:3699–3707. doi:10.1021/ac800112r
14. Dungchai W, Chailapakul O, Henry CS (2010) Use of multiple colorimetric indicators for paper-based microfluidic devices. Anal Chim Acta 674:227–233
15. Li X, Zwanenburg P, Liu X (2013) Magnetic timing valves for fluid control in paper-based microfluidics. Lab Chip 13:2609–2614
16. Zang DJ, Ge L, Yan M, Song XR, Yu JH (2012) Electrochemical immunoassay on a 3D microfluidic paper-based device. Chem Commun 48:4683–4685. doi:10.1039/C2cc16958d
17. Li X, Zhao C, Liu X (2015) A paper-based microfluidic biosensor integrating zinc oxide nanowires for electrochemical glucose detection. Microsyst Nanoeng 1, article number: 15014, 7pp
18. Dungchai W, Chailapakul O, Henry CS (2009) Electrochemical detection for paper-based microfluidics. Anal Chem 81:5821–5826
19. Jagadeesan KK, Kumar S, Sumana G (2012) Application of conducting paper for selective detection of troponin. Electrochem Commun 20:71–74
20. Zan X et al (2013) Freestanding graphene paper decorated with 2D-assembly of Au@Pt nanoparticles as flexible biosensors to monitor live cell secretion of nitric oxide. Biosens Bioelectron 49:71–78
21. Zhang Y et al (2014) Flexible paper-based ZnO nanorod light-emitting diodes induced multiplexed photoelectrochemical immunoassay. Chem Commun 50:1417–1419
22. Liu B, Du D, Hua X, Yu XY, Lin Y (2014) Paper-based electrochemical biosensors: from test strips to paper-based microfluidics. Electroanalysis 26:1214–1223
23. Maxwell EJ, Mazzeo AD, Whitesides GM (2013) Paper-based electroanalytical devices for accessible diagnostic testing. MRS Bull 38:309–314
24. Nie Z, Deiss F, Liu XY, Akbulut O, Whitesides GM (2010) Integration of paper-based microfluidic devices with commercial electrochemical readers. Lab Chip 10:3163–3169. doi:10.1039/c0lc00237b
25. Xu J et al (2012) Sensitive point-of-care monitoring of HIV related DNA sequences with a personal glucometer. Chem Commun 48:10733–10735
26. Su J et al (2013) Sensitive detection of copper (II) by a commercial glucometer using click chemistry. Biosens Bioelectron 45:219–222
27. Fu L et al (2013) Portable and quantitative monitoring of heavy metal ions using DNAzyme-capped mesoporous silica nanoparticles with a glucometer readout. J Mater Chem B 1:6123–6128
28. Rowe AA et al (2011) CheapStat: an open-source, "do-it-yourself" potentiostat for analytical and educational applications. PLoS One 6:e23783
29. Carrilho E, Martinez AW, Whitesides GM (2009) Understanding wax printing: a simple micropatterning process for paper-based microfluidics. Anal Chem 81:7091–7095. doi:10.1021/ac901071p
30. Bard AJ, Faulkner LR (1980) Electrochemical methods: fundamentals and applications, vol 2. Wiley, New York
31. Zhao C, Thuo MM, Liu X (2013) A microfluidic paper-based electrochemical biosensor array for multiplexed detection of metabolic biomarkers. Sci Technol Adv Mater 14:054402
32. Pohanka M, Skládal P (2008) Electrochemical biosensors—principles and applications. J Appl Biomed 6:57–64

Chapter 11
Droplet Microfluidics for Screening of Surface-Marker and Secretory Protein Expression

Pooja Sabhachandani, Saheli Sarkar, and Tania Konry

Abstract Current techniques of biomarker detection employ population-based studies, which fail to provide information on cellular and molecular heterogeneity and variances in functional behavior, which is important for diseased state characterization. Additionally, low concentration of samples in biological fluids creates hurdles in detection using these methods. Thus, improvement in detection techniques of clinical biomarkers has become imperative to halt disease progression at the earliest possible stage. Droplet microfluidics has become quite popular as a diagnostic platform due to its ability to sensitively detect low volume of clinical samples in a rapid and high-throughput manner. These advantages make the droplets an ideal platform for low-abundance biomarker detection. This chapter focuses on various techniques for detection of surface and secreted protein biomarkers using droplet microfluidics platform.

Keywords Droplet microfluidics • Diagnostics • Single cell analysis • Surface-marker proteins • Secretory proteins

11.1 Introduction

Since the initial emergence of micro Total Analysis Systems (μTAS) [1], the last two decades have seen diverse applications of "lab-on-a-chip" (LOC) devices in biomedical research. This system has shown great potential not only in sample manipulation and handling but also created a reputation for itself in being able to integrate multiparametric analysis in single platforms. LOC microfluidic devices have found a major niche in various facets of biochemical research including drug discovery [2, 3], diagnostics [4–8], single cell analysis [9–11], single molecule

P. Sabhachandani, M.S. • S. Sarkar, Ph.D. • T. Konry, Ph.D. (✉)
Department of Pharmaceutical Sciences, Northeastern University, 360 Huntington Avenue, 140 The Fenway, Boston, MA 02115, USA
e-mail: sabhachandani.p@husky.neu.edu; s.sarkar@neu.edu; t.konry@neu.edu

© Springer International Publishing Switzerland 2016

C. Lu, S.S. Verbridge (eds.), *Microfluidic Methods for Molecular Biology*,
DOI 10.1007/978-3-319-30019-1_11

detection [12–14], chemical synthesis [15, 16], and other biological and chemical assays [17–20].

These technologies manipulating fluid flow at micron and submicron level enjoys considerable benefits when compared to conventional assay techniques. One of its most prominent advantages is reduction of sample and reagent quantities, which is of consequence when dealing with low quantities of samples available for analysis. Also, reduced volumes ensure faster reaction rates, which results in reduced reaction time [21].

Droplet microfluidics is a distinct class of microfluidic technologies that deals with compartmentalization of reagents into monodisperse droplets using immiscible liquid phases driven through microchannels. This system, in addition to the advantages presented by other microfluidic technologies, is capable of conducting biological assays in droplet micro-reactors in a massively parallel fashion with high throughput and precise control over reaction conditions [22, 23]. Introduced for biochemical applications in 2001 [24], the complexity of assays performed in droplet microfluidic platforms has since increased with impressive tools being employed to study complex, multistep reactions [25]. The recently developed droplet platforms are capable of performing a wide variety of droplet manipulations like merging [26], mixing [27], splitting [28], reagent addition [29], and sorting [30] as per the assay requirements.

Generation of femtoliter to nanoliter volume droplets using microfluidic tools has exhibited the capability to isolate single molecules/cells into these microscopic bioreactors in a highly reproducible manner. In comparison to traditional protein detection assays, nano-compartmentalization reduces reagent volumes to 10,000 times the original. Constricted volumes ensure fast reagent mixing, which accelerates the reaction kinetics and reduces the detection time from hours to minutes [31]. Also, due to the immiscible nature of the continuous phase and the dispersed phase, any kind of diffusive loss of the secreted proteins is minimized. Thus constricted volume in droplets allows for ultra-sensitive detection of minute concentrations of secreted proteins and low-abundant surface-marker proteins, which is imperative for diseased state diagnostics and has thus gained considerable momentum in the field of biomolecule and biomarker detection. These advantages make it one of the most extensively researched platforms for studying myriad of biological applications involving genomic, proteomic, and metabolomics detection.

Explorations of cellular and molecular diversity are important factors to be considered for improvement of diagnostic and therapeutic techniques. Analysis of clinical biomarkers for diseased states sheds light on disease progression, cellular mechanisms, and pathways at play, which can help develop more efficient and targeted identification and therapeutic platforms.

Genomics biomarker characterization using droplet-based PCR (polymerase chain reaction) techniques has been widely researched in the past few years [22, 32, 33]. These methods are effective, but require complicated microchips capable of efficiently executing multiple processes involved in PCRs including thermocycling steps [22]. Although DNA and proteins can both be used as biomarkers for disease characterization, anomalies in mRNA expression and subsequent protein generation can only be accounted for during protein biomarker detection [34]. Also, according to the widely accepted paradigm, one gene can contain code for many proteins, and these proteins may undergo alternative post-translational modifications relevant for certain pathophysiological indicators of diseased states. Thus proteins, being the ones that directly execute the biological processes, are more effective biomarkers compared to DNA [35].

Current protein biomarker detection techniques provide information based on population-averaged studies like ELISA and western blot, which are unsuccessful in providing key information regarding heterogeneous molecular variations and differences in functional behavior between individual cells or molecules. Also, isolation and high-throughput analysis of single cells and single molecules is difficult with traditional methods, as they require high sample volumes for efficient detection [36]. Apart from heterogeneity concerns, serum concentrations of many biomarkers are extremely low at the commencement of the disease and are progressively amplified. Traditional immunoassay techniques for diseased state biomarker detection need protein concentrations above 10^{-12} M for conclusive detection since signal generated from these low concentrations will diffuse into large assay volumes (minimum 50 μl) and will provide a very weak, virtually undetectable signal compared to background [37, 38]. Another consequence of traditional methods due to sample dilution is longer incubation times which leads to delayed detection. Thus, the need for improved assays for detection of protein biomarkers has become imperative to halt disease progression at the earliest possible stage.

In this chapter, we will focus on recent advances made in the field of biomarker detection in droplet microfluidics, with emphasis on screening of secreted and surface marker proteins as disease biomarkers.

11.2 Droplet Device and Experimental Design Considerations for Biochemical Applications

Droplet microfluidic devices used biological assays that are typically made up of glass-polydimethylsiloxane (PDMS) by soft lithography method [39]. PDMS is an inert elastomer, which is nontoxic to cells and gas permeable. Other materials such as silica, resin, glass, polymethyl methacrylate (PMMA), and polystyrene have been used occasionally in instances where PDMS may be adversely affected like in incidences where organic solvent has to be employed [40, 41].

The effectiveness of droplet-microfluidics system relies on its principle to generate monodisperse droplets of uniform size and volume. The most common techniques of droplet formation rely on fluid shearing in T-junction and flow-focusing methods [23]. These channel design variations have been employed to generate water-in-oil (W/O) or oil-in-water (O/W) emulsions as well as complex multiple phase emulsions [42]. For stable droplet formation, the fluid flow into the microfluidic device must be steady, to maintain uniformity in terms of size, contents, and reagent volume. Fluid phases can be driven by passive techniques like syringe pumps (flow-controlled), pressure regulators, vacuum, or centrifugal forces [43]. Other methods to generate droplets include Dielectrophoresis (DEP) and Electrowetting on dielectric (EWOD). These methods are comparatively low-throughput compared to non-electrohydrodynamic methods.

Dynamic properties of droplet formation like size and rate of formation are determined by dimensionless constant capillary number. This quantifies the relationship between viscosity, velocity of flow, and interfacial tension between the two liquid phases [44]. The critical capillary number provides the break-off point for the droplets and varies with varying device geometry, and droplet generation phenomenon such as dripping and jetting are based on the increasing values of the critical capillary number respectively [45, 46]. Once the reagents have been compartmentalized, there is a need for stable docking of droplets for variable periods of time depending on the kinetic and dynamic assessment needs of applications. On-chip docking for spatiotemporal monitoring of preselected droplet over time and off-chip incubation and processing of droplets has both been utilized extensively. On-chip droplet incubation systems have been developed bypass traps, vertical posts, side wells, and top wells [47–50]. Droplet traps ensure stable droplet retention but may also be employed to release droplets from docking sites upon flow of continuous phase for further droplet processing if required [51]. Large droplet docking arrays called Dropspots were designed where droplet-docking sites were connected by constricted channels [52]. A highly sophisticated and programmable droplet docking and merging microchip was designed by Hansen et al. to study phenotypic and genotypic characteristics of bacterial cells by manipulating wetting properties in a controlled fashion [53]. A recent publication reported the development of a 6000-site droplet-trapping array combined with an automated image analysis program for single cell viability studies [54].

Another important parameter is the emulsification system for droplet formation in biochemical assays (i.e. continuous phase, dispersed aqueous phase, and surfactant). The continuous phase and the surfactant must be biocompatible to ensure inertness at the oil–water interface, which is of utmost importance for the protection of droplet contents. Mineral oils have been widely used for droplet generation but their application is limited in reactions with incubation times ranging from minutes to a few hours [55]. For long-term cell survival in droplets, emulsification systems containing fluorinated oils have been most successful owning to their inherent property that permits gas exchange [52, 56]. The primary function of surfactant in the emulsion system is to prevent droplet coalescence and promote stability of droplet micro-reactors. For hydrocarbon oils like mineral oil, standard surfactants such as span and tween have been routinely employed. For fluorinated oils, the range of surfactants is limited since fluorosurfactants like PFPE (perfluoropolyethers) have questionable biocompatibility. Chemical modifications to generate biocompatible derivatives of PFPE and other fluorosurfactants have been successful and commonly employed in droplet-based biochemical assays [57, 58].

11.3 Detection of Cell-Surface Proteins

Cell-surface proteins have been established as an important class of biomarkers, which function as useful and clinically relevant indicators of disease state, for example, Her2 in breast cancer [59].

Traditional methods of detection are centered on principle of flow cytometry where cells containing biomarker cell-surface proteins are sorted from complex cell populations. Although high-throughput, flow cytometry fails to offer sensitivity when detection is aimed at cells expressing low copy number cell-surface marker protein, which maybe important for early disease prognosis and detection [60]. It is also detrimental to cells since they are exposed to high shear and voltage, which may provide faulty results when dealing with fragile stem cells and make post-processing and further use of these samples impossible. Droplet microfluidics offers advantages over flow cytometry on these accounts and thus researchers are looking for novel methods for detection of cell-surface proteins using droplet platforms.

One of the first instances of utilization of cell-surface protein for identification and sorting of rare progenitor cells from cell population in droplet microfluidics was described by Srisa-Art et al. [50]. Periosteal cells from periosteum tissue were extracted and used as cell source due to its widely heterogeneous nature consisting of phenotypically distinct populations of mesenchymal and hematopoietic cell types, including a small subset of stem cell-like cells, at various stages of differentiation. Also, its therapeutic potential in musculoskeletal tissue repair and regeneration made it a lucrative candidate to be studied. Fluorescence detection of cell-surface proteins, alkaline phosphatase (ALP), and Stro-1 was targeted to differentiate cells based on stages of maturation and lineage commitment, where Stro-1 is cell-surface marker for mesenchymal stem cells with osteogenic potential and ALP is an established surface marker for cells to be differentiated into osteoblasts. They targeted multiplexed fluorescence detection of the two cell-surface markers, and three phenotypically distinct cell populations were identified based on differences in expression these surface proteins.

Cells emitting coincident fluorescent signals for ALP and Stro-1 were identified as pre-osteoblast cells and the other cells were classified as either progenitor or osteoblast cells depending on the emitted fluorescence. The comparison between data obtained from flow cytometry was identical when compared to expression levels obtained by droplet microfluidic platform. They were able to effectively identify subtle changes in cell-surface marker expression levels in a donor-dependent manner. Thus, they paved the way for use of droplet microfluidics for cell sorting and phenotypic classification based on cell-surface marker proteins.

Joensson and colleagues used enzyme amplification as a tool for detection of low copy number cell-surface marker proteins in single cells (Fig. 11.1A). They first labeled the surface proteins with specific biotinylated antibody, which was conjugated to streptavidin-coupled β-galactosidase. Subsequently, the cells labeled with fluorescent markers called Qdots (Quantum dots) for identification. The enzyme-labeled cells were injected in microfluidic device and co-encapsulated in droplets

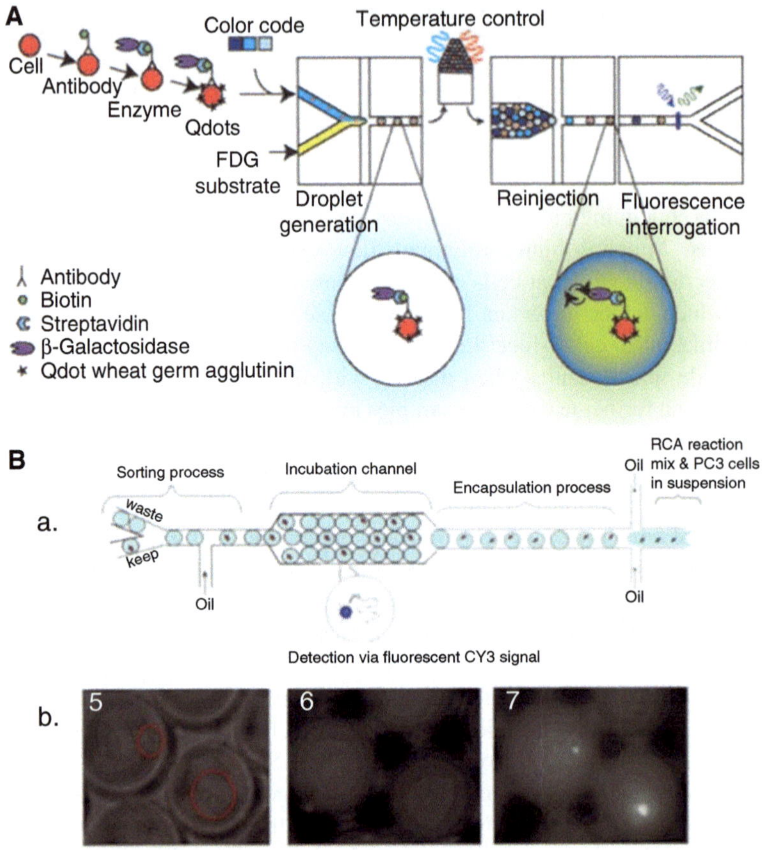

Fig. 11.1 Droplet microfluidics for the detection of cell-surface marker proteins. (**A**) Schematic of detection of low-abundant surface markers CD19 and CCR5 using enzyme amplification as described by Joensson et al. [61]. (**B**) Illustration of the RCA strategy for detection of low-abundance cell-surface marker protein EpCAM on human prostate cancer cell [62]. (**a**) Schematic of the droplet generation and sorting device, (**b**) *5*, encapsulated cells in droplets; *6*, fluorescence image at time 0; *7*, images of fluorescence expression of cell-surface marker protein after 1.5 h of incubation with RCA reagents (copyright—Wiley-VCH Verlag GmbH & Co. KGaA. Reproduced from Refs. [61] and [62] respectively with permission)

with fluorogenic substrate. They multiplexed the detection of two biomarkers on a human monocytic cell line: CCR5, a low-abundance cell-surface receptor in HIV-1 infection and a B-cell lineage marker, CD19. They also devised an optical color-coding scheme in droplets to allow for concurrent analysis of multiple samples. A comparative FACS study was conducted which validated the droplet system as more accomplished in terms of signal resolution [61].

An interesting approach was to exploit the Rolling circle amplification technique (RCA) for the detection of low copy number cell-surface markers in cancer, as described by Konry et al. [62]. RCA is a simple, isothermal method combining specific antigen–antibody interaction with DNA amplification. Briefly, the process involves binding of DNA tags to low-abundant expressed cell-surface marker proteins via specific antigen–antibody complexes. These DNA tags serve as scaffolds for directed RCA assembly to bind and subsequently produce million of DNA mini circle replicas called RCA amplicons. These replicas are constructed from fluorescently labeled nucleotides, which allow for direct detection of these amplified surface markers via optical methods. This method allows for a 1000-fold amplification of the surface marker to allow for single-target molecule detection. Konry et al. performed proof-of-principle study with amplification of epithelial cell adhesion molecule (EpCAM) on PC3 human prostate cancer cell line (Fig. 11.1B). EpCAM is of consequence in the detection of circulating tumor cells (CTCs) but its expression is below detection limit for conventional flow cytometry platforms. Detectable fluorescence signal could be visualized in less than 2 h, where local clusters of amplicons were represented as unambiguous bright dots on cell surfaces. Single molecule detection on single cells, as described above, is important for detection of response of heterogeneous cell populations toward therapeutics for efficacy screening. Although all the methods described here show tremendous advancement in the field of cell-surface biomarker detection, improvements in terms of throughput, robustness, and ease of experimental setup would make it a preferred system over FACS for personalized and point-of-care diagnostics applicable in clinics.

11.4 Detection of Secreted Proteins

Secretomes are another important class of biomarkers for diagnostic and therapeutic characterization since they help comprehend the effect of inter-, intra-, and extra-cellular environment on intracellular functions. Traditional, well-established secretome analysis techniques have low limits of detection, which is of consequence when the detection is aimed at low-abundance secreted proteins. Also, these methods detect bulk behavior, which masks the responses of cellular subpopulations, and lack the capability to dynamically detect changes in the secretory profiles of proteins [63]. Thus, detection of picomolar concentrations of secreted proteins and characterizing the secretory behavior of these atypical cellular populations has become of increasing importance. Droplet microfluidics provides a valuable tool for secretome detection based on these shortcomings. Apart from the abovementioned functions, it also has the capability to monitor dynamic changes in phenotypic behavior of these subcellular populations, which together with secretome profiles provides a complete picture of multiple cellular processes concurrently in a single platform.

11.4.1 Cellular Encapsulation and Protein Detection in Droplets

In 2008, Huebner et al. described a static droplet microarray platform for bacterial cell isolation in individual droplets and measure time-dependent enzymatic activity in picoliter volumes [51]. Since then, the droplet platform has been used extensively for single cell isolation and protein detection in bacterial infection [52, 64], cancers [65, 66], hybridoma screening [30, 67], and immunology [68–70].

Shim et al. developed a droplet microfluidic platform for compartmentalizing single bacteria in droplets and measuring secretory enzyme activity of individual clones in a time-dependent fashion [49]. Individual *E. coli* clones, with a capability to secrete Alkaline Phosphatase (AP), were co-encapsulated in droplets, along with fluorogenic substrates to measure and quantify AP activity of individual clones. They were able to successfully distinguish between the wild-type clone from mutant clone by quantifying the rate of AP release and their semiautomated system was able to accrue data from 4000 droplets over a total experimental time frame of 20 h. Gelled microdroplets have also been used to encapsulate individual bacterial cells and measure enzyme secretion [64]. This method is preferred over aqueous droplets when (1) the assay workflow requires multiple reagents need to be diffused into the droplets; (2) the cells need to be incubated for an extended period of time where the hydrogel matrix would allow for nutrient and gas exchange to maintain cellular health; (3) recovery of microdroplets from the continuous oil phase can be easily achieved with centrifugation and washing and; (4) subsequent post-processing of microdroplets with encapsulated cells by FACS is more robust due to increased droplet stability.

Cytokines are an important class of secretory biomarker proteins involved in various cellular processes such as growth, metabolism, lineage progression, and intercellular signaling [71, 72]. Changes in cytokine secretion profiles are important indicators of pathophysiological conditions such as inflammation, sepsis, trauma, and cancers [72]. The secretory profiles of cytokines may vary depending on disease progression, site of inflammation or infection, and cellular heterogeneity. Thus, compartmentalizing cytokine producing cells in picoliter droplets to measure dynamic cytokine secretion is of great consequence. Konry et al. encapsulated primary regulatory T cells (CD4+/CD25+) in droplets along with functionalized microspheres-based sensors and fluorescently labeled reporter antibody for IL-10 detection [68]. The droplets were incubated for 1 h and optical measurements based on change in fluorescent intensity on microbead surface were correlated with quantity of cytokine released (Fig. 11.2A).

They performed subsequent studies with simultaneous detection of surface marker CD86 on dendritic cells and secreted cytokine, IL-6 upon dendritic cell–T cell immune synapse (IS) formation in droplets [69]. This system facilitated the dynamic and simultaneous detection of cell-surface markers and secreted cytokines, with concurrent monitoring of phenotypical changes upon IS formation while accounting for heterogeneity in the cell population. This work, in the future, could

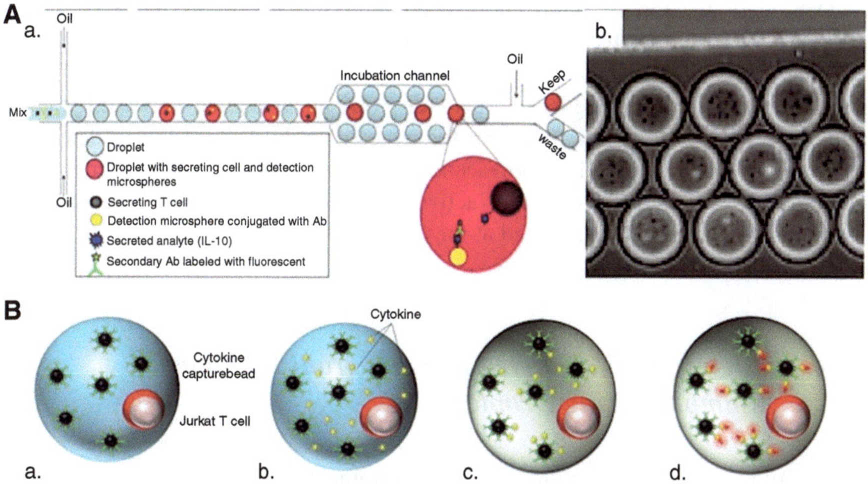

Fig. 11.2 Single cell secretome analysis using droplet microfluidics. (**A**) Droplet microfluidic-based detection of IL-10 cytokine from T cells [68]. (**a**) Schematic illustration of device used for droplet generation, incubation, and fluorescence-based sorting, (**b**) fluorescence expression due to IL-10 detection in droplets (reproduced from Ref. [68] with permission from Elsevier). (**B**) Schematic overview of detection of cytokines from single cells entrapped in agarose droplets [70]. (**a**) Single-cell and cytokine detection beads co-encapsulation in a agarose hydrogel droplet, (**b**) secretion and (**c**) capture of cytokine on detection bead, (**d**) detection of captured cytokines using fluorescently labeled reporter antibody (reproduced from Ref. [70] with permission from The Royal Society of Chemistry (RSC))

be used for correlation of secretory profiles with phenotypical changes, which occur concurrently in many cellular processes like cell division, apoptosis, and necrosis.

In a similar approach, agarose hydrogel droplets were utilized to detect cytokine release profiles from single cells (Fig. 11.2B). Briefly, activated jurkat T cells were encapsulated in a droplet together with functionalized microspheres for cytokine detection. The droplets were incubated off chip for 18–24 h for cytokine production and capture by detection microbeads. The agarose droplets were then polymerized and incubated with reporter antibody for detection of IL-2, IFN-γ, and TNF-α secretions. They monitored over 7000 cell-containing agarose beads for fluorescence intensity via flow cytometry [70].

Antibodies, produced by immune cell innately, are important diagnostic and therapeutic tools. Therapeutic monoclonal antibodies are the most researched class of therapeutics today, with a special emphasis in the field of cancer immunotherapy. These monoclonal antibodies are conventionally prepared by hybridoma cell technology, which involve complicated processes of clonal expansion and cell immortalization to obtain detectable quantity of the antibody. As a result, only a few thousand antibody-producing clones can be screened. Droplet microfluidic platform seems promising for screening of hybridoma cells on a single cell level.

Debs et al. developed a microfluidic platform to screen 300,000 hybridoma cells in a day to sort for cells producing antibodies to inhibit angiotensin-converting enzyme 1 (ACE-1) from a heterogeneous hybridoma cell population [67]. In a similar study, single mouse hybridoma cells were encapsulated in 50-pL droplets with streptavidin-coated microspheres for secreted antibody capture and detection. The droplet also contained a fluorescent reporter antibody for fluorescence measurement to distinguish between active and non-active cells. The droplets containing active cells were sorted at the rate of 200 Hz [30].

11.4.2 Detection of Secreted Proteins from Biological Samples Using Droplets

Apart from single cell capture and its secretome analysis, droplet microfluidic platform has been used for detection of low-abundance protein biomarkers from various biological samples using numerous strategies, some of which are highlighted below.

Mazutis et al. described a method for in vitro transcription and translation for cotA laccase gene in droplets, using a series of three devices performing six on-chip and two off-chip operations, the end point being the kinetic catalytic activity of the secreted laccase enzyme found in many microorganisms including fungi [25].

Chen et al. proposed the use of Matrix metalloproteinases (MMPs) from cellular supernatant as biomarkers as they are key operators in extracellular matrix and play a significant role in tumor progression and metastasis [66]. They developed a single microfluidic device with a biomolecule concentrator and a droplet generator for enhancing the detection sensitivity of the low-abundance MMPs. For biomolecule concentration, they utilized electrokinetic force opposing the pressure-driven flow to concentrate the low-abundance MMPs at the droplet formation junction, where they are mixed with the substrate and co-encapsulated in a droplet. This method resulted in decrease of reaction time by tenfold using about 25 μl of cellular supernatant for detection. They also developed an integrated platform with droplet generation, picoinjection, and droplet barcoding capabilities for multiplexed detection of multiple MMPs and their activity from 20 μl of peritoneal fluid obtained from endometriosis patients (Fig. 11.3) [73].

Another group developed a digitalized droplet-based ELISA platform using bead-based sensors for the detection of Prostate-specific antigen (PSA). The detection sensitivity was about 1000-fold higher compared to standard ELISA detection platforms and concentrations as low as 46 fM could be detected by compartmentalizing them in these femtoliter droplets [65]. Similar bead-based immunosensors were used for the detection of anti-tetanus immunoglobulin (IgG) from spiked human sera as a proof-of-concept study using one-step co-encapsulation in a simple droplet microfluidic device [74].

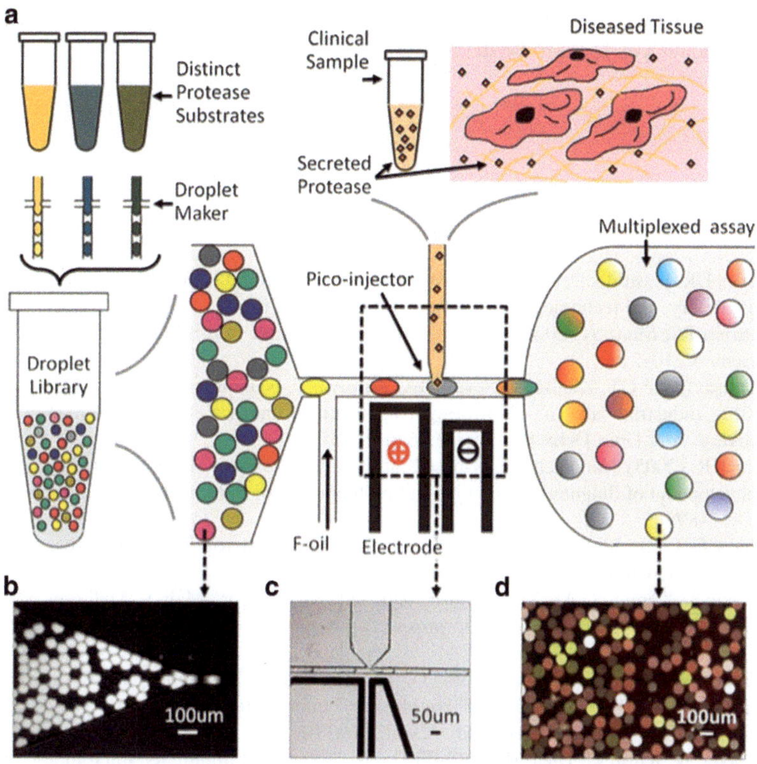

Fig. 11.3 Droplet-microfluidic-based detection of proteins from clinical samples. (**a**) Droplet-based workflow for detection of MMPs from peritoneal fluid [73]. Various protease substrates and inhibitors are used to generate droplet libraries (**b**); Clinical sample containing MMPs are introduced into the droplets via picoinjector (**c**); multiplexed identification of various reactions are detected using fluorescence determination (**d**) (reproduced from Ref. [73] with permission from American Chemical Society)

11.5 Conclusion and Future Perspectives

Use of droplet microfluidics for protein biomarker characterization has become one of the most powerful tools in diagnostics due to the various advantages the platform provides. Measurement of surface and secreted proteins on single molecule and single cell resolution has been rendered achievable by use of this technology by providing lower detection limits along with dynamic monitoring of phenotypic and secretory profiles with superior sensitivity. Although many techniques and methods for protein detection using droplet microfluidics have been discussed in this chapter, translation into highly robust and fully automated systems still remains a challenge. Also renovation of these technologies for point-of-care personalized diagnostics in clinics is important, with special focus on developing compact and portable systems. Despite the current predicaments, droplet microfluidics offers enough incentives for it to be developed into real-world applications for biomarker detection.

References

1. Reyes DR, Iossifidis D, Auroux PA, Manz A (2002) Micro total analysis systems. 1. Introduction, theory, and technology. Anal Chem 74(12):2623–2636
2. Dittrich PS, Manz A (2006) Lab-on-a-chip: microfluidics in drug discovery. Nat Rev Drug Discov 5:210–218
3. Vladisavljević GT, Khalid N, Neves MA, Kuroiwa T, Nakajima M, Uemura K, Kobayashi I (2013) Industrial lab-on-a-chip: design, applications and scale-up for drug discovery and delivery. Adv Drug Deliv Rev 65(11):1626–1663
4. Jain KK (2005) Nanotechnology-based lab-on-a-chip devices. In: Fuchs J, Podda M (eds) Encyclopedia of diagnostic genomics and proteomics, vol 2. Marcel Dekkar Inc., New York, pp 891–895
5. Chin CD, Linder V, Sia SK (2012) Commercialization of microfluidic point-of-care diagnostic devices. Lab Chip 12(12):2118–2134
6. Kumar S, Kumar S, Ali M, Anand P, Agrawal VV, John R, Maji S, Malhotra BD (2013) Microfluidic-integrated biosensors: prospects for point-of-care diagnostics. Biotechnol J 8 (11):1267–1279
7. Watkins NN, Hassan U, Damhorst G, Ni H, Vaid A, Rodriguez W, Bashir R (2013) Microfluidic CD4+ and CD8+ T lymphocyte counters for point-of-care HIV diagnostics using whole blood. Sci Transl Med 5(214):214ra170
8. Weaver W, Kittur H, Dhar M, Di Carlo D (2014) Research highlights: microfluidic point-of-care diagnostics. Lab Chip 14(12):1962–1965
9. Hochstetter A, Stellamanns E, Deshpande S, Uppaluri S, Engstler M, Pfohl T (2015) Microfluidics-based single cell analysis reveals drug-dependent motility changes in trypanosomes. Lab Chip 15(8):1961–1968
10. White AK, VanInsberghe M, Petriv I, Hamidi M, Sikorski D, Marra MA, Piret J, Aparicio S, Hansen CL (2011) High-throughput microfluidic single-cell RT-qPCR. Proc Natl Acad Sci U S A 108(34):13999–14004
11. Van den Brink FT, Gool E, Frimat JP, Bomer J, van den Berg A, Le Gac S (2011) Parallel single-cell analysis microfluidic platform. Electrophoresis 32(22):3094–3100
12. Witters D, Knez K, Ceyssens F, Puers R, Lammertyn J (2013) Digital microfluidics-enabled single-molecule detection by printing and sealing single magnetic beads in femtoliter droplets. Lab Chip 13(11):2047–2054
13. Kühnemund M, Witters D, Nilsson M, Lammertyn J (2014) Circle-to-circle amplification on a digital microfluidic chip for amplified single molecule detection. Lab Chip 14(16):2983–2992
14. Wunderlich B, Nettels D, Benke S, Clark J, Weidner S, Hofmann H, Pfeil F, Schuler B (2013) Microfluidic mixer designed for performing single-molecule kinetics with confocal detection on timescales from milliseconds to minutes. Nat Protoc 8(8):1459–1474
15. Keng PY, Chen S, Ding H, Sadeghi S, Shah GJ, Dooraghi A, Phelps M, Satyamurthy N, Chatziioannou A, Van Dam RM (2012) Micro-chemical synthesis of molecular probes on an electronic microfluidic device. Proc Natl Acad Sci U S A 109(3):690–695
16. Sebastian Cabeza V, Kuhn S, Kulkarni AA, Jensen KF (2012) Size-controlled flow synthesis of gold nanoparticles using a segmented flow microfluidic platform. Langmuir 28 (17):7007–7013
17. Hansen C, Quake SR (2003) Microfluidics in structural biology: smaller, faster em leader better. Curr Opin Struct Biol 13(5):538–544

18. Chovan T, Guttman A (2002) Microfabricated devices in biotechnology and biochemical processing. Trends Biotechnol 20(3):116–122
19. Huynh T, Sun B, Li L, Nichols KP, Koyner JL, Ismagilov RF (2013) Chemical analog-to-digital signal conversion based on robust threshold chemistry and its evaluation in the context of microfluidics-based quantitative assays. J Am Chem Soc 135(39):14775–14783
20. Sun J, Xianyu Y, Jiang X (2014) Point-of-care biochemical assays using gold nanoparticle-implemented microfluidics. Chem Soc Rev 43(17):6239–6253
21. Whitesides GM (2006) The origins and the future of microfluidics. Nature 442(7101):368–373
22. i Solvas X (2011) Droplet microfluidics: recent developments and future applications. Chem Commun 47(7):1936–1942
23. Teh SY, Lin R, Hung LH, Lee AP (2008) Droplet microfluidics. Lab Chip 8(2):198–220
24. Thorsen T, Roberts RW, Arnold FH, Quake SR (2001) Dynamic pattern formation in a vesicle-generating microfluidic device. Phys Rev Lett 86:4163–4166
25. Mazutis L, Baret JC, Treacy P, Skhiri Y, Aragh AF, Ryckelynck M, Taly V, Griffiths AD (2009) Multi-step microfluidic droplet processing: kinetic analysis of an in vitro translated enzyme. Lab Chip 9(20):2902–2908
26. Niu X, Gulati S, Edel JB, deMello AJ (2008) Pillar-induced droplet merging in microfluidic circuits. Lab Chip 8(11):1837–1841
27. Song H, Bringer MR, Tice JD, Gerdts CJ, Ismagilov RF (2003) Experimental test of scaling of mixing by chaotic advection in droplets moving through microfluidic channels. Appl Phys Lett 83:4664–4666
28. Link DR, Anna SL, Weitz DA, Stone HA (2004) Geometrically mediated breakup of drops in microfluidic devices. Phys Rev Lett 92(5):054503
29. Sato T, Kawai K, Kanai M, Shoji S (2009) Development of an all-fluoroplastic microfluidic device applied as a nanoliter sample injector. Jpn J Appl Phys 48(6S):06FJ03
30. Mazutis L, Gilbert J, Ung WL, Weitz DA, Griffiths AD, Heyman JA (2013) Single cell analysis and sorting using droplet-based microfluidics. Nat Protoc 8:870–891
31. Guo MT, Rotem A, Heyman JA, Weitz DA (2012) Droplet microfluidics for high-throughput biological assays. Lab Chip 12(12):2146–2155
32. Didelot A, Kotsopoulos SK, Lupo A, Pekin D, Li X, Atochin I, Srinivasan P, Zhong Q, Olson J, Link DR, Laurent-Puig P, Blons H, Hutchinson JB, Taly V (2013) Multiplex picoliter-droplet digital PCR for quantitative assessment of DNA integrity in clinical samples. Clin Chem 59 (5):815–823
33. Zanoli LM, Licciardello M, D'Agata R, Lantano C, Calabretta A, Corradini R, Marchelli R, Spoto G (2013) Peptide nucleic acid molecular beacons for the detection of PCR amplicons in droplet-based microfluidic devices. Anal Bioanal Chem 405(2–3):615–624
34. Huber LA (2002) Preface: proteomics and genomics technologies. J Mammary Gland Biol Neoplasia 7:357–358
35. Sahab ZJ, Semaan SM, Sang QA (2007) Methodology and applications of disease biomarker identification in human serum. Biomark Insights 2:21–43
36. Liu Y, Singh AK (2013) Microfluidic platforms for single-cell protein analysis. J Lab Autom 18(6):446–454
37. Giljohann DA, Mirkin CA (2009) Drivers of biodiagnostic development. Nature 462 (7272):461–464
38. Rissin DM, Walt DR (2006) Digital concentration readout of single enzyme molecules using femtoliter arrays and Poisson statistics. Nano Lett 6(3):520–523
39. Duffy DC, McDonald JC, Schueller OJ, Whitesides GM (1998) Rapid prototyping of microfluidic systems in poly (dimethylsiloxane). Anal Chem 70(23):4974–4984
40. Subramanian B, Kim N, Lee W, Spivak DA, Nikitopoulos DE, McCarley RL, Soper SA (2011) Surface modification of droplet polymeric microfluidic devices for the stable and continuous generation of aqueous droplets. Langmuir 27(12):7949–7957
41. Sackmann EK, Fulton AL, Beebe DJ (2014) The present and future role of microfluidics in biomedical research. Nature 507(7491):181–189

42. Okushima S, Nisisako T, Torii T, Higuchi T (2004) Controlled production of monodisperse double emulsions by two-step droplet breakup in microfluidic devices. Langmuir 20 (23):9905–9908
43. Haeberle S, Zengerle R (2007) Microfluidic platforms for lab-on-a-chip applications. Lab Chip 7(9):1094–1110
44. Anna SL, Mayer HC (2006) Microscale tipstreaming in a microfluidic flow focusing device. Phys Fluids 18(12):121512
45. Gu H, Duits MH, Mugele F (2011) Droplets formation and merging in two-phase flow microfluidics. Int J Mol Sci 12(4):2572–2597
46. Fu T, Wu Y, Ma Y, Li HZ (2012) Droplet formation and breakup dynamics in microfluidic flow-focusing devices: from dripping to jetting. Chem Eng Sci 84:207–217
47. Boukellal G, Hertel D, Valette R, Münstedt H, Agassant JF (2008) Investigation of LDPE converging flows using fieldwise measurements techniques. Int J Mater Form 1(1):687–690
48. Shi J, Mao X, Ahmed D, Colletti A, Huang TJ (2008) Focusing microparticles in a microfluidic channel with standing surface acoustic waves (SSAW). Lab Chip 8(2):221–223
49. Shim JU, Olguin LF, Whyte G, Scott D, Babtie A, Abell C, Huck WTS, Hollfelder F (2009) Simultaneous determination of gene expression and enzymatic activity in individual bacterial cells in microdroplet compartments. J Am Chem Soc 131(42):15251–15256
50. Srisa-Art M, DeMello AJ, Edel JB (2010) High-efficiency single-molecule detection within trapped aqueous microdroplets. J Phys Chem B 114(48):15766–15772
51. Huebner A, Bratton D, Whyte G, Yang M, Abell C, Hollfelder F (2009) Static microdroplet arrays: a microfluidic device for droplet trapping, incubation and release for enzymatic and cell-based assays. Lab Chip 9(5):692–698
52. Schmitz CH, Rowat AC, Köster S, Weitz DA (2009) Dropspots: a picoliter array in a microfluidic device. Lab Chip 9(1):44–49
53. Leung K, Zahn H, Leaver T, Konwar KM, Hanson NW, Pagé AP, Chien-Chi L, Chain PS, Hallam SJ, Hansen CL (2012) A programmable droplet-based microfluidic device applied to multiparameter analysis of single microbes and microbial communities. Proc Natl Acad Sci U S A 109(20):7665–7670
54. Khorshidi MA, Rajeswari PKP, Wählby C, Joensson HN, Svahn HA (2014) Automated analysis of dynamic behavior of single cells in picoliter droplets. Lab Chip 14(5):931–937
55. Aharoni A, Amitai G, Bernath K, Magdassi S, Tawfik DS (2005) High-throughput screening of enzyme libraries: thiolactonases evolved by fluorescence-activated sorting of single cells in emulsion compartments. Chem Biol 12(12):1281–1289
56. Clausell-Tormos J, Lieber D, Baret J-C, El-Harrak A, Miller OJ, Frenz L, Blouwolff J, Humphry KJ, Koester S, Duan H, Holtze C, Weitz DA, Griffiths AD, Merten CA (2008) Droplet-based microfluidic platforms for the encapsulation and screening of mammalian cells and multicellular organisms. Chem Biol 15(5):427–437
57. Roach LS, Song H, Ismagilov RF (2005) Controlling nonspecific protein adsorption in a plug-based microfluidic system by controlling interfacial chemistry using fluorous-phase surfactants. Anal Chem 77(3):785–796
58. Jeanneaux F, Krafft MP, Rábai J, Riess JG (1998) Perfluoroalkylated amphiphiles with a morpholinophosphate or a dimorpholinophosphate polar head group. New J Chem 22 (6):609–613
59. Ross JS, Fletcher JA, Bloom KJ, Linette GP, Stec J, Symmans WF, Pusztai L, Hortobagyi GN (2004) Targeted therapy in breast cancer the HER-2/neu gene and protein. Mol Cell Proteomics 3(4):379–398
60. Newman JR, Ghaemmaghami S, Ihmels J, Breslow DK, Noble M, DeRisi JL, Weissman JS (2006) Single-cell proteomic analysis of S. cerevisiae reveals the architecture of biological noise. Nature 441(7095):840–846
61. Joensson HN, Samuels ML, Brouzes ER, Medkova M, Uhlén M, Link DR, Andersson-Svahn H (2009) Detection and analysis of low-abundance cell-surface biomarkers using enzymatic amplification in microfluidic droplets. Angew Chem Int Ed 121(14):2556–2559

62. Konry T, Smolina I, Yarmush JM, Irimia D, Yarmush ML (2011) Ultrasensitive detection of low-abundance surface-marker protein using isothermal rolling circle amplification in a microfluidic nanoliter platform. Small 7(3):395–400
63. Walt DR (2012) Optical methods for single molecule detection and analysis. Anal Chem 85 (3):1258–1263
64. Hosokawa M, Hoshino Y, Nishikawa Y, Hirose T, Yoon DH, Mori T, Sekiguchi T, Shoji S, Takeyama H (2015) Droplet-based microfluidics for high-throughput screening of a metagenomic library for isolation of microbial enzymes. Biosens Bioelectron 67:379–385
65. Shim JU, Ranasinghe RT, Smith CA, Ibrahim SM, Hollfelder F, Huck WT, Klenerman D, Abell C (2013) Ultrarapid generation of femtoliter microfluidic droplets for single-molecule-counting immunoassays. ACS Nano 7(7):5955–5964
66. Chen CH, Sarkar A, Song YA, Miller MA, Kim SJ, Griffith LG, Lauffenburger DA, Han J (2011) Enhancing protease activity assay in droplet-based microfluidics using a biomolecule concentrator. J Am Chem Soc 133(27):10368–10371
67. El Debs B, Utharala R, Balyasnikova IV, Griffiths AD, Merten CA (2012) Functional single-cell hybridoma screening using droplet-based microfluidics. Proc Natl Acad Sci U S A 109 (29):11570–11575
68. Konry T, Dominguez-Villar M, Baecher-Allan C, Hafler DA, Yarmush ML (2011) Droplet-based microfluidic platforms for single T cell secretion analysis of IL-10 cytokine. Biosens Bioelectron 26(5):2707–2710
69. Konry T, Golberg A, Yarmush M (2013) Live single cell functional phenotyping in droplet nano-liter reactors. Sci Rep 3:3179
70. Chokkalingam V, Tel J, Wimmers F, Liu X, Semenov S, Thiele J, Figdor CG, Huck WT (2013) Probing cellular heterogeneity in cytokine-secreting immune cells using droplet-based microfluidics. Lab Chip 13(24):4740–4744
71. Borish LC, Steinke JW (2003) Cytokines and chemokines. J Allergy Clin Immunol 111(2): S460–S475
72. Steinke JW, Borish L (2006) Cytokines and chemokines. J Allergy Clin Immunol 117(2): S441–S445
73. Chen CH, Miller MA, Sarkar A, Beste MT, Isaacson KB, Lauffenburger DA, Griffith LG, Han J (2012) Multiplexed protease activity assay for low-volume clinical samples using droplet-based microfluidics and its application to endometriosis. J Am Chem Soc 135(5):1645–1648
74. Golberg A, Yarmush ML, Konry T (2013) Picoliter droplet microfluidic immunosorbent platform for point-of-care diagnostics of tetanus. Microchim Acta 180(9–10):855–860

Chapter 12
Wire-guided Droplet Manipulation for Molecular Biology

Dustin K. Harshman and Jeong-Yeol Yoon

Abstract Wire-guided droplet manipulation (WDM) is a simple method of manipulating liquid droplets in a hydrophobic environment to conduct experiments, reactions, and assays. In WDM, a wire (or needle tip) manipulates microliter-sized liquid droplets within an immiscible liquid or on a hydrophobic surface. The attributes of this format for liquid handling address some of the challenges facing the use of conventional techniques. Specifically, WDM provides solutions for development of automated, sample-to-answer, point-of-care systems with potential applications in medicine, life science research, forensics, veterinary diagnostics, and disease control.

The widespread applicability of this technique is due to its inherent simplicity stemming from the attractive force between the droplet and the wire. The physics of this interaction will be explained in this chapter. WDM can be applied to standard protocols and is easily reprogrammable for different liquid handling applications. Dilution, mixing, centrifugation, and thermocycling have all been automated by WDM (You and Yoon, J Biol Eng 6:15, 2012). If desired, the principles of droplet manipulation can be easily integrated into the common scientific automation strategy, using commercially available robotic pipetting systems. WDM is automatable, reprogrammable, easy to use, and robust. These are essential features of rapid, all-in-one, sample-to-answer systems to be used at the point-of-care.

The applications of WDM within molecular biology that have been demonstrated include DNA extraction (lysing, precipitation, washing, and rehydration), nanoparticle surface deposition for fabrication of a protein nanoarray, immunoassay, PCR thermocycling, and real-time PCR.

Keywords Wire-guided droplet manipulation • Electrowetting • Superhydrophobic • Sessile droplet • Contact angle • Work of adhesion • Electrospinning • Pendant droplet • Polymerase chain reaction (PCR) • Rapid thermocycling • Sample preparation • PCR inhibition • Protein nanoarray •

D.K. Harshman, Ph.D. • J.-Y. Yoon, Ph.D. (✉)
Biomedical Engineering Graduate Interdisciplinary Program and Department of Agricultural and Biosystems Engineering, The University of Arizona, 1177 East 4th Street, Room 403, Tucson, AZ 85721-0038, USA
e-mail: dustin.harshman@gmail.com; jyyoon@email.arizona.edu

© Springer International Publishing Switzerland 2016
C. Lu, S.S. Verbridge (eds.), *Microfluidic Methods for Molecular Biology*,
DOI 10.1007/978-3-319-30019-1_12

Particle immunoassay • Endpoint detection • Real-time quantification • Interfacial effects • Water-in-oil emulsion

12.1 Droplet Manipulation in Molecular Biology

Liquid droplets are used in molecular biology to take advantage of their potential as discrete reaction vessels [1]. Droplets can be manipulated in microfluidic channels [2] or by surface-based manipulations [3]. These manipulation strategies strive to conduct droplet operations such as actuation, fission, fusion, mixing, and immobilization. While droplets are isolated from other droplets and from their environment, they do interact with their environment at an interface. For this reason, the choice of droplet milieu and an understanding of interfacial effects are critical for the success of these technologies.

Various surface-based droplet manipulation methods have been demonstrated and provide a particularly promising opportunity to develop miniaturized diagnostic systems for point-of-care applications [3]. These methods include electrowetting-on-dielectric (EWOD) [4, 5], magnetofluidics [6, 7], magnetofluidics with surface energy trapping [8, 9], electrospun fiber mats [10], light-induced dielectrophoresis [11], optical trapping [12], pneumatic actuation on a thin superhydrophobic membrane [13], actuation by vibration [14], heater controlled surface tension gradients known as thermocapillary forces [15], and wire-guided droplet manipulation (WDM) [1, 16–18]. The high degree of interest in developing new droplet manipulation technologies follows from their importance in advancing molecular biology and its applications.

12.2 The Physics of Wire-guided Droplet Manipulation (WDM)

WDM operates on the principles of interfacial tension (γ; mJ/m^2 = mN/m), which is defined as the interface energy (mJ) per interface area (m^2). When a liquid droplet rests on a surface as illustrated in Fig. 12.1, it is known as a sessile droplet.

The contact angle (θ) between a liquid and a surface can be described by Young's equation (Eq. 12.1). The contact angle is measured at the three-phase borderline, where an equilibrium is established between the surface tensions between liquid–vapor (γ_{LV}; approximated to γ_L since vapor can be approximated as vacuum), surface–vapor (γ_{SV}; approximated to γ_S), and surface–liquid (γ_{SL}).

$$\gamma_{SL} = \gamma_s - \gamma_L \cos\theta \tag{12.1}$$

The work of adhesion is central to WDM [16] and can be described by the Dupré equation (Eq. 12.2) and the Young-Dupré equation (Eq. 12.3) [19]. The work of

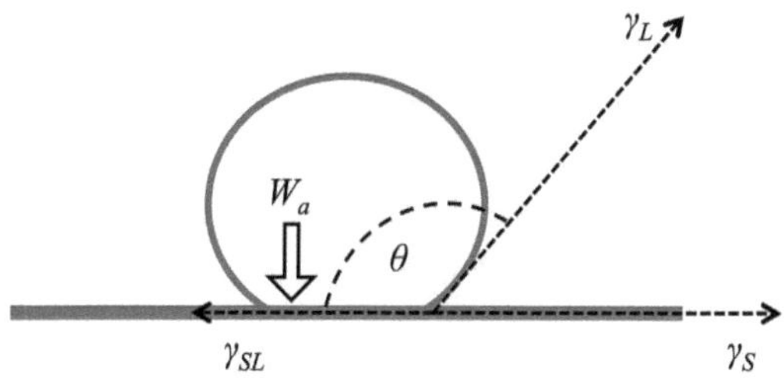

Fig. 12.1 Illustration of a liquid droplet resting on a solid surface, known as a sessile droplet. The contact angle (θ), the interfacial tension between liquid–vapor (γ_{LV}; approximated to γ_L), the interfacial tension between surface–vapor (γ_{SV}; approximated to γ_S), and the interfacial tension between surface–liquid (γ_{SL}) are represented

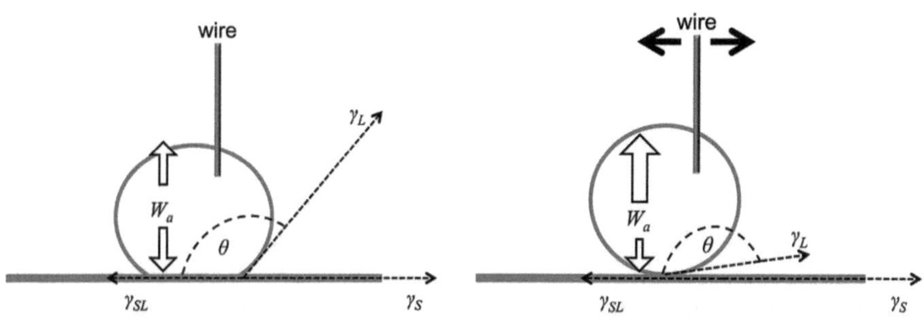

Fig. 12.2 Illustration of the work of adhesion of a liquid droplet to a surface and to a metal wire guide. In the case shown, the contact angle is sufficiently large so that the magnitude of the work of adhesion between the droplet and the wire is larger than the work of adhesion between the droplet and the surface. In this situation, the wire can be used to move the droplet across the surface

adhesion is inversely related to the contact angle: surfaces with higher contact angles have lower works of adhesion (Fig. 12.1).

$$W_a = \gamma_s + \gamma_L - \gamma_{SL} \tag{12.2}$$

$$W_a = \gamma_L(1 + \cos\theta) \tag{12.3}$$

In the case of WDM, the work of adhesion of a liquid droplet to a surface must be overcome by the work of adhesion of the liquid droplet to a wire (Fig. 12.2). Surfaces with the highest contact angles (i.e., superhydrophobic surfaces) are ideal for most applications of WDM since they minimize the work of adhesion of the liquid droplet to the surface. In one unique example of droplet manipulation demonstrated by Nicolini et al. [10], a surface of electrospun PCL fibers is used to achieve droplet immobility for assays that are impervious to external vibration. This surface has an apparent hydrophobicity ($\theta = 119°$) due to air pockets between the fibers, while maintaining a local hydrophilicity for droplet adhesion. Even though droplets bead up on this surface, the work of adhesion is sufficiently high to resist droplet movement.

In an extreme case, droplet contact with the surface is lost completely and the work of adhesion of the liquid droplet to the surface becomes zero. This situation,

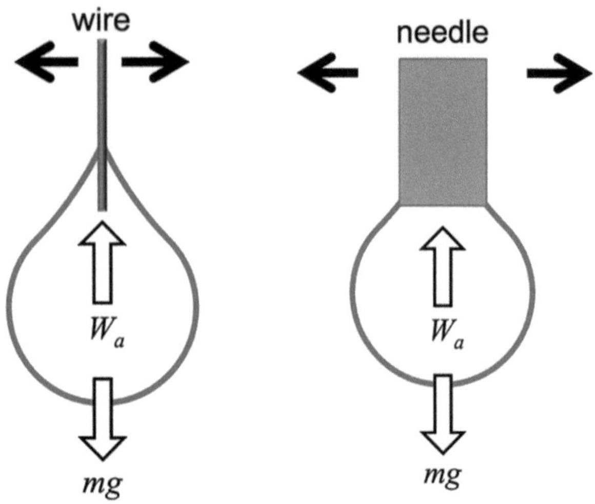

Fig. 12.3 Illustration of pendant droplets with the upward acting work of adhesion (W_a) of the droplet to a wire or to a needle and the weight of the droplet (mg) acting downward

illustrated in Fig. 12.3, is known as a pendant droplet and can be described by the Young-Laplace equation (Eq. 12.4). The Young-Laplace equation relates interfacial tension to droplet shape and shows that the pressure difference across the interface (Δp) is equal to balance of the effective weight of the droplet (ρgh) and the interfacial tension (γ) times the radii of curvature (R_1 and R_2). The radii of curvature can be determined and used to calculate the interfacial tension by computer-aided image processing from images of the pendant droplet from the side view [20].

For the case of a pendant droplet, the maximum size of a droplet that can be manipulated by WDM using a round tip is described by Tate's Law (Eq. 12.5) [21]. Tate's Law describes the maximum weight of a liquid pendant droplet (mg) with interfacial tension (γ) that will not fall from the tip with radius (r).

$$\Delta p = \rho gh - \gamma\left(\frac{1}{R_1} + \frac{1}{R_2}\right) \tag{12.4}$$

$$mg = 2\pi r\gamma \tag{12.5}$$

During the following discussion of applications, we will see the roles that these physical principles play in the practical uses and advantageous attributes of WDM.

12.3 Applications of Wire-guided Droplet Manipulation

12.3.1 Electrowetting

An early use of WDM and precursor to its invention is the use of electrowetting to study biomolecular adsorption to a surface such as a microfluidic chip [22]. In this application, sessile droplets of protein solutions rested on a hydrophobic dielectric surface and a voltage was applied across the surface and a platinum wire in contact with the top of the droplet. This electrical potential causes the contact angle of the droplet to change according to the Young-Lippmann equation (Eq. 12.6).

$$cos\,\theta_v = cos\,\theta_0 + \frac{1}{2}\frac{\varepsilon\varepsilon_0}{\gamma_{LV}d}V^2 \tag{12.6}$$

where θ_0 and θ_v are the contact angles before and after applying voltage, ε and ε_0 are the dielectric constants for the insulating layer of the electrode surface and the surrounding environment (air or oil), d is the thickness of insulating layer, and V is the applied voltage. Application of a potential can also cause biomolecules to adsorb to the surface through electrostatic attractions changing the contact angle over time and negatively affecting the performance of EWOD-based devices [22]. Additionally, it is known that proteins adsorb at water–oil interfaces [23], which can cause complications in EWOD-based devices, but we will see that WDM has been used to leverage this interfacial protein adsorption toward sample purification, PCR efficiency enhancement, and sequestration of PCR inhibitors.

12.3.2 *Particle Immunoassay*

It has been demonstrated that WDM could be used for repeated movement and merging of 5–20 μL droplets on a superhydrophobic surface [16]. This is possible because the work of adhesion of a 10 μL droplet to a superhydrophobic surface ($\theta = 155°$) is 14 nJ, which is overcome by the work of adhesion to a metal wire ($\theta = 10°$) of 57 nJ. In this format, the liquid droplets have minimal contact with solid surfaces, and biofouling is therefore minimized. A three-axis manipulator was programmed to move droplets, merge a droplet containing the target with a droplet containing antibody-conjugated particles, mix the merged droplet by mechanical vibration of the wire, and position the droplet at a detection site (Fig. 12.4). The particle immunoagglutination assay was conducted by WDM and achieved detection of bovine viral diarrhea virus (BVDV) with a detection limit of 2.5 $TCID_{50}$/ml ($TCID_{50} = 50\,\%$ of tissue culture infectious dose) and detection of *E. coli* with a detection limit of 85 CFU/ml.

12.3.3 *Protein Nanoarray*

In order to create a high-density protein array, nanowells were created by electron beam lithography (EBL) on a polymethyl methacrylate (PMMA)-coated surface and 100 nm protein-conjugated beads were deposited into the nanowells by vibrated-wire-guided droplet manipulation (Fig. 12.5) [24]. The beads were suspended in a 0.5 μL droplet and passed over the surface three times to achieve optimal array saturation. The wires were vibrated to provide necessary energy to the protein-conjugated beads to be deposited and patterned into the nanowells. The linear movement of a droplet over a nanoarray surface functioned as washing steps.

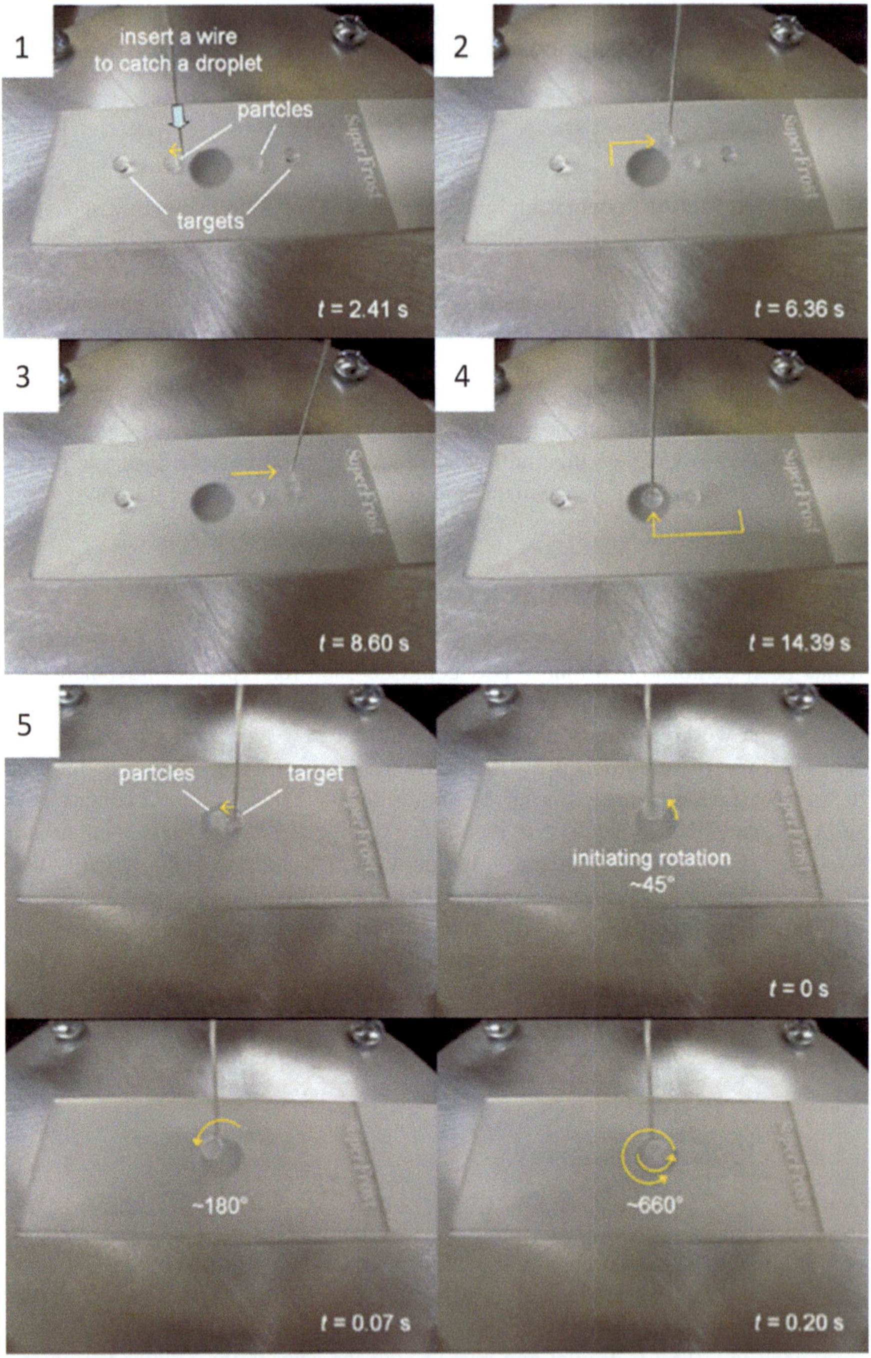

Fig. 12.4 Snapshots from preprogrammed wire-guided droplet manipulations including (**1**) catching a 10 μL droplet, (**2**) linearly moving a 10 μL droplet, (**3**) merging two 10 μL droplets, and (**4**) linearly moving a 20 μL droplet. (**5**) Rapidly mixing the merged droplet by mechanical vibration of the wire. (Images taken from [16], with permission, © Yoon and You)

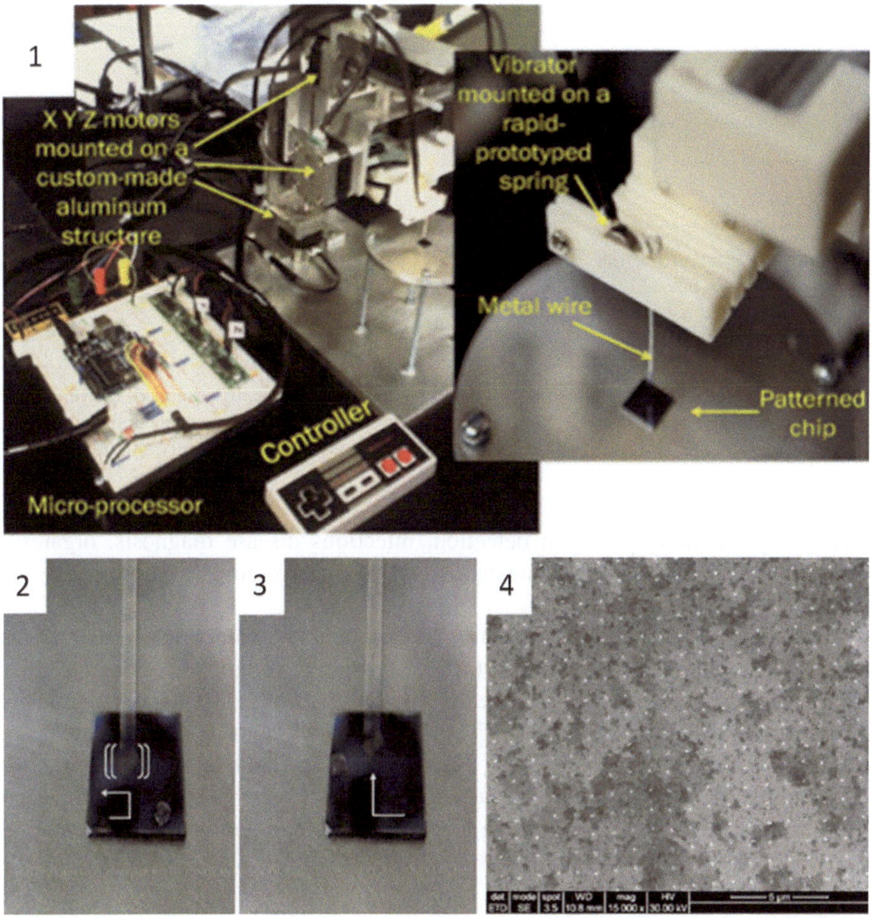

Fig. 12.5 (**1**) Three-axis manipulator with vibration motor mounted on the wire for droplet vibration. (**2**) Droplets containing a bead suspension on a polymethyl methacrylate (PMMA) substrate with a nanometer well array created by electron beam lithography (EBL). The droplets are vibrated to produce high-density arrays of 100 nm beads in nanowells. (**3**) The vibrated droplet is moved across the surface. (**4**) SEM image of the high-density nanowell array with 100 nm beads. (Images taken from Ref. [24], with permission, © Springer)

12.4 Polymerase Chain Reaction (PCR)

12.4.1 PCR Theory

The polymerase chain reaction (PCR) is a molecular biology technique to specifically amplify DNA fragments enzymatically [25]. PCR is a technology that is used ubiquitously across the life sciences and in medical diagnostics. Some of its

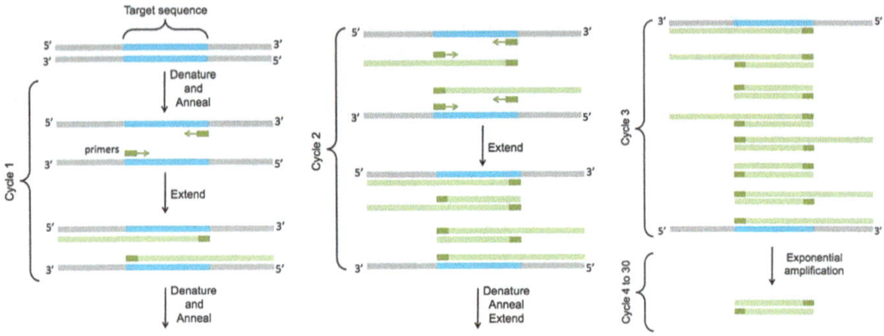

Fig. 12.6 Schematic illustration of the polymerase chain reaction (PCR). In cycle 1, the target sequence is denatured, primers anneal, and the sequence is enzymatically extended. In cycle 2, the process continues and the first fragments of known length are generated. In cycle 3, the amplicons begin to accumulate, and in subsequent cycles the amplicons are exponentially amplified

applications include pathogen detection, infectious disease diagnosis, organism genotyping, genetic mutation detection, DNA fingerprinting for forensics, and cloning.

PCR is a cyclical reaction with three phases: double-stranded DNA (dsDNA) is denatured to single-stranded DNA (ssDNA), primers (oligonucleotides typically 20–30 base pairs or bp in length) are annealed to their target sequences, and DNA polymerases extend the amplicon (Fig. 12.6). A forward and a reverse primer are designed to be complementary to the DNA template and flank the target sequence to be amplified. The extension products resulting from the hybridization of each primer are complementary, meaning that it is possible to double the concentration of the amplified product every reaction cycle.

12.4.2 Rapid Thermocycling

PCR thermocycling is the process of heating and cooling the reaction mixture: denaturation (90–98 °C), annealing (50–70 °C), and extension (70–80 °C) (Fig. 12.7). At temperatures above 90 °C, there is enough thermal energy to disrupt the hydrogen bonding interactions between conjugate base pairs of the two strands. While most proteins will also be denatured at temperatures exceeding 90 °C, the DNA polymerase used in modern PCR, *Taq* polymerase, is thermally stable [26]. *Taq* polymerase was isolated from a thermophilic bacterium, *Thermus aquaticus*, discovered in thermal springs [27]. The thermal stability of *Taq* polymerase enables thermal cycles to proceed continuously without the need to replenish the enzyme.

PCR is widely used and many of its applications are highly urgent, so it is highly desirable to decrease the reaction time. Thermocycling of the reaction mixture

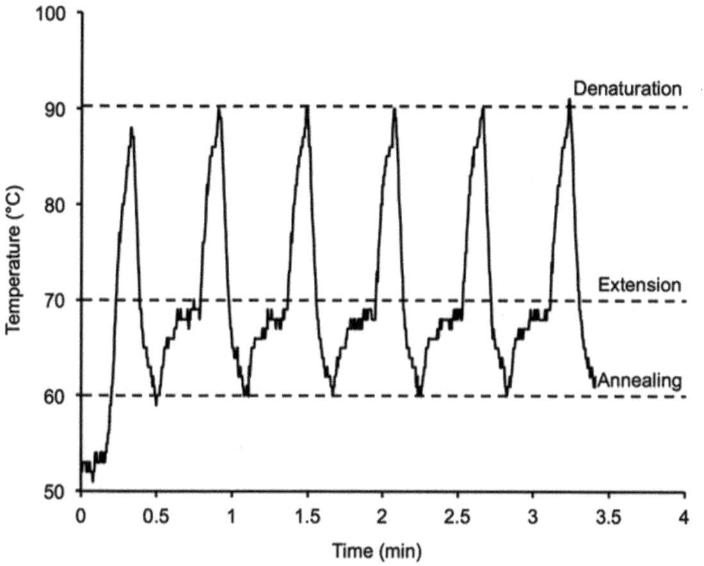

Fig. 12.7 Representative temperature profile for the first five cycles of a PCR. The phases of the reaction are denaturation at 90 °C, annealing at 60 °C, and extension at 72 °C

between three different temperatures is the most time-consuming process. Conventional PCR methods require two to three minutes per cycle, due to limitations in conductive heating and cooling. To increase the speed of DNA analysis by PCR, many researchers have focused on developing rapid thermocycling technologies to achieve thermal cycle times as low as 10 s [28–30]. There are several limiting factors in achieving such extreme thermocycling speeds. The primary obstacle has been the need for instrumentation capable of rapid heat transfer. This limitation can be eased somewhat by designing PCR primers with higher melting temperatures to allow for annealing and extension at the same temperature (65–75 °C) and decreasing the magnitude of temperature change required for thermocycling. DNA denaturation and primer annealing take place in less than 1 s [31]. Since the time required for annealing is partially limited by diffusion, minimum annealing times can be ensured by the use of high primer concentrations. The enzymatic elongation of the target sequence is limited by *Taq* polymerase activity, which has an extension rate of 60–120 bp per second at 72 °C [32]. With these limitations in mind, many different strategies have been employed to achieve rapid thermocycling.

Conventional thermocyclers use a thermoelectric heating element for active heating and cooling of a metal heater block that makes contact to plastic tubes containing 10–100 μL reactions. Heat transfer is made by conduction through the walls of the plastic tube. Thin walled tubes are important for minimize heat transfer resistance in conventional thermocycling. Many innovative strategies have been conceived for more efficient heat transfer and amplification at accelerated speeds.

In WDM, a 5–10 μL pendant droplet is submerged in heated oil and continuously moved so heat transfer occurs by convection [17, 18]. A computational fluid dynamics study showed that thermocycling in this manner can theoretically be achieved at 13 s/cycle and amplification was actually demonstrated at 13.7 s/cycle using an automated 3-axis computer numerical controlled (CNC) system for droplet positioning [17]. PCR amplification with WDM, using a circular robotic positioning system, was demonstrated at the quicker speed of 8.67 s/cycle [18].

12.4.3 Sample Preparation

It is important not to overlook the time requirement for sample preparation prior to PCR thermocycling. PCR can be inhibited in several ways including incomplete cell lysis for DNA extraction, interference with DNA capture, DNA degradation, and inhibition of polymerase activity [33]. Sources of PCR inhibitors come from sample matrices such as bodily fluids, food samples, wastewater samples, and culture media [33, 34]. Specifically, blood and tissue samples contain inhibitors such as cellular debris, bile salts, heparin, and proteins (especially hemoglobin, lactoferrin, and bilirubin) [35].

Typically, PCR workflows have obligatory DNA extraction procedures to purify DNA and remove any PCR inhibitors. Sample preparation, including DNA extraction, has been previously demonstrated using WDM (Fig. 12.8) [1]. Serial dilution, centrifugation, and DNA extraction were demonstrated for extracting genomic DNA from *E. coli* culture using WDM on a hydrophobic surface. A syringe needle was utilized to add/remove reagents and buffer solutions, rotate the droplet at high rpm (through vibrating the needle), and transport/merge droplets.

12.4.4 Inhibition Relief

Since DNA extraction procedures are not perfectly efficient, with carryover reagents inhibiting PCR [34], direct amplification from cell lysates has been attempted. This direct amplification may also benefit from the reduced time and effort for sample preparation. To this end, oil overlays are used, initially to prevent evaporation [36], but later used to enable this direct amplification from cell lysates. The observed increase in specificity and yield results from the action of oil overlays to segregate detergents used for cell lysis at the oil–water interface. This segregation minimizes their inhibitory effects and disruption of primer annealing specificity. While proteins can act as an inhibitor, the most abundant blood protein, albumin, has been shown to facilitate amplification in the presence of common inhibitors [37].

In different systems and at different concentrations, the same substance can be either an inhibitor or a facilitator [33]. A good example of this duality is

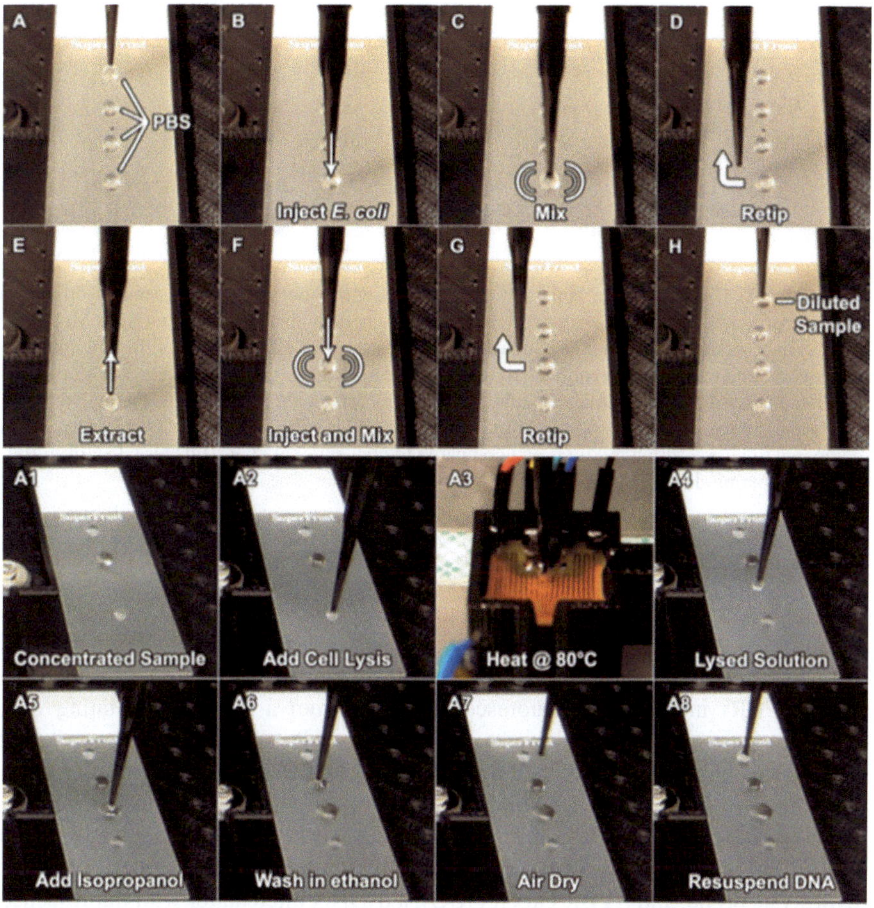

Fig. 12.8 (**a–h**) Images from a protocol for serial dilution of *E. coli* by WDM including droplet splitting, moving, and mixing. (**a1–a8**) Images from a DNA extraction protocol by WDM including cell lysis, heat incubation, washing, drying, and resuspension. (Images taken from Ref. [1], with permission, © You and Yoon)

biomolecular adsorption, which can pose a challenge to microfluidic PCR strategies and limit PCR efficiency [28]. In contrast, in WDM, biomolecular adsorption is used advantageously to separate contaminant molecules at the oil–water interface [18]. Simultaneously, the adsorption of contaminants to the oil–water interface competitively prevents adsorption of critical PCR components (Fig. 12.9). While many efforts have been made toward rapid thermocycling, rapid PCR workflows with minimal need for sample purification must also be developed to decrease real-world sample-to-answer times. This is achieved by WDM in which it has been demonstrated that PCR efficiency is actually enhanced by the presence of tissue contaminants [38].

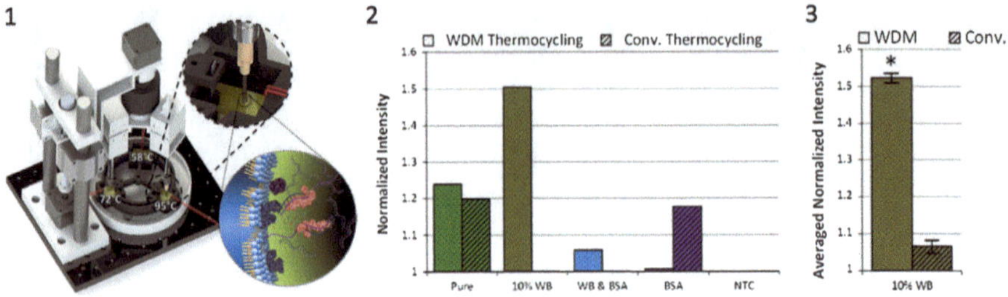

Fig. 12.9 (**1**) Circular WDM apparatus for rapid PCR thermocycling. The pendant droplet is moved continuously in the heated oil baths maintained at discrete temperatures. Molecular partitioning occurs at the oil–water interface by self-assembly. (**2**) PCR with different contaminants was conducted by conventional and WDM thermocycling, gel band intensities were quantified and normalized to compare the inhibitory effects between the two thermocycling methods. (**3**) In the presence of 10 % whole blood (WB), the WDM PCR efficiency is enhanced when compared to pure reaction mixtures and compared to conventional PCR. (Images taken from Ref. [18], with permission, © Elsevier)

12.4.5 Real-Time PCR (qPCR)

Quantifying the extent of PCR amplification had been limited to endpoint analysis [39], but in 1992, Higuchi et al. [40] described a method for real-time quantitative PCR (qPCR) in which the fluorescence signal from a DNA intercalating dye, ethidium bromide, was monitored each thermal cycle. Subsequently, many other chemistries have been developed and used for qPCR including nonspecific DNA labeling dyes [41], adjacent hybridization probes with resonance energy transfer [41], primer-based fluorophores [42], molecular beacons [43], and hydrolysis probes [44]. The most common qPCR chemistries are TaqMan probes, a type of hydrolysis probe, and SYBR Green I (SG), a dsDNA intercalator with fluorescence 1000 times greater when complexed with dsDNA [45] but not ssDNA.

An endpoint detection strategy has been demonstrated on a WDM PCR platform where the droplet is immersed in silicone or coconut oil and rests on a heated, hydrophobic surface [46]. The droplet is thermocycled by moving it over three different heaters, merged with a droplet of SG, and positioned for fluorescence detection by a smartphone-based fluorescence microscope (Fig. 12.10). This system has the added benefit of being shock and vibration resistant because the droplet is in contact with the surface and the thermocouple. The use of coconut oil makes this device particularly portable. These unique features mean this surface-heated WDM PCR platform can be used in the field or in a moving vehicle.

The exponential amplification of PCR, as monitored in real time, can be described mathematically [40, 47] by Eq. (12.7), where E is the reaction efficiency, C_n is the number of cycles, N_0 is the initial number of target sequences, and N_n is the number of amplicons after n cycles.

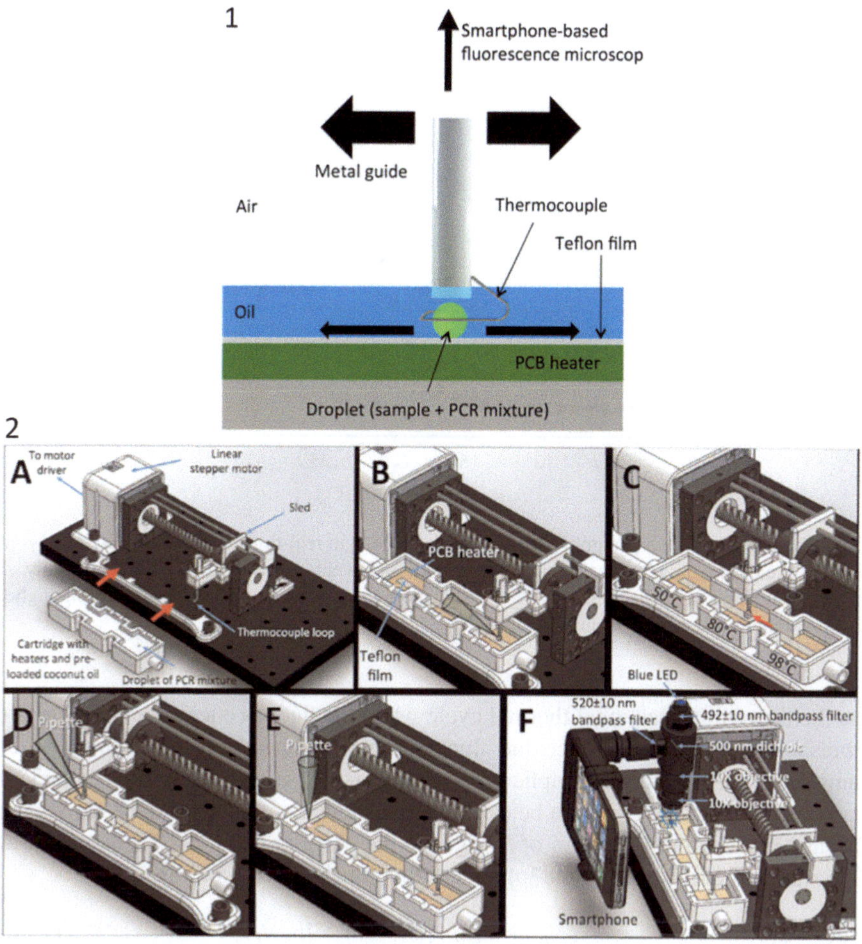

Fig. 12.10 (**1**) Schematic of the droplet resting on the surface and manipulated by the thermocouple wire-guide, which also provides droplet temperature feedback. (**2**) Schematic illustrating the operation of the surface-heated WDM PCR device for endpoint detection. A disposable cartridge is prefilled with oil and is loaded to the device platform. Heating commences, melting the coconut oil, and the preloaded PCR mixture becomes available to be merged with the target droplet. A thermocouple loop acts as the wire-guide during thermocycling across the different temperature regions. After thermocycling, the droplet is merged with a droplet of SYBR Green I (SG) and is positioned at the detection site and dislodged from the thermocouple loop with a pipette tip. Endpoint detection of the amplified DNA is achieved by a smartphone-based fluorescence microscope. (Images taken from Ref. [46], with permission, © Elsevier)

$$N_n = N_0(E + 1)^{C_n} \tag{12.7}$$

For fluorescence-based monitoring of PCR amplification, a threshold (F_t) is chosen to calculate the corresponding threshold cycle (C_t). The threshold cycle is defined

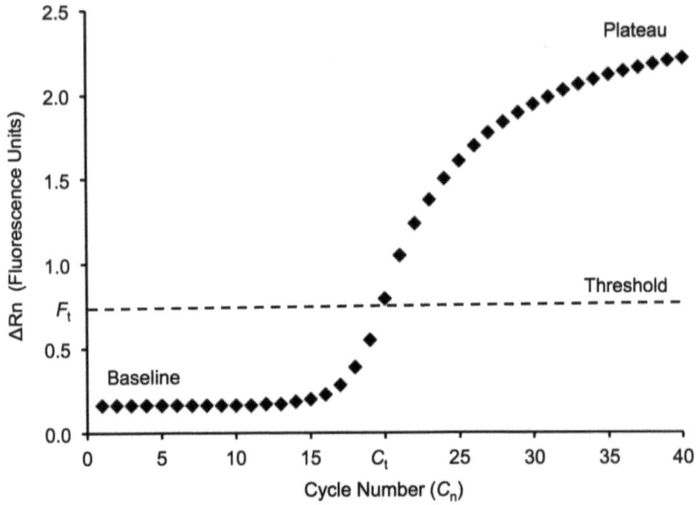

Fig. 12.11 Representative amplification curve monitored in real-time using SYBR Green I. From 0 to 15 cycles, the fluorescence signal is at baseline, and eventually begins to rise exponentially. A fluorescence threshold (F_t) is set such that a threshold cycle (C_t) can be calculated. During higher cycles, an amplification plateau is reached

as the theoretical fractional thermal cycle when the fluorescence signal rises to the threshold (Fig. 12.11). Since the fluorescence is dependent on the number of amplicons, the number of amplicons at C_t (N_t) is the same for any N_0. A standard curve for a given reaction can be established by calculating the C_t value for known N_0 and plotting the logarithm of N_0 against the corresponding C_t. The equation of the standard curve follows the log–linear relationship in Eq. (12.8).

$$logN_0 = -C_t log(E + 1) + logN_t \tag{12.8}$$

While qPCR detection modalities have been limited primarily to fluorescence, qPCR has been achieved by WDM by monitoring interfacial effects [38]. This methodology is termed *droplet-on-thermocouple silhouette real-time PCR* (DOTS qPCR), and it monitors the size of the droplet over the course of thermocycling. DOTS qPCR was invented based on the observation that PCR amplification in the presence of SYBR Green I (SG) by droplet-on-thermocouple thermocycling is accompanied by a change in droplet volume and a decrease in interfacial tension. It was also observed that water-in-oil emulsion droplets were forming in the oil phase (Fig. 12.12).

This WDM qPCR approach is sensitive to changes to the molecular composition of the droplet over the course of thermocycling, and detection at early PCR thermal cycles is achieved (Fig. 12.12). The droplet height is monitored by a smartphone camera every five thermal cycles, and the change in droplet height is used for amplification detection. Real-time quantification of bacterial genomic DNA in the

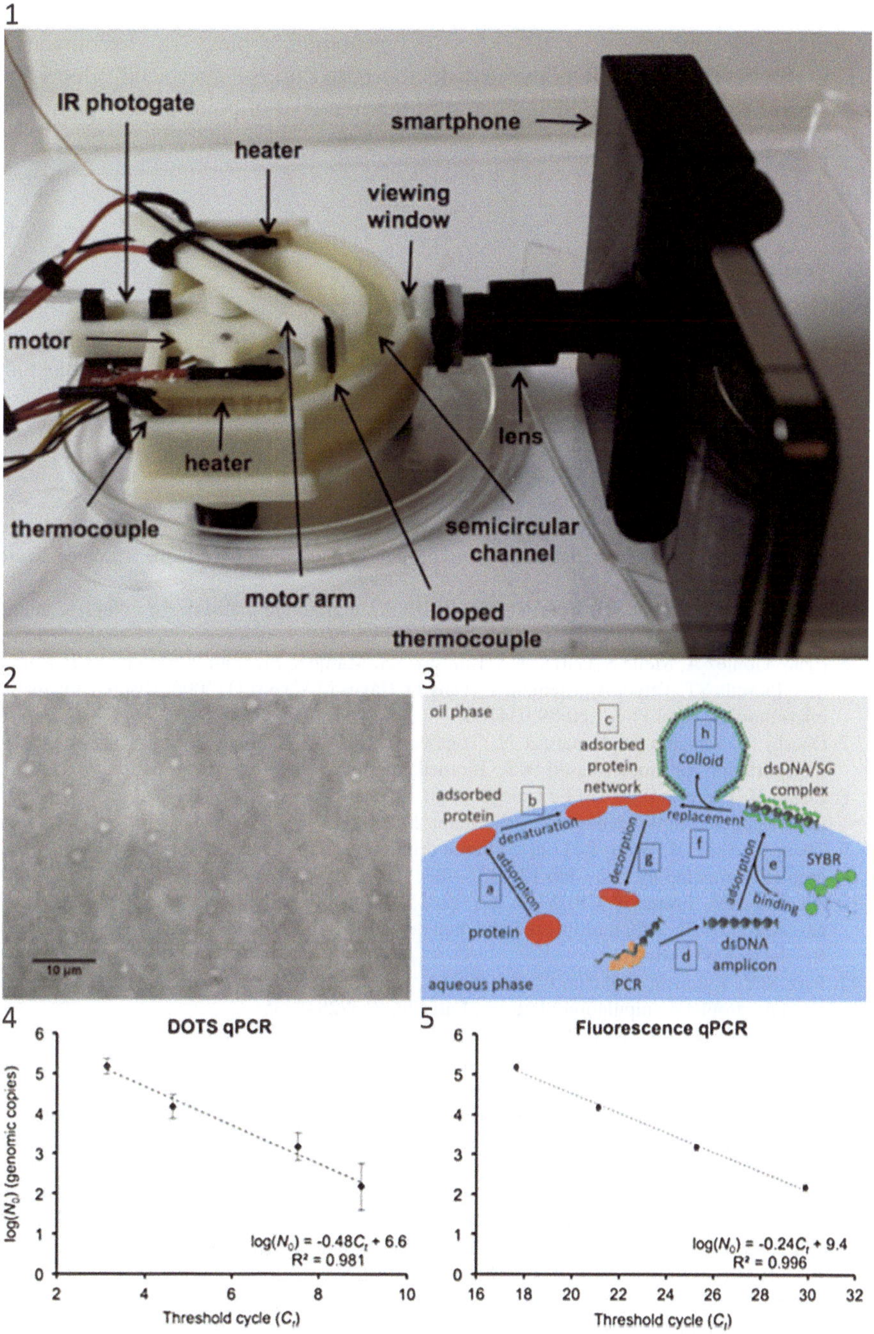

Fig. 12.12 (**1**) DOTS qPCR device for real-time PCR by WDM monitoring interfacial effects. (**2**) Water-in-oil emulsion droplets found in the oil phase after PCR amplification in the presence of SYBR Green I (SG). (**3**) Schematic depiction of inhibition relief and droplet volume decrease by colloid formation. (**4**) Standard curve for DOTS qPCR showing quantification is achieved during early thermocycles. (**5**) Standard curve for fluorescence PCR for comparison. (Images taken from Ref. [38], with permission, © Harshman et al.)

concentration range, 1.5×10^2–1.5×10^5 genomic copies, is demonstrated (Fig. 12.12). DOTS qPCR is the first demonstrated use of interfacial effects for sensing of PCR amplification and could further the goal of point-of-care pathogen detection.

References

1. You DJ, Yoon J-Y (2012) Droplet centrifugation, droplet DNA extraction, and rapid droplet thermocycling for simpler and faster PCR assay using wire-guided manipulations. J Biol Eng 6:15
2. Schneider T, Kreutz J, Chiu DT (2013) The potential impact of droplet microfluidics in biology. Anal Chem 85:3476–3482
3. Zec H, Shin DJ, Wang T-H (2014) Novel droplet platforms for the detection of disease biomarkers. Expert Rev Mol Diagn 14:787–801
4. Cho SK, Moon H, Kim C-J (2003) Creating, transporting, cutting, and merging liquid droplets by electrowetting-based actuation for digital microfluidic circuits. J Microelectromech Syst 12:70–80
5. Barbulovic-Nad I, Au SH, Wheeler AR (2010) A microfluidic platform for complete mammalian cell culture. Lab Chip 10:1536–1542
6. Egatz-Gómez A, Melle S, García AA, Lindsay SA, Márquez M, Domínguez-García P, Rubio MA, Picraux ST, Taraci JL, Clement T, Yang D, Hayes MA, Gust D (2006) Discrete magnetic microfluidics. Appl Phys Lett 89:034106
7. Ohashi T, Kuyama H, Hanafusa N, Togawa Y (2007) A simple device using magnetic transportation for droplet-based PCR. Biomed Microdevices 9:695–702
8. Zhang Y, Wang T-H (2013) Full-range magnetic manipulation of droplets via surface energy traps enables complex bioassays. Adv Mater 25:2903–2908
9. Zhang Y, Shin DJ, Wang T-H (2013) Serial dilution via surface energy trap-assisted magnetic droplet manipulation. Lab Chip 13:4827–4831
10. Nicolini AM, Fronczek CF, Yoon J-Y (2015) Droplet-based immunoassay on a "sticky" nanofibrous surface for multiplexed and dual detection of bacteria using smartphones. Biosens Bioelectron 67:560–569
11. Park S-Y, Kalim S, Callahan C, Teitell MA, Chiou EPY (2009) A light-induced dielectrophoretic droplet manipulation platform. Lab Chip 9:3228–3235
12. He M, Edgar JS, Jeffries GDM, Lorenz RM, Shelby JP, Chiu DT (2005) Selective encapsulation of single cells and subcellular organelles into picoliter- and femtoliter-volume droplets. Anal Chem 77:5214–5219
13. Huang CJ, Fang WF, Ke MS, Chou HYE, Yang JT (2014) A biocompatible open-surface droplet manipulation platform for detection of multi-nucleotide polymorphism. Lab Chip 14:2057–2062
14. Daniel S, Chaudhury MK (2005) Vibration-actuated drop motion on surfaces for batch microfluidic processes. Langmuir 4:4240–4248
15. Darhuber AA, Valentino JP, Davis JM, Troian SM, Wagner S (2003) Microfluidic actuation by modulation of surface stresses. Appl Phys Lett 82:657
16. Yoon J-Y, You DJ (2008) Backscattering particle immunoassays in wire-guide droplet manipulations. J Biol Eng 2:15
17. You DJ, Tran PL, Kwon H-J, Patel D, Yoon J-Y (2011) Very quick reverse transcription polymerase chain reaction for detecting 2009 H1N1 influenza A using wire-guide droplet manipulations. Faraday Discuss 149:159–170

18. Harshman DK, Reyes R, Park TS, You DJ, Song J-Y, Yoon J-Y (2014) Enhanced nucleic acid amplification with blood in situ by wire-guided droplet manipulation (WDM). Biosens Bioelectron 53:167–174
19. de Gennes PG (1985) Wetting: statics and dynamics. Rev Mod Phys 57:827–863
20. Song B, Springer J (1996) Determination of interfacial tension from the profile of a pendant drop using computer-aided image processing. J Colloid Interface Sci 184:64–76
21. Tate T (1864) On the magnitude of a drop of liquid formed under different circumstances. Philos Mag Ser 4(181):176–180
22. Yoon J-Y, Garrell RL (2003) Preventing biomolecular adsorption in electrowetting-based biofluidic chips. Anal Chem 75:5097–5102
23. Bartzoka V, Brook MA, McDermott MR (1998) Protein–silicone interactions: how compatible are the two species? Langmuir 14:1887–1891
24. Tran PL, Gamboa JR, You DJ, Yoon J-Y (2010) FRET detection of Octamer-4 on a protein nanoarray made by size-dependent self-assembly. Anal Bioanal Chem 398:759–768
25. Mullis K, Faloona F, Scharf S, Saiki R, Horn G, Erlich H (1986) Specific enzymatic amplification of DNA in vitro: the polymerase chain reaction. Cold Spring Harb Symp Quant Biol 51:263–273
26. Chien A, Edgar DB, Trela JM (1976) Deoxyribonucleic acid polymerase from the extreme thermophile Thermus aquaticus. J Bacteriol 127:1550–1557
27. Brock T, Freeze H (1969) Thermus aquaticus gen. n. and sp. n., a non-sporulating extreme thermophile. J Bacteriol 98:289–297
28. Zhang C, Xing D (2007) Miniaturized PCR chips for nucleic acid amplification and analysis: latest advances and future trends. Nucleic Acids Res 35:4223–4237
29. Roper MG, Easley CJ, Landers JP (2015) Advances in polymerase chain reaction on microfluidic chips. Anal Chem 77:3887–3893
30. Farrar JS, Wittwer CT (2015) Extreme PCR: efficient and specific DNA amplification in 15–60 seconds. Clin Chem 61:145–153
31. Wittwer CT, Chris G, Garling J (1990) Minimizing the time required for DNA amplification by efficient heat transfer to small samples. Anal Biochem 331:328–331
32. Innis MA, Myambo KB, Gelfand DH, Brow MAD (1988) DNA sequencing with Thermus aquaticus DNA polymerase and direct sequencing of polymerase chain reaction-amplified DNA. Proc Natl Acad Sci U S A 85:9436–9440
33. Wilson IG (1997) Inhibition and facilitation of nucleic acid amplification. Appl Environ Microbiol 63:3741–3751
34. Rossen L, Norskov P, Hoimstrom K, Rasmussen OF (1992) Inhibition of PCR by components of food samples, microbial diagnostic assays and DNA-extraction solutions. Int J Food Microbiol 17:37–45
35. Ra P, Al-soud WA, Jo LJ, Rådstro P (2001) Purification and characterization of PCR-inhibitory components in blood cells. J Clin Microbiol 39:485–493
36. Katzman M (1993) Use of oil overlays in "oil-free" PCR technology. Biotechniques 14:36–40
37. Kreader CA (1996) Relief of amplification inhibition in PCR with bovine serum albumin or T4 gene 32 protein. Appl Environ Microbiol 62:1102–1106
38. Harshman DK, Rao BM, McLain JE, Watts GS, Yoon J-Y (2015) Innovative qPCR using interfacial effects to enable low threshold cycle detection and inhibition. Sci Adv 1:e1400061
39. Becker A, Reith A, Kadenbach B (1996) A quantitative method of determining initial amounts of DNA by polymerase chain reaction cycle titration using digital imaging and a novel DNA stain. Anal Biochem 237:204–207
40. Higuchi R, Dollinger G, Walsh PS, Griffith R (1992) Simultaneous amplification and detection of specific DNA sequences. Nat Biotechnol 10:413–417
41. Wittwer CT, Herrmann MG, Moss AA, Rasmussen RP (1997) Continuous fluorescent monitoring of rapid cycle DNA amplification. Biotechniques 22:130–138

42. Gašparič MB, Tengs T, La Paz JL, Holst-Jensen A, Pla M, Esteve T, Žek J, Gruden K (2010) Comparison of nine different real-time PCR chemistries for qualitative and quantitative applications in GMO detection. Anal Bioanal Chem 396:2023–2029
43. Tyagi S, Kramer FR (1996) Molecular beacons: probes that fluoresce upon hybridization. Nat Biotechnol 14:303–308
44. Heid CA, Stevens J, Livak KJ, Williams PM (1996) Real time quantitative PCR. Genome Res 6:986–994
45. Dragan AI, Pavlovic R, McGivney JB, Casas-Finet JR, Bishop ES, Strouse RJ, Schenerman MA, Geddes CD (2012) SYBR Green I: fluorescence properties and interaction with DNA. J Fluoresc 22:1189–1199
46. Angus SV, Cho S, Harshman DK, Song J-Y, Yoon J-Y (2015) A portable, shock-proof, surface-heated droplet PCR system for Escherichia coli detection. Biosens Bioelectron 74:360–368
47. Rutledge RG, Côté C (2003) Mathematics of quantitative kinetic PCR and the application of standard curves. Nucleic Acids Res 31:e93

Chapter 13
Mechanical and Electrical Principles for Separation of Rare Cells

Elisa M. Wasson, Temple A. Douglas, and Rafael V. Davalos

Abstract Early detection of disease has long been a goal of many research projects and public health initiatives, as prevention from disease advancement is one of the most effective cures. Many diseases that are difficult to treat in late stage such as melanoma, pancreatic cancer, and highly metastatic tumors are treatable with a much better prognosis in early stage. One method of interest in early cancer detection has been the isolation of circulating tumor cells (CTCs) from the blood. By isolating CTCs, we can determine the presence of cancer from only a blood sample rather than requiring multiple screening modalities be performed on a patient. In addition, cancers without effective screening modalities, such as pancreatic cancer, may produce CTCs while the patient is asymptomatic, allowing for a much earlier start to treatment and improved prognosis [1]. In this chapter, we present several approaches that employ microfluidics to separate rare cells of interest. We start with a basic overview of fluid dynamics that includes scaling analysis and several solutions to the Navier-Stokes equations. We then introduce mechanical and electrical separation techniques along with an overview of their respective theories and several examples of each.

Keywords Inertial microfluidics • Dielectrophoresis • DEP • Circulating tumor cells (CTCs)

13.1 Introduction and Brief Description of Chapter

Early detection of disease has long been a goal of many research projects and public health initiatives, as prevention from disease advancement is one of the most effective cures. Many diseases that are difficult to treat in late stage such as

E.M. Wasson (✉) • T.A. Douglas • R.V. Davalos, Ph.D.
Bioelectromechanical Systems Lab, Department of Mechanical Engineering, Virginia
Tech—Wake Forest University, School of Biomedical Engineering and Sciences,
330 Kelly Hall (MC0298), Blacksburg, VA 24061, USA
e-mail: ewasson1@vt.edu; tadougla@vt.edu; davalos@vt.edu

© Springer International Publishing Switzerland 2016

C. Lu, S.S. Verbridge (eds.), *Microfluidic Methods for Molecular Biology*,
DOI 10.1007/978-3-319-30019-1_13

melanoma, pancreatic cancer, and highly metastatic tumors are treatable with a much better prognosis in early stage. One method of interest in early cancer detection has been the isolation of circulating tumor cells (CTCs) from the blood. By isolating CTCs, we can determine the presence of cancer from only a blood sample rather than requiring multiple screening modalities be performed on a patient. In addition, cancers without effective screening modalities such as pancreatic cancer may produce CTCs while the patient is asymptomatic, allowing for much earlier start to treatment and improved prognosis [1].

One of the key challenges in isolating CTCs is their rarity. The presence of CTCs in the blood is thought to be at concentration 1–10 cells in 10 mL of whole blood [2]. Without some method to specifically isolate these cells from a sample, their low concentrations make them almost impossible to detect. While many successful microfluidic devices use biomarker-specific analysis to detect the presence of cancer, the focus of this chapter will be microfluidic techniques that provide label-free isolation and enrichment with high cell viability, allowing them to be cultured for further analysis and potential targeted treatment.

This chapter focuses on the isolation of rare mammalian cells as opposed to viruses and bacteria. First, a brief overview of microfluidic theory will be given including scaling laws, several analytical solutions to the Navier-Stokes equations, dimensionless parameters used in microfluidics, as well as Stokes law and drag force. Next, particle dynamics will be discussed using a force balance approach for both passive and active methods. Mechanical (passive) and electrical (active) methods of separation will be discussed in depth, giving several examples and their limitations for each technique. Implications for the future of microfluidic technologies to separate rare cell populations are also discussed.

13.2 Basic Physics in Microfluidics

Microfluidic devices allow researchers to replicate conditions found inside the body in order to investigate mammalian cells in their natural environment. Devices typically have length scales and dimensions that are comparable to fluid flow systems in the body such as arteries (0.1–10 mm), alveoli in the lungs (200 μm), as well as sinusoids (30–40 μm) and lobules in the liver (500–1000 μm). Parameters such as geometry, shear stress, flow rate, oxygen concentration, and pulsatile flow can be precisely controlled and modified to replicate conditions found in vivo using an in vitro platform. These capabilities of microfluidic devices give researchers the power to investigate the cell's role in diseases such as cancer [3–5], malaria [6–8], and HIV [9–11]. Microfluidic devices can also take advantage of physical phenomena that occur at the microscale, lending to cell separation techniques that have the potential to lower the cost, time, resources, reagents, and lab infrastructure necessary to operate many of the current clinical separation technologies [12–16].

Therefore, when separating and analyzing mammalian cells using microfluidic devices, it is necessary to have a firm grasp of the physics and forces that dominate at the microscale.

13.2.1 The Continuum Hypothesis

Particle spacing and scale play an important role in determining how to model a fluidic system. Consideration should be given to what physics must be captured in order to scale the system accordingly. It is important to choose the relevant characteristic length scale that describes the phenomena of interest within the system. For example, if you are interested in particle dynamics, the diameter of the particle may serve as the best characteristic length. On the other hand, if you are interested in the fluid flow in a microfluidic device, the channel diameter should be chosen as the characteristic length. The physics of a system may have different influence at varying scales. If there is a nanofluidic channel with a characteristic length scale defined as the channel diameter, on the order of a few nanometers, the particles in the fluid experience forces differently than particles in a flowing river with a characteristic length scale defined as the width, in kilometers. Intermolecular forces, such as van der Waals attraction and Brownian motion, influence the flow in a way that can be considered negligible in the flowing river. This difference is due to the continuum assumption, which is defined using the dimensionless Knudsen number:

$$K_n = \frac{\lambda}{L} \tag{13.1}$$

For microfluidics, the Knudsen number is the ratio between the intermolecular distance of water (λ) and the characteristic length scale of the system (L) which is usually the diameter of the device channel. If $K_n < 0.01$, the continuum hypothesis is valid [17]. Water has an intermolecular distance of 0.3 nm [18], meaning that a water molecule must travel this distance before colliding with its neighbor. Therefore, if the characteristic length scale is on the same order as the intermolecular distance, like the nanofluidic channel, the fluid can no longer be considered continuous and there may be a point in the fluid that does not have a finite value defined for properties such as density and velocity. These physical properties are then considered discontinuous and the continuum assumption breaks down. On the contrary, if the characteristic length scale of the system is much larger than that of the particle spacing, it can be assumed that the fluid is continuous and all physical properties are averaged over many molecules. While the elements of fluid being analyzed are large enough to consider mean representations of molecular effects, they are also small enough to allow them to still be considered differential elements. This allows for spatial and temporal variations of fluid properties to be investigated using continuum mechanics. Microfluidic devices are typically tens to hundreds of

Table 13.1 Knudsen values for different flow regimes

$K_n < 0.01$	Continuum model, non-slip flow
$0.01 \leq K_n < 1$	Continuum model, slip-flow
$1 \leq K_n < 10$	Transition region
$K_n \geq 10$	Free-molecular region

microns in size resulting in $K_n \ll 0.01$ allowing continuum mechanics to be used. A nanofluidic device, on the other hand, with a diameter of 10 nm will have a Knudsen number of 0.3 which is greater than 0.01. This leads to slip-flow where the molecules of the fluid may collide with channel walls before they interact with each other, therefore inducing discontinuities in properties (i.e., density, velocity) at the fluid-solid surface.

Systems with a Knudsen number higher than one no longer satisfy the continuum assumption. At this scale, complex collisions occur and intermolecular forces between particles must be considered. In order to analyze a non-continuum system, deterministic or statistical molecular models must be used which require extensive computational resources [19–21]. Non-continuum models are complicated and difficult to solve whereas continuum is not only easier to compute, but well characterized and more predictable [17]. As summarized in Table 13.1, the Knudsen number, along with several other parameters discussed in a later section, can be a powerful tool in classifying the flow of the system and determining the correct modeling approach needed to solve the problem. Depending on the scale of the system, the relevant physics at each regime will vary, demanding a closer look at the effects scaling can have in microfluidics.

13.2.2 Scaling Laws

As demonstrated by the continuum hypothesis, taking a macroscopic approach to define a system can lead to a drastically different outcome than applying a microscopic approach to an identical system. This is apparent when administering drugs to patients of different sizes. A man who is 6 feet tall and weighs 200 lb is typically going to need a larger dose of the same drug than that of a small boy who weighs 70 lb. This is due to differences in their size and metabolism. The same is true for other physical phenomena. At the macroscale, volumetric forces such as gravity and inertia typically dominate the physics that occur on a day-to-day basis. If we were to apply these macroscopic principles to objects at the microscale, it will lead to detrimental errors in our calculations. In order to avoid this, it is necessary to develop a method to scale physical parameters and gain a more accurate understanding. This is known as scaling analysis and can be classified as either geometric or dynamic [22].

13.2.2.1 Geometric Scaling

Geometric scaling relates directly to the size of an object. Table 13.2 lists scaling laws for length, area, volume, weight, and mass. Despite the simplicity of these laws, they provide immense insight for dynamic systems involving rules that are more complex.

One example of geometric scaling is to compare a human to the Brazilian pygmy gecko. This 4 cm long reptile resides in the Amazon jungle, and in order to survive the large amount of rainfall, has adapted to its climate by developing the ability to float on water while standing. Its small stature and hydrophobic skin help to repel water and prevent its body from breaking the surface tension of puddles on the jungle floor [23]. If we approximate the small gecko with a cube having a side length of 4 cm, we can calculate both its surface area and volume to be 96 cm^2 and 64 cm^3, respectively. The ratio of its surface area to volume is found to be 1.5. This is obviously oversimplifying the analysis, but for demonstrating purposes shows how drastically size can affect the behavior of an object at different scales. If a human attempts to walk on water, the outcome is not as favorable as that of the gecko. This is due to the fact that an average male who is 170 cm tall has a surface area-to-volume ratio of 0.035. In order for a person to have a chance of walking on water, they would need to have a surface area 42 times larger than what it currently is. From Table 13.2, we can see that surface area is a second-order scaling law whereas volume is a third-order law. If the length of the man were to double, his surface area would quadruple and his volume would octuple. Therefore, for larger objects, volume will always increase faster than surface area and lead to a smaller ratio of the two. At the microscale, the surface area-to-volume ratio is large and hence surface forces and effects dominate over volumetric forces. Sometimes it is useful to create a visual representation of scaling parameters in order to gain an intuition for how small objects will behave. Figure 13.1 shows the surface area-to-volume ratio for a cell, the nematode *Caenorhabditis elegans* (*C. elegans*), the Brazilian pygmy gecko, and an average-sized man.

As described previously, the pygmy gecko has a surface area-to-volume ratio 42 times larger than the average-sized man, therefore making it possible for it to stand on water without sinking. The adult nematode, *C. elegans*, is on average 1 mm long and has a surface area-to-volume ratio of 60. This nematode is capable of swimming in highly viscous fluids due to its small size and the undulations it undertakes to propel itself forward. Finally, a single cell with an average diameter of 10 μm has a surface area-to-volume ratio of 6000! Clearly, a single cell experiences forces differently than an average-sized man at the macroscale. Scaling laws of these different forces are discussed in the next section to understand just how much of an influence macroscale forces have at the microscale.

Table 13.2 Geometric scaling laws

Name	Formula	Order
Length (L)	$L \sim l^1$	1
Area (A)	$A \sim l^2$	2
Volume/weight/mass (V)	$V \sim l^3$	3

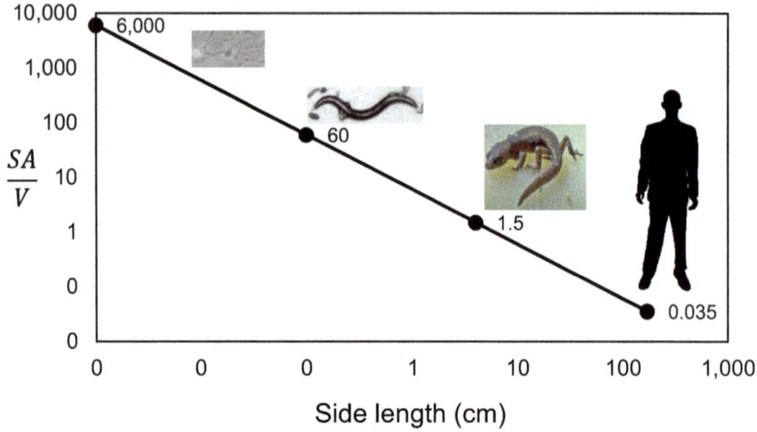

Fig. 13.1 Surface area-to-volume ratio for different organisms

13.2.2.2 Dynamic Scaling

In dynamic scaling, the laws of motion come into play and forces need to be scaled in order to understand their behavior at the microscale. Starting with Newton's second law, we can determine the scaling laws for both the acceleration of an object (a) and the time (t) it takes to get from point A to B in terms of force (F), mass (m), and length scale (l) [24]. Based on Newton's second law:

$$a = \frac{F}{m} \sim Fl^{-3} \tag{13.2}$$

Taking this a step further to determine the scaling law for the time (t) it takes an object to move from one point to another, we can denote displacement as l and use the following kinematic equation:

$$l = v_o t + \frac{1}{2} a t^2 \tag{13.3}$$

Assuming that initial velocity, v_o, is equal to zero, we are left with an equation for time:

$$t = \left(\frac{2l}{a}\right)^{0.5} \sim l^2\, F^{-0.5} \tag{13.4}$$

Depending on the force causing motion of the object, time and acceleration will scale differently. Table 13.3 summarizes the scaling laws for various forces and parameters that play a key role in microfluidics. For example, if we were interested in the effect a volumetric force, such as gravity, $\left(F_g = \rho V g\right)$ has on the time it takes

Table 13.3 Dynamic scaling laws

Name	Formula	Variables	Scaling	Order
Gravity/inertia	V	V—volumetric force	l^3	3
Surface tension	$F = 2\pi\sigma r$	σ—surface tension r—contact radius	l^1	1
Reynolds number	$Re = \frac{\rho VD}{\eta}$	ρ—density V—velocity D—diameter η—dynamic viscosity	l^2	2
Viscous forces	$F = A\eta \frac{du}{dx}$	A—area η—viscosity dx—distance from body u—velocity	l^{1+u}	$1 + u$
Magnetic force	$F = iL\mu_0 \left(\frac{n}{L}\right) i$	i—current L—length of wire n—# turns in wire μ_0—permeability	l^4	4
Kinetic energy	$F = \frac{1}{2} mv^2$	m—Mass v—velocity	l^3	3
Resistance	$R = \rho \frac{L}{A}$	ρ—density L—length A—area	l^{-1}	−1
Capacitance	$C = \varepsilon \frac{A}{x}$	A—area ε—permittivity x—distance between source and ground	l^1	1
Electric field	$E = \frac{V}{x}$	V—electric potential x—distance between source and ground	l^{-1}	−1
Electrostatic	$F = \frac{\varepsilon_o A (\Delta V)^2}{2d^2}$	A—area ΔV—potential d—distance between source and ground ε_o—permittivity of free space	l^2	2

an object to move from one place to another, force would scale as l^3 and time would scale as $l^{0.5}$. Most forces tend to scale as order two; therefore microscale objects tend to accelerate faster than macroscale objects [24]. As objects become smaller ($l \downarrow$), the time it takes for them to get from one point to another also decreases. For a force that scales as second order, the object will have a time scale of the first order. Therefore, if one object is ten times larger than another object, it will take ten times as long to travel from one point to another.

It is important to keep in mind that these scaling laws can be applied using the length scale of the particle being investigated, or as an alternative may be applied using the characteristic length of the channel itself. When using scaling analysis, it is necessary to be aware of the physics you are interested in capturing, what is being scaled, and then scale the system accordingly. As a result, the scaling laws found in Table 13.3 may vary slightly depending on what is being scaled.

13.2.3 Fluid Dynamics

13.2.3.1 Navier-Stokes Equations

Since most microfluidic devices have a Knudsen number less than 0.01, continuum mechanics may be used and the Navier-Stokes equations become the governing equations. The Navier-Stokes equations consist of three nonlinear, partial differential equations with five unknown parameters where the velocity vector is denoted by $\vec{u}$ and has xyz components, p is pressure, and ρ is density. The constants g and η are defined as gravity and dynamic viscosity, respectively. In order to solve for all of the parameters, we need a fifth equation to fully define the system. We can assume that the flow is incompressible and therefore define density, ρ, to be constant, making it spatially uniform. This fifth equation of state, along with conservation of momentum (continuity), now provides a complete description of fluid flow for incompressible, Newtonian fluids.

Mass conservation (continuity):

$$\nabla \cdot \vec{u} = 0 \tag{13.5}$$

Navier-Stokes:

$$\rho \frac{\partial \vec{u}}{\partial t} + \rho \vec{u} \cdot \nabla \vec{u} = -\nabla p + \eta \nabla^2 \vec{u} + \rho \vec{g} \tag{13.6}$$

The first term in the Navier-Stokes equations is the unsteady, local inertial term. The second term is the convective inertial term and is responsible for nonlinear behavior in the fluid flow. The right-hand side of Eq. (13.6) contains both body forces (gravity) and surface forces (shear stress and pressure) of the fluid. It is possible to solve the Navier-Stokes equations analytically for different conditions by making some basic assumptions about the flow or geometry of the system and applying the no-slip boundary condition.

No-slip boundary:

$$\vec{u} = 0 \tag{13.7}$$

This equation states that the velocity of the fluid at a motionless wall is zero. This holds true for most devices at the microscale because the length scale of the system is generally larger than the intermolecular spacing and continuum applies. In the following section, we discuss the solutions for different geometries that are common in microfluidic devices.

13.2.3.2 Analytical Solutions for Different Geometries

Most microfluidic channels have cross-sectional areas that are either circular or rectangular due to photolithography and etching techniques used to fabricate the devices. Solutions to the Navier-Stokes equations for these two geometries can be found analytically using integration or through series expansion methods. As seen in Table 13.4, a channel with a circular cross section of radius R is often approximated as a cylindrical tube with length L and its velocity equation is solved for analytically using cylindrical coordinates. The Navier-Stokes equations in cylindrical coordinates can be found in Munson, Young Okiishi, and Huebsch's "Fundamentals of Fluid Mechanics" [25]. Assuming steady, incompressible, laminar, fully developed, and axially symmetric flow and applying the no-slip and symmetry boundary conditions lead to the following equation that describes the velocity profile.

Velocity profile (cylinder):

$$u_z(r) = \frac{1}{4\eta}\frac{\partial p}{\partial z}\left(r^2 - R^2\right) \tag{13.8}$$

Here pressure is defined as p and η is the viscosity of the fluid. We can also define the pressure gradient and calculate flow rate by integrating the velocity over the differential area. Note that the pressure gradient is negative because pressure decreases in the direction of flow.

Pressure drop (cylinder):

$$-\frac{\partial p}{\partial z} = \frac{\Delta p}{L} \tag{13.9}$$

Volumetric flow rate (cylinder):

$$Q = \int_0^R u_z(r)\, dA = 2\pi \int_0^R u_z(r)\, r dr = \frac{\pi R^4 \Delta p}{8\eta L} \tag{13.10}$$

This results in the familiar Hagen-Poiseuille equation whose step-by-step derivation can be found in Munson's "Fundamental Fluid Mechanics" [25]. We can see that for a given pressure drop, flow rate scales as l^4 since the flow rate is proportional to the radius to the fourth power. As a channel gets smaller, the flow rate will decrease drastically, therefore making it difficult to drive fluids in microchannels using conventional pressure drop methods. High pressures are necessary to obtain high flow rates due to the fluidic resistance of the channels. Table 13.4 summarizes the resistance of different shaped cross-section microchannels which can be determined using an analogous form of Ohm's law. In this equation, voltage is represented by pressure drop (Δp), current is represented by flowrate (Q), and resistance (R) is defined as fluidic resistance. Techniques such as electrokinetic

Table 13.4 Flow conditions for various channel geometries

	Parallel plate	Cylindrical	Rectangular
Geometry	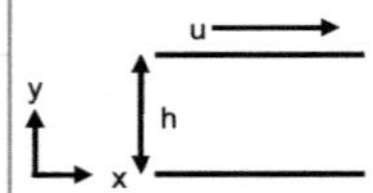	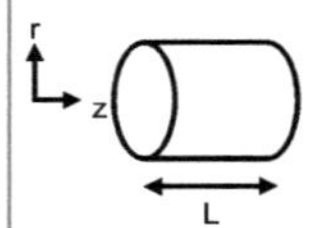	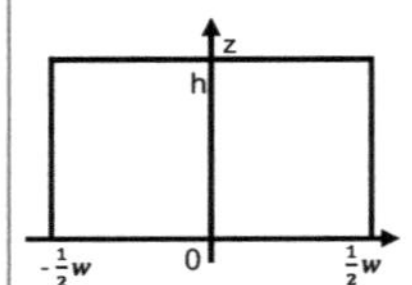
Velocity profile	$u_x(y) = \frac{h^2}{2\mu}\frac{\Delta p}{L}\left[\left(\frac{y}{h}\right) - \left(\frac{y}{h}\right)^2\right]$	$u_z(r) = \frac{1}{4\eta}\frac{\Delta p}{L}(R^2 - r^2)$	$u_x(y,z) = \frac{4h^2\Delta p}{\pi^3\eta L}\sum_{n,\mathrm{odd}}^{\infty}\frac{1}{n^3}\left[1 - \frac{\cosh\left(n\pi\frac{y}{h}\right)}{\cosh\left(n\pi\frac{w}{2h}\right)}\right]\sin\left(n\pi\frac{z}{h}\right)$
Flow rate	$Q = \frac{h^3\Delta p}{12\eta L}$	$Q = \frac{\pi R^4\Delta p}{8\eta L}$	$Q = \frac{h^3 w\Delta p}{12\eta L}\left[1 - 0.630\frac{h}{w}\right]$
Fluidic resistance	$R_h = \frac{12\eta L}{h^3}$	$R_h = \frac{8\eta L}{\pi R^4}$	$R_h = \frac{12\eta L}{(1 - 0.63)\left(\frac{h}{w}\right)h^3 w}$

flow where fluid is driven by the movement of molecules in an electric field either by electroosmosis and electrophoresis are commonly used in microfluidic devices for this reason [26].

Due to standard fabrication techniques, many microchannels are not circular but instead have a rectangular cross-sectional area with width w and height h. The solution for a rectangular channel is a bit more involved than a circular channel because it is a two-dimensional, elliptic partial differential equation. Therefore, integration cannot be used to solve directly for the velocity profile, but instead a Fourier series expansion is used. Assuming steady, laminar, fully developed, and incompressible flow leads to a reduced form of the Navier-Stokes equations. Performing a Fourier series expansion and applying the no-slip boundary conditions at each surface yield the equation for the velocity profile derived in full by Bruus' "Theoretical Microfluidics" [18].

Velocity profile (rectangular):

$$u_x(y,z) = \frac{4h^2\Delta p}{\pi^3\eta L}\sum_{n,\,\mathrm{odd}}^{\infty}\frac{1}{n^3}\left[1-\frac{\cosh\left(n\pi\frac{y}{h}\right)}{\cosh\left(n\pi\frac{w}{2h}\right)}\right]\sin\left(n\pi\frac{z}{h}\right) \tag{13.11}$$

In order to find an equation for the flow rate, we need to integrate the velocity equation over the area of the channel. This results in a complicated solution, but we are able to simplify this even further by assuming that our channel is going to be very wide and flat where the width is much larger than the height $\left(\frac{h}{w}\rightarrow 0\right)$ [18].

Flow rate (rectangular):

$$Q \approx \frac{h^3 w\Delta p}{12\eta L}\left[1-0.630\frac{h}{w}\right] \tag{13.12}$$

Flow in a microchannel can also be approximated as a 2D flow where we can model the channel walls as two infinitely long parallel plates with a height, h, separating them. In this instance, we have pressure-driven flow and consider the same assumptions. Flow is steady, incompressible, fully developed, and laminar, therefore reducing the Navier-Stokes equations significantly and leaving a balance between pressure and shear stress. Solving through integration and applying the no-slip boundary condition [25] at the top $(y=h)$ and bottom walls $(y=0)$, we can obtain the equation for the velocity profile, flow rate, and pressure drop in the channel.

Velocity profile (parallel plate):

$$u_x(y) = \frac{h^2}{2\mu}\frac{dp}{dx}\left[\left(\frac{y}{h}\right)^2-\left(\frac{y}{h}\right)\right] \tag{13.13}$$

Pressure drop (parallel plate):

$$-\frac{dp}{dx} = \frac{\Delta p}{L} \tag{13.14}$$

Flow rate (parallel plate):

$$Q = \frac{h^3 \Delta p}{12\eta L} \tag{13.15}$$

Table 13.4 summarizes the results found by solving the Navier-Stokes equations for a cylinder, rectangular, and parallel plate channel. The geometry of the system is defined as well as the velocity profile, flow rate, and fluidic resistance equations.

13.2.3.3 Nondimensionalization and Stokes Flow

In order to gain a better understanding of the Navier-Stokes equation and deduce which forces become important at the microscale, it is necessary to nondimensionalize the equation. As with any engineering system, nondimensionalization provides a method to gain a better intuition for what is happening in the system. If viscous forces are large, how large are they? In relation to what? These are all questions that can be answered by taking a nondimensional approach. To start our nondimensionalization, we need to pick characteristic values in order to form the correct dimensionless variables, which will be represented with an asterisk.

Here, we choose a length scale of l to nondimensionalize our spatial variables, a characteristic velocity, U, to nondimensionalize our velocity terms and a time that scales as velocity over distance $\left(t^* = \frac{tU}{l}\right)$. We choose to use viscous pressure $\left(p^* = \frac{pl}{\eta U}\right)$ instead of dynamic pressure $\left(p = \frac{P}{\rho U^2}\right)$ since, through our scale analysis, we have determined that inertial forces will become negligible at small scales in comparison to viscous forces which will dominate [27, 28]. For this reason, we also assume that body forces such as gravity are negligible. If we were to choose dynamic pressure and follow through with the nondimensionalization, the pressure term will be eliminated from the equation. Yet we know that pressure does in fact play a role at the microscale and this would result in an inaccurate solution. Substituting these dimensionless variables into the Navier-Stokes equations and simplifying gives the following equation:

$$Re_{\mathrm{c}}{*}\left(\frac{\partial \vec{u}^*}{\partial t^*} + \vec{u}^* \cdot \nabla^* \vec{u}^*\right) = -\ \nabla^* p^* + \nabla^{*2} \vec{u}^* \tag{13.16}$$

The dimensionless number that is present, Re_{c}, is the channel Reynolds number and is one of the most widely used parameters in the microfluidics field. The channel Reynolds number is the ratio of inertial forces to viscous forces and is defined as

$$Re_{\mathrm{c}} = \frac{\text{inertial forces}}{\text{viscous forces}} = \frac{\rho U l}{\eta} = \frac{\rho U D_{\mathrm{h}}}{\eta} \tag{13.17}$$

where l is the characteristic length and is usually defined as the hydraulic diameter of the channel which is dependent on area, A, and perimeter, P $\left(D_{\mathrm{h}} = \frac{4A}{P}\right)$, when analyzing fluid flow. Since microfluidic devices have relatively small diameters (10–100 μm) and low velocities (1 μm/s–1 cm/s), the Reynolds number is typically between 10^{-4} and 1 for most devices, making the majority of flow laminar and viscous dominated [29]. In this case, it is possible to simplify the governing equation even further. Taking the limit of Reynolds number as it approaches zero results in the following simplified momentum equations in three dimensions:

$$0 = -\nabla^{*} p^{*} + \nabla^{*2} \vec{u}^{*} \tag{13.18}$$

This is known as Stokes flow for low Reynolds number ($Re_{\mathrm{c}} \ll 1$). The governing equation has been simplified greatly, eliminating the convective inertial term and nonlinearity of the system. Typical boundary conditions are no-slip at the wall interface, but nonlinearities may still be introduced through boundary conditions and surface interactions found at small scales. If we were analyzing a nanofluidic channel for instance, the slip boundary condition may need to be taken into account.

13.2.3.4 Dimensionless Parameters

13.2.3.4.1 Particle Reynolds Number

While channel Reynolds number is important when analyzing flow conditions that occur in a microchannel, particle Reynolds number is a crucial dimensionless number when analyzing particle dynamics in the channel. In many of the passive and active separation techniques that will be discussed in the following sections, particle Reynolds number defines the way forces influence motion of particles:

$$Re_{\mathrm{p}} = Re_{\mathrm{c}} \frac{a^2}{{D_{\mathrm{h}}}^2} = \frac{\rho u_{\mathrm{f}} a^2}{\eta D_{\mathrm{h}}} \tag{13.19}$$

Here a is particle radius, ρ is density, u_{f} is fluid velocity, η is viscosity, and D_{h} is the hydraulic diameter of the channel. When $Re_{\mathrm{p}} \ll 1$, viscous forces dominate particle motion whereas when $Re_{\mathrm{p}} \gg 1$, inertial forces dominate [30, 31]. The influence of these inertial forces is discussed in Sect. 13.4.

13.2.3.4.2 Blockage Ratio

The size of the particle in relation to the channel dimensions will also play a role in flow/particle interactions [30]:

$$\kappa = \frac{a}{D_{\mathrm{h}}} \tag{13.20}$$

Here a is particle radius and D_h is hydraulic diameter of the channel. This will again have an influence on the migration of particles when both Reynolds numbers are greater than one and inertial forces dominate flow.

13.2.3.4.3 Dean Number

Some microfluidic devices have Reynolds numbers close to or greater than one that cannot be approximated using Stokes flow. This is a result of induced inertial effects caused by certain channel geometries and flow conditions. If a microfluidic device has a curved channel, where the radius of curvature (r) is larger than that of the channel radius (R), fluid with a higher velocity in the center of the channel will be carried to the outer wall due to an induced centrifugal force on the fluid. Fluid at the walls, which is usually stagnant, will now recirculate inward to the middle of the channel causing two symmetric vortices that rotate in opposite directions to form. This secondary flow phenomenon is known as Dean flow. The magnitude of Dean flow can be characterized by the dimensionless Dean number which was first developed by Dean in 1928 [32] and investigated further by Berger et al. in 1983 [33]:

$$De = Re_c\left(\frac{D_h}{2r}\right)^{1/2} \tag{13.21}$$

The Dean number depends on channel Reynolds number (Re_c), hydraulic diameter (D_h), and radius of curvature (r). As the Dean number increases, the vortices will shift their centers closer to the outside wall and a boundary layer will develop. While this secondary force can help to increase mixing at the microscale, it can also be tuned to separate heterogeneous populations of cells. Several examples of this application are discussed in Sect. 13.3.1.3.

13.2.3.4.4 Péclet Number

Since most microfluidic devices have a low Reynolds number and Stokes flow is applied, inertial forces are negligible and mixing is very difficult since diffusion becomes the main method of transport. Mixing in a microfluidic device can require very large time scales, which can be either desired or undesired depending on the application. In the case of devices used for rapid detection of viruses such as HIV, it is imperative that mixing of reagents and samples takes place in a timely manner. The gold standard in detection, known as an ELISA (enzyme-linked immunoassay), requires several mixing and washing steps. Therefore, in order for a microfluidic device to be successful in detecting the virus, it must be capable of replicating some of these steps. On the other hand, for devices whose main objective is to separate rare cells, it is imperative to control mixing. Keeping subpopulations of cells separate is crucial to the success of the devices. One way to determine the amount of mixing that will occur is to use the ratio of convection to diffusion [34], which is described using the dimensionless Péclet number (*Pe*):

$$Pe = \frac{\text{convection}}{\text{diffusion}} = \frac{Uw}{D} \tag{13.22}$$

where width (w), velocity (U), and the diffusion coefficient (D) comprise the equation. Since the *Pe* number depends on the diffusion coefficient of objects, this alone can be used to separate cells of different sizes and diffusion constants. According to Stokes-Einstein relation, the diffusion coefficient of a particle is inversely proportional to its size

$$D = \frac{kT}{6\pi\eta a} \sim \frac{1}{a} \tag{13.23}$$

where k is Boltzmann's constant, T is temperature, η is dynamic viscosity, and a is the radius of the particle. Therefore, since large objects typically have smaller diffusion coefficients and diffuse slower than small objects, this phenomenon can be used to separate cells of different sizes. In 1996, Brody et al. developed a simple H-filter capable of separating molecules such as solute ions ($a = 0.1$ nm) from viruses (100 nm) and mammalian cells (10 μm) [35]. A small particle such as a solute ion has a high *Pe* number characterized by a large diffusion coefficient and small diameter, whereas a mammalian cell has a large *Pe*. This results in the small particles diffusing to one side of the channel while the larger particles are restricted to the opposite side. The mean square distance (width) the particles will travel in the microchannel can be defined as

$$w = \sqrt{2Dt} \tag{13.24}$$

where w is width, D is the diffusion coefficient of the particle in the media, and t is the time it takes the particle to reach the other side. It is possible to determine the time it takes particles to diffuse a certain distance using this equation:

$$t \sim \frac{w^2}{D} \tag{13.25}$$

13.2.4 Particle Dynamics

Now that we have gone into detail about the fluid dynamics of the system, it is important to look at what is happening to the cells as they are flowing down the length of the channel. We will do so using a force balance analysis on a single cell in order to determine which forces dominate and how they play a role in separation. For simplification purposes, a cell is assumed to take the shape of a sphere. Realistically, many of the cells we are interested in separating, such as red blood cells (RBCs) and cancer cells, are not spherical in shape and have the ability to deform. In order to generalize the balance of forces for many different cases and to simplify calculations, a sphere will be sufficient [27, 28]. The cell will be considered much smaller than the channel it is traveling in as a way to reduce any local effects that may be induced in the overall flow. This is a realistic scenario since cells typically have diameters of 8–10 μm whereas channel width and height are usually 100 μm. First, we will derive Stokes drag and expand from there.

13.2.4.1 Stokes Drag

If we revisit Stokes laws to describe a particle of radius a moving in a fluid, we can solve for the radial and colatitudinal velocities, in spherical coordinates, analytically as demonstrated by Kirby's "Micro- and Nanoscale Fluid Mechanics" [27].

Radial velocity:

$$u_r = u_c \cos\theta\left(1 - \frac{3}{2}\frac{a}{r} + \frac{1}{2}\frac{a^3}{r^3}\right) \tag{13.26}$$

Colatitudinal velocity:

$$u_\theta = -u_c \sin\theta\left(1 - \frac{3a}{4r} - \frac{1}{4}\frac{a^3}{r^3}\right) \tag{13.27}$$

where u_c is the velocity of the cell and a is the radius of the cell. Following Kirby [27] we can use these equations to calculate the pressure and viscous shear stress associated with Stokes flow around a sphere.

Pressure:

$$\Delta p = -\eta u_c \frac{3}{2}\frac{a}{r^2}\cos\theta \tag{13.28}$$

Shear stress:

$$\tau = -\frac{3}{2}\frac{a}{r^2}\eta u_c \sin\theta \tag{13.29}$$

These become important when we are attempting to calculate the total force of the fluid on the sphere, which, in this case, is the drag force. In order to do so, we must add all of the surface stresses and integrate over the area of the region. Therefore, we add the normal stress (pressure) and tangential stress (shear stress) together and integrate over the area of the sphere. Doing so results in the Stokes drag equation where u_c is velocity of the cell, η is dynamic viscosity of the medium, and a is the radius of the cell:

$$F_{\text{drag}} = 6\pi\eta a u_c \tag{13.30}$$

If the cell happens to be rotating in the fluid due to the application of external forces, the drag due to the torque on the cell associated with its angular rotation, ω, is defined as

$$T = 8\pi\eta a^3\omega \tag{13.31}$$

13.2.4.2 Cell in Moving Fluid

Microfluidic devices used to separate rare cell types use continuous flow to their advantage, capitalizing on the balance of forces that occur and manipulating them until separation is achieved. Since we now know Stokes drag, we can use it to determine the force balance on a cell as it is flowing in a microfluidic device, whether the fluid is transporting it alone, as shown in Fig. 13.2, or under the influence of an external force. Since the velocity of the fluid is carrying the cell down the channel, the drag force the cell experiences will increase if cell acceleration increases due to the application of an external force. Mechanical and electrical forces such as inertial lift and dielectrophoresis (DEP) that cause this acceleration to occur are discussed in depth in Sect. 13.4.

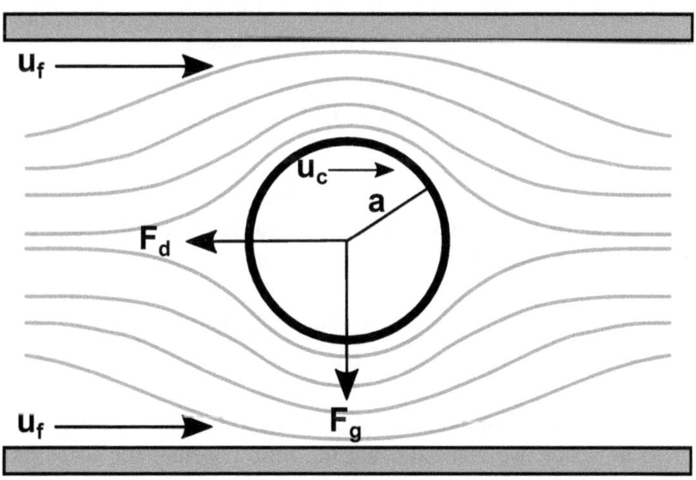

Fig. 13.2 Balance of forces on a cell in viscous flow

For now, we start with deriving the velocity of a cell under the influence of some arbitrary external force using Newton's second law as the foundation shown below:

$$F = m\frac{d\vec{u}_{\mathrm{c}}}{dt} \tag{13.32}$$

Knowing that the cell will experience some drag as it is being carried with the fluid, we need to define a suitable equation to describe the force. We can use the following relation where $\vec{u}_{\mathrm{c}}$ is the velocity of the cell and f is the friction factor which is dependent on the shape of the cell under investigation. Taking into account the fact that the cell will be moving at a different velocity $(\vec{u}_{\mathrm{c}})$ than the fluid $(\vec{u}_{\mathrm{f}})$, our drag force is defined in terms of the relative velocity of the particle:

$$F_{\mathrm{d}} = -f(\vec{u}_{\mathrm{c}} - \vec{u}_{\mathrm{f}}) \tag{13.33}$$

For the sake of simplification, we will assume that the cell is spherical and therefore using Stokes law, we are able to define the friction factor as

$$f = 6\pi\eta a \tag{13.34}$$

where again η is dynamic viscosity. Using these equations, we can expand Newton's second law:

$$F_{\mathrm{ext}} - f(\vec{u}_{\mathrm{c}} - \vec{u}_{\mathrm{f}}) = m\frac{d\vec{u}_{\mathrm{c}}}{dt} \tag{13.35}$$

Dividing by mass and rearranging the equation yields a linear, first-order differential equation that can be easily solved by multiplying by an appropriate integrating factor and integrating the equation. This gives us the velocity of the cell in terms of the external force, drag force, and velocity of the fluid:

$$\vec{u}_{\mathrm{c}} = \frac{F_{\mathrm{ext}}}{f} + \vec{u}_{\mathrm{f}}\left(1 - e^{-\frac{f}{m}t}\right) \tag{13.36}$$

With the equation above, we can see that the exponential term is due to the acceleration of the cell under the applied external force and the time constant for acceleration is defined as follows:

$$\tau = \frac{m}{f} = \frac{\rho\frac{4}{3}\pi a^3}{6\pi\eta a} = \frac{2\rho a^2}{9\eta} \tag{13.37}$$

For a typical cell, that has a diameter of 10 μm and the same density (ρ) and fluid viscosity (η) as water, we can calculate the time constant for acceleration to be 0.02 μs. Therefore, for times that are much larger than the characteristic time for acceleration (τ), we can take the limit of the velocity equation as $\tau \rightarrow 0$ and find that we are left with the terminal velocity of the cell, which is independent of time:

$$\vec{u}_c = \frac{F_{ext}}{f} + \vec{u}_f \tag{13.38}$$

This means that the cell is no longer accelerating and has reached a constant velocity. From Eq. (13.38), we see that this velocity is a combination of the fluid's velocity and the velocity caused by the external force. Therefore, since the characteristic time constant for acceleration of a cell with a diameter of 10 μm $\left(\tau \sim 10^{-8}\,\mathrm{s}\right)$ is usually much smaller than what can be observed experimentally $(t \sim 0.03\,\mathrm{s})$ [36], the cell can be assumed to have reached its terminal velocity, which may be approximated by Eq. (13.38) in any further modeling.

13.2.4.3 Gravity

The main external force that a cell experiences while suspended in a fluid is gravity. It is possible to perform a force balance on the cell using Stokes' law.

Figure 13.3 shows that gravity is pulling the cell down in the channel causing sedimentation while the drag force is acting upwards against gravity. The force due to gravity is described with the equation below:

$$F_g = V_c(\rho_c - \rho_f)g \tag{13.39}$$

where the volume of the cell is denoted as V_c, g is the gravitational constant, ρ_c is the density of the cell, and ρ_f is the density of the fluid. When gravity is balanced with drag force, the equation becomes

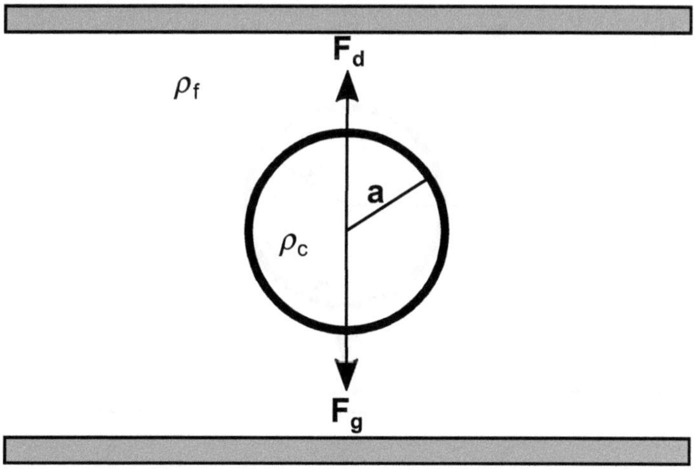

Fig. 13.3 Sedimentation of a cell in viscous flow

$$f^{*}u_{\mathrm{g}} = V_{\mathrm{c}}(\rho_{\mathrm{c}} - \rho_{\mathrm{f}})g \tag{13.40}$$

where u_{g} is the velocity of the cell due to gravity and f is Stokes drag on the cell. Substituting both the equation for volume of a sphere and Stokes drag (Eq. 13.30) results in the final equation for velocity of a cell in a gravitational field:

$$u_{\mathrm{g}} = \frac{2}{9}\frac{a^2(\rho_{\mathrm{c}} - \rho_{\mathrm{f}})g}{\eta} \tag{13.41}$$

13.2.4.4 Magnus Effect

In the analysis of the above situation, it was assumed that the cell in question was simply translating in the flow and not rotating due to viscous effects. This is an ideal case where Reynolds number is assumed zero and Stokes flow is applied, but this is not always what occurs in actual flow. In microfluidic devices used to separate rare cells, it is possible for these cells to experience rotation $Re \sim 1$ in the fluid flow as shown in Fig. 13.4 [37].

This may drastically affect flow streamlines around the cell and as a result affect the influence that drag force, pressure, and viscous shear stress have on cell motion. During rotation, the no-slip condition still applies and causes the fluid on the surface of the cell to move along with it, where instead of simply translating the fluid also begins to rotate [38, 39]. On the top side of the cell, the fluid follows the rotation of the cell because of the no-slip boundary condition, and begins to flow in the same direction that the cell is rotating. On the bottom side of the cell, fluid continues to flow downstream opposing the rotating flow at the top of the cell. This causes an

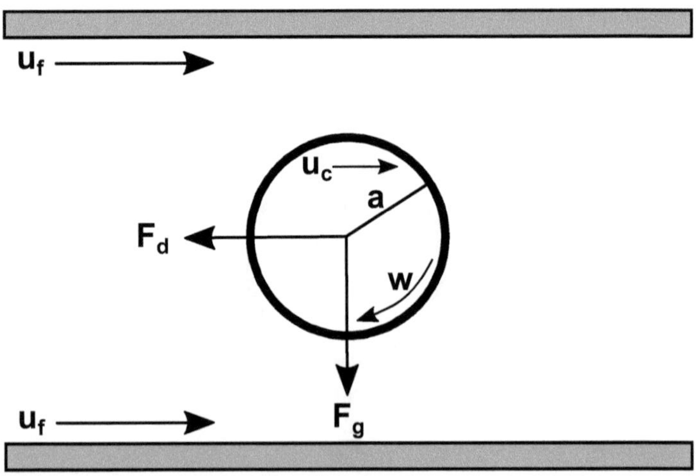

Fig. 13.4 Rotation of a cell in viscous flow

imbalance of pressures at the surface, therefore causing the flow of fluid on the bottom of the cell to stagnate and the flow on top to curve downward along with the cell. Due to Newton's third law, a force equal and opposite is created in a direction normal to the cell's axis of rotation, therefore creating a lift force that pushes the cell upward. This is known as the *Magnus effect* or *Magnus force* [40]:

$$F_{\text{mag}} = 4\pi a^3 \rho_{\text{f}} \left[\vec{\omega} \times (\vec{u}_{\text{c}} - \vec{u}_{\text{f}}) \right] \tag{13.42}$$

where ω is the angular velocity of the cell, ρ_{f} is the density of the fluid and other constants are defined in Sect. 13.2.4.2.

13.2.4.5 Buoyancy Due to Thermal Effects

When applying an electric field to drive flow or separate cells from a heterogeneous population using dielectrophoresis (DEP), thermal gradients may develop because of joule heating in the channel, causing the density of the fluid to change as well. This gradient in density results in a buoyant force on the cell causing lift:

$$F_{\text{b}} = \Delta\rho_{\text{f}} V_{\text{c}} g = \frac{\partial \rho_{\text{f}}}{\partial T} V_{\text{c}} \Delta T g \tag{13.43}$$

where T is temperature and g is the gravitational acceleration on the particle. Buoyancy can usually be neglected in many mechanical separation techniques due to the fact that heating is not generated in the device. In electrical separation techniques, buoyancy is also usually neglected since DEP force dominates the system, therefore minimizing the effect buoyancy has on the cell.

13.3 Mechanical Methods of Rare Cell Isolation

If we revisit the basic force balance on a cell found in Sect. 13.2, we can see that there may be an external force acting on the cell as described in Sect. 13.2.4.2. In terms of rare cell isolation, these forces may be passive or active. Passive forces are inherent in the fluid flow as a consequence of channel geometry or cell behavior, whereas active forces require the application of an external force such as an electric field. Here, we present several examples of passive separation techniques that involve mechanical forces acting on the particle. Specifically, inertia and secondary Dean flow will be discussed.

13.3.1 Inertial Forces

As discussed in Sect. 13.2.3.3, most devices have a Reynolds number much less than one, but in the case of a channel with a width of 100 μm and fluid velocity of 1 cm/s, it is possible to have a Reynolds number of one. If channel dimensions or fluid velocity happen to be larger, the Reynolds number can be greater than one, allowing inertial effects to enter the equation. Inertial effects were first documented and studied by Segré and Silberberg in 1961 [41]. In their investigation of laminar flow in a cylindrical pipe, they found that spherical particles initially in a uniform distribution, position themselves in an annulus ~0.6 times the radius of the pipe. This phenomenon was coined the *tubular pinch* effect and the points where particles migrated became known as the equilibrium positions. This was also studied and characterized for channels with rectangular cross sections and it was found that four equilibrium positions exist and are symmetric along the cross section of the channel [42]. In order to explain this phenomenon, several studies [30, 43–46] determined that this lateral migration of particles occurs due to inertial lift forces acting on the particles in conjunction with Stokes drag. These inertial forces are shown in Fig. 13.5. The wall repulsion force (F_{wL}) is caused by the fact that movement of the particle creates an axisymmetric vorticity wake distribution which is disrupted by the presence of the nearby wall [39]. This induces a velocity, which acts to push the particle away from the wall. The shear gradient lift force (F_{SL}), due to the parabolic velocity profile, acts to push particles away from the centerline.

These forces balance one another at certain locations in the channel, therefore establishing an equilibrium. Channel geometry, particle size, and particle Reynolds number can have a significant effect on how these inertial lift forces influence the motion of the particle. Initially, studies indicated that as long as the blockage

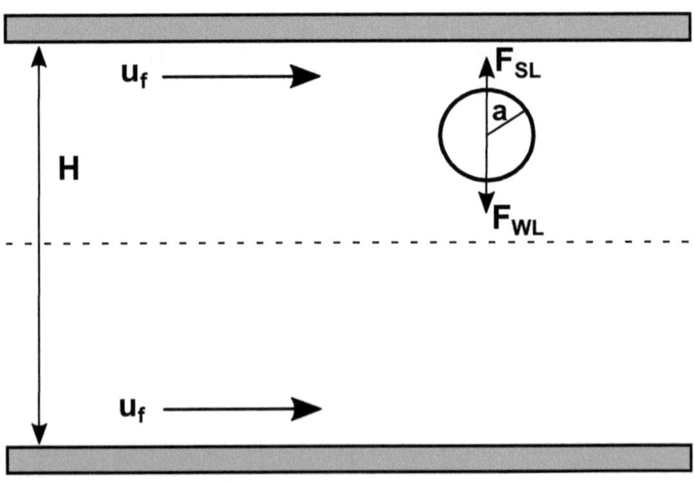

Fig. 13.5 Inertial lift forces on a cell

ratio (κ) and particle Reynolds number (Re_p) were much less than one, the particle will not influence flow conditions and lift force will be uniformly distributed along the channel [30]. The lift force can then be described by $F_L = \left(\rho U_f^2 a^4\right)/H^2$ where ρ is the density of the fluid, U is the velocity of the fluid, a is the radius of the particle, and H is the characteristic length of the channel. Di Carlo et al. [31, 47] further investigated the lift force in confined flows and determined that depending on the aspect ratio of the channel, particle position, and Reynolds number, the lift force can vary. If the blockage ratio approaches one, the lift force may act on the particle differently depending on its position in the channel. Near the wall, it can be described as the following:

$$F_{Lw} = \frac{f_c \rho {U_f}^2 a^6}{H^4} \tag{13.44}$$

whereas at the centerline, it is described by

$$F_{Ls} = \frac{f_c \rho {U_f}^2 a^3}{H} \tag{13.45}$$

where f_c is the dimensionless lift coefficient, which is dependent on distance from the channel wall and channel Reynolds number [48]. A higher channel Reynolds number, Re_c, was found to decrease lift force near the wall ($F_{Lw} \downarrow$) and increase lift force near the centerline ($F_{Ls} \uparrow$). Therefore, as Reynolds number increases, the particles in the channel are pushed toward the walls even more.

Since aspect ratio of the channel, particle size, and the Reynolds number of the channel can influence the lift forces that the particle experience, these parameters can be varied in order to isolate and capture cells of interest [31, 48, 49]. Several devices that apply inertial separation techniques are discussed in the following sections.

13.3.1.1 Straight Microchannels

As aspect ratio (AR) becomes larger $\left(\frac{h}{w} \sim 2\right)$, the number of equilibrium points along the channel cross section will decrease due to the fact that the shear gradient lift force is weakened. In a straight, rectangular cross section, particles will focus at the center of the longer, side channels. Many researchers have taken advantage of this phenomenon and used it to focus and subsequently separate cells of interest. Since inertial lift force scales with particle radius, larger particles will migrate faster than smaller particles to their respective equilibrium positions. Therefore, larger cells will separate from smaller cells and can be collected using bifurcating outlets. Bhagat et al. [50, 51] used this technique to design a microfluidic device capable of continuously filtering 1.9 μm polystyrene beads from a mixture containing 590 nm polystyrene beads [50, 51]. Hur et al. [52] also successfully filtered adrenal cortical progenitor cells from heterogeneous tissue samples using a similar design.

Recently, Wang et al. [53] focused particles of different sizes into a single stream and were able to perform on-chip flow cytometry at a throughput of 850 cells/s. The device consists of a low AR channel, which causes particles to migrate to equilibrium positions at the center of the top and bottom channel walls. This single channel is then bifurcated into two high AR channels with different hydrodynamic resistances (Fig. 13.6a). This difference in channel resistance causes the flow to separate asymmetrically in accordance with the resistance values and all particles begin to trace the inner wall of channel one. Due to the high AR, after a certain length, the particles no longer form two equilibrium positions, but migrate to a single focal position at the center of the channel. Using numerical and experimental techniques, the optimal resistance ratio (R_1/R_2) between the two bifurcating channels and channel Reynolds number ($Re_c = 40$) were found. A Reynolds number any lower than 40 would have resulted in a decrease of inertial lift forces and any higher would result in more equilibrium positions, therefore lowering efficiency of the device. Efficiency of separation with these optimized parameters was ~99 %. Initially, fluorescent beads 15 μm in diameter were tested, but the device proved to be effective on different sized beads and cells. Microbeads with a $\geq$ 15μm diameter obtained an efficiency >99 %, microbeads with a $\leq$10 μm diameter obtained an efficiency of 55 %, and mouse fibroblasts ($a \sim 17$ μm) obtained an efficiency near 100 %. Incorporating flow cytometry into the device using a laser to count cells downstream led to a relatively high throughput for both beads (2200 beads/s) and fibroblasts (850 cells/s) at a flow rate of 0.15 mL/min. Recovery rate was ~99 % although no post-separation analysis was performed to confirm this. Although throughput for this device is higher than many microfluidic separation techniques, it is still not comparable to gold standards such as fluorescence-activated cell sorting (FACS). Inertial forces can also cause cell lysis, potentially affecting the phenotype and gene expression of cells; yet this was not investigated. Despite the high separation efficiency and relatively high throughput, straight channels operate at a low flow rate and Reynolds number, therefore limiting the amount of cells they can process at one time. The devices also do not offer multiple stages of separation that may lead to a higher outlet purity than is achieved in a single channel. These limitations have been resolved by combining inertial focusing with microscale vortices, which is discussed in the next section.

13.3.1.2 Microscale Vortices

Designing symmetric side reservoirs along a straight, high AR channel induces the formation of vortices and sheath flow (Fig. 13.6b). Due to the sharp expansion of the channel width at each reservoir, the wall-induced lift force becomes small and shear gradient dominates, pushing particles into the vortex. The lift velocity, U_L, scales as the fluid velocity squared multiplied by the particle radius ($U_f^2 a$).

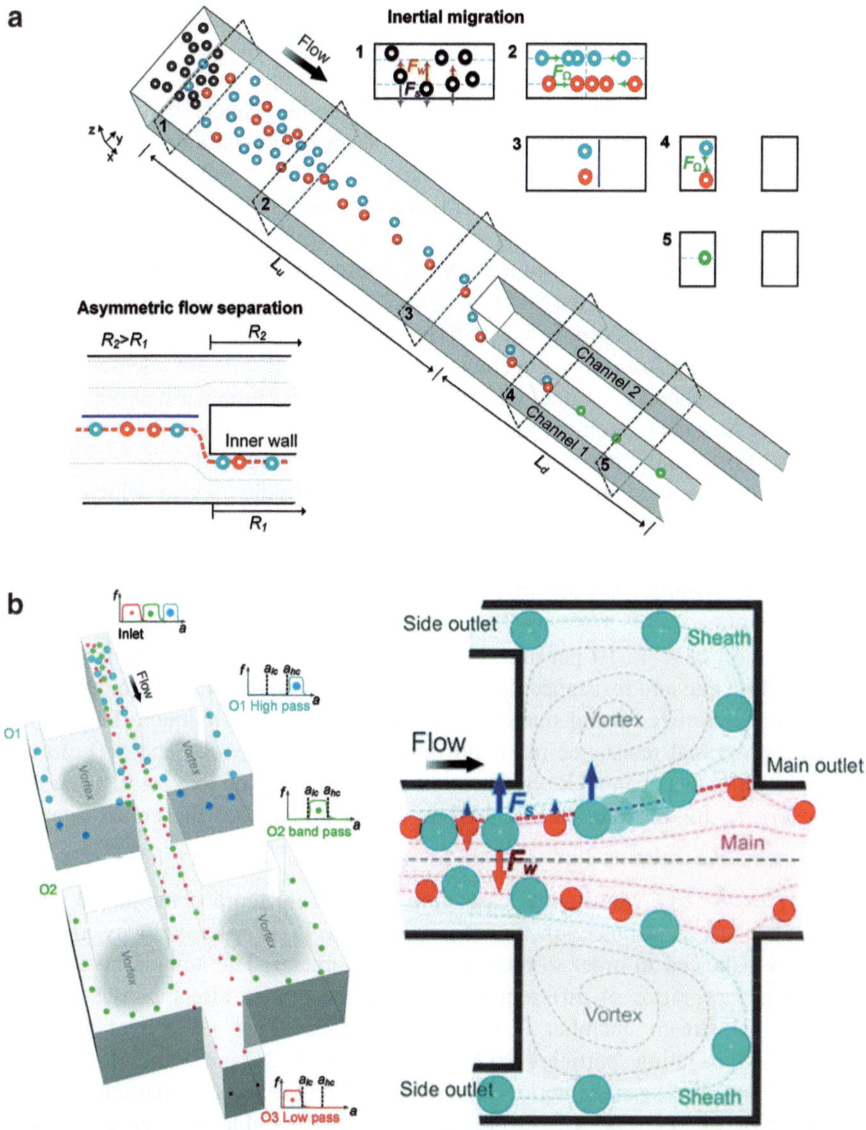

Fig. 13.6 Inertial microfluidic devices for cell separation. (**a**) A single low-aspect-ratio channel is bifurcated into two, high-aspect-ratio channels with different hydrodynamic resistances causing cells to become focused along the inner wall (adapted from Ref. 53 with permission of The Royal Society of Chemistry). (**b**) A continuous, multimodal separation device that uses microscale vortices to trap cells of different sizes (adapted from Ref. 59 with permission of The Royal Society of Chemistry)

Therefore, larger particles migrate across streamlines faster and quickly become entrained in the vortex and sheath flow, while small cells remain in the center channel flow. Several researchers have used this technique to separate polystyrene beads with diameters ranging from 6 to 20 μm [54], cancer cells from blood [55, 56], and pleural effusions [57]. Despite successful separation at a relatively high throughput, these devices had an average efficiency ~60 %, which is much lower than conventional techniques. This may be because the trapping mechanism is not continuous, therefore leading to a threshold number of cells and particles that are allowed to stay entrained in the vortex. If this threshold number is exceeded, cells may circulate out of the vortex and back into the main flow. Moreover, once the entire sample is processed, and the important cells captured, the extraneous cells need to be washed from the main channel before isolated cells are released. During this process, isolated cells may leave the vortex and become washed out with the main flow. As a result, continuous separation methods were developed to minimize this loss of efficiency experienced during the washing step [58].

Recently Wang et al. [59] developed a microfluidic device which uses a two-stage, continuous vortex design to enhance purity, resolution, and efficiency of separation using only one inlet. The device consists of a straight, high AR channel with symmetric reservoirs on each side of the channel (Fig. 13.6b). First, polymer microspheres were used as a proof of concept to test the device at a concentration of 4×10^4per mL. These microspheres ranged in size from 10 to 27 μm. Flow rate and resistance ratio between the side reservoirs and main channel (r/R) were optimized using numerical and experimental techniques. The authors found the optimal resistance ratio and flow rate to be $r/R = 5.4$ and 500 μL/min, corresponding to a channel Reynolds number of 110. This led to a cutoff separation diameter, a_c, for the particles of 14 μm. It was found that the cutoff diameter has a sensitive relationship to both resistance ratio $\left(a_c \sim (r/R)^2\right)$ and fluid velocity ($a_c \sim U_f$). Multimodal cascading is therefore more complicated than it seems because resistance of the channels as well as inlet velocity must be precisely balanced in order to optimize cutoff diameter. The authors investigated how resistance ratios of different branches of the device affect particle separation. The upstream chamber (O1) has a resistance ratio (R_1/R_c) of 4 that is capable of isolating particles with a diameter of 21 μm. The downstream chamber has a resistance ratio (R_3/R_4) of 4.9 and separated particles with a diameter of 11 μm. The second branch (O2) was then able to separate particles in between the two ranges at 18.5 and 15 μm. The two-stage design allows for higher separation efficiency. Not only does it prevent loss of cells during a washing step, but also further refines the heterogeneous sample that is injected into the device. Separation efficiency was found to be >90 % for particles with a size difference of ~3 μm and >70 % for particles with a size difference of 1 μm.

Many of the devices mentioned above that are used to separate cells require long channel lengths in order to see inertial forces have an effect on the cells in the fluid. This is because both designs depend heavily on the resistance ratio of two channels, which is mainly controlled by their lengths. Therefore, in order to increase throughput and efficiency, it is necessary to connect devices in series or parallel leading to increased channel length and longer processing times.

In order to shorten this length and focus cells more quickly, researchers began to curve their microchannels and capitalize on a secondary flow that resulted, known as Dean flow, which is discussed in the next section.

13.3.1.3 Curved Microchannels

As mentioned briefly in Sect. 13.2.3.4, Dean flow occurs when a microchannel is not straight, but instead has a curved geometry. Due to this curve in the geometry, the flow experiences centrifugal effects causing fluid at the midline of the channel cross section to travel outwards, while stagnant fluid at the top and bottom of the channel travels inward, satisfying conservation of mass [32, 48, 60]. As a result, counter-rotating vortices form in the fluid that act in conjunction with inertial lift forces, changing the equilibrium positions of particles. The two dimensionless numbers that characterize this secondary flow are the Dean number and curvature ratio. In Sect. 13.2.3.4 the Dean number is defined as

$$De = Re_c \left(\frac{D_h}{2r}\right)^{1/2} \tag{13.46}$$

where r is the radius of curvature, D_h is the hydraulic diameter and the channel Reynolds number (Re_c) and curvature ratio $\left(\delta = \frac{D_h}{2r}\right)$ affect how the secondary flow will influence particle dynamics. Since Dean flow affects inertial forces that the particles experience in the channel, it causes a drag force that is in the same direction of the secondary flow. The Dean drag force can be defined as

$$F_{De} \sim \frac{\rho U_f{}^2 a D_h{}^2}{2r} \tag{13.47}$$

where ρ is the density, u_f is the flow velocity, a is the particle radius, D_h is the hydraulic diameter, and r is the radius of curvature. Particle position in the channel is dependent upon the balance of inertial and Dean drag forces, which act in superposition. If Dean drag (F_{De}) is comparable to inertial lift (F_L), these forces have the ability to reduce the number of equilibrium position channel, therefore ordering particles in the channel. If F_{De} is much larger than F_L, particles will simply mix due to the dominant centrifugal motion. On the other hand, if F_{De} is less than F_L, only inertial forces will act on the particle. Therefore, in order to determine which forces will affect the fluid and influence particle dynamics, it is important to calculate the ratio of Dean drag and lift forces as defined by Matas et al. and later Di Carlo et al. [48, 61]:

$$\frac{F_{\mathrm{L}}}{F_{De}} \sim \frac{2ra^2}{{D_{\mathrm{h}}}^3} \tag{13.48}$$

Many research groups have capitalized on this phenomena using expansion/contraction designs [62–64] and spiral microchannels [65–70].

Reese et al. [71] recently demonstrated that combining straight and asymmetrically curved microchannels in series provides quicker focusing and higher efficiency. Single- and multi-stage device configurations were tested as a means to increase sample purity and efficiency. The authors tested several device configurations. The first to be tested was a straight, low AR channel with an expansion near the outlet (a), second was a high AR asymmetrically curved channel followed by a straight channel expansion (b), and third was a high AR straight channel followed by a channel expansion (c) (Fig. 13.7a). Polystyrene beads with a diameter of 9.9 μm were injected into the devices at a concentration of 2×10^6 beads per mL and velocity of 0.5 m/s $\left(Re_{\mathrm{p}} = 1.5\right)$. As expected, in devices (a) and (c), particles were focused at their equilibrium positions along the centers of the top and side walls, respectively. At the channel expansion, particles in device (a) became unfocused and occupied 20 % of the channel area. Device (b) performed better with particles occupying about 5 % of the channel and device (c) performed the best with particles only occupying 2 % of the channel area. This is because Dean flow focuses the particles into a single stream that traces the inside wall of the curvature. When particles exit into the expansion, the average fluid velocity decreases, in turn also decreasing interparticle spacing. The fact that the asymmetrically curved channel reduced the number of equilibrium positions to one therefore leads to better focusing in the expansion. Next, a four-stage device was tested to determine if combining more focusing techniques in series would lead to an increase in efficiency. As seen in Fig. 13.7b, a wide straight channel was followed by a wide asymmetrically curved section, a straight high AR section, and finally a channel expansion with three outlets for collection. Section 1 was used to focus particles into two vertical equilibrium positions centered along the channel width. Section 2 caused migration of the particles to one side of the microchannel and section 3 functioned to space the particles and laterally align them before entering the expansion. Outlets 1 and 2 served to collect the concentrated solution while the remaining outlet collected any of the remaining bulk sample. Solutions from outlets 1 and 2 were re-injected into the device and processed three more times. By pass number four, a 25-fold increase in the concentration of particles was reached and a total of $\sim 50 \times 10^6$ beads were collected.

While the device was successful in increasing concentration of particles with each pass through the four stages, separation of different sized particles was not investigated. This is a significant limitation in terms of rare cell isolation, but may serve beneficial in other sample-handling processes such as washing steps that require centrifugation for complete removal of reagents.

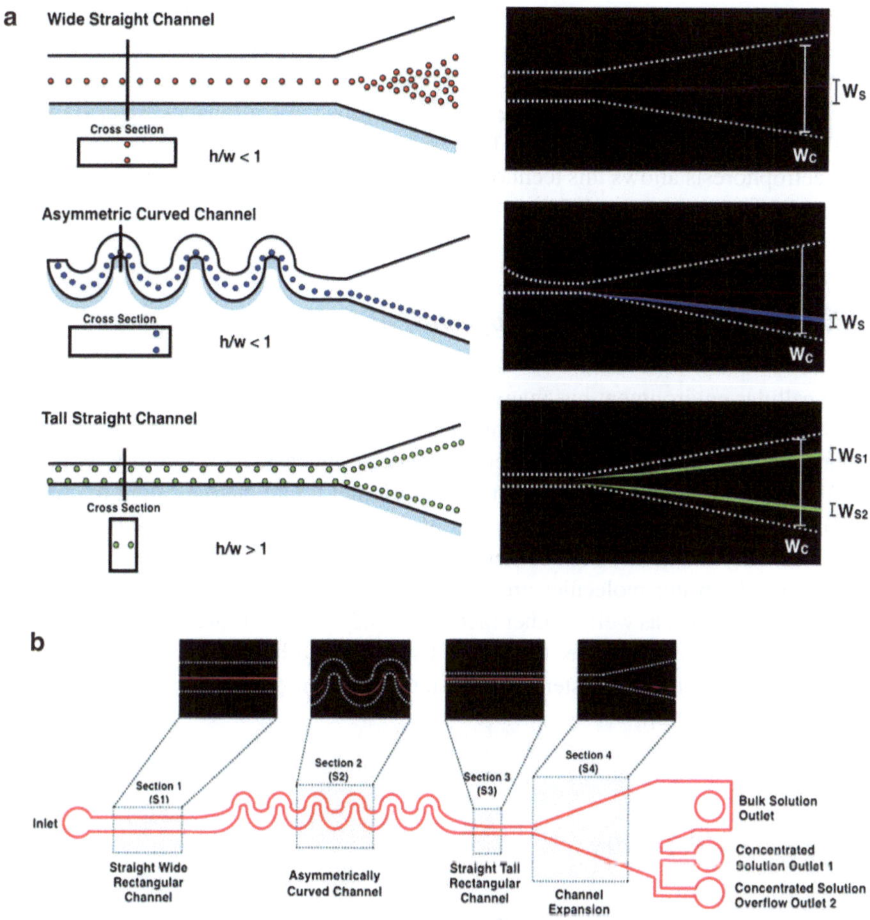

Fig. 13.7 Four-stage curved device for cell focusing. (**a**) Three channel designs used to optimize cell focusing (adapted from Ref. 71 with permission of The Royal Society of Chemistry). (**b**) Schematic of entire device showing the optimized design (adapted from Ref. 71 with permission of The Royal Society of Chemistry)

13.4 Electrical Methods of Rare Cell Isolation

Dielectrophoresis (DEP) is the motion of a particle due to its polarization in a nonuniform electric field. Using this technique, particles can be separated in solution. Different types of cells in particular, but also DNA, and proteins, have been separated via dielectrophoresis based on their intrinsic polarizability [72–75]. The application of microfluidic chips has been useful in, as they have been utilized to design systems with low Reynolds number regimes and high electric field gradients. The high electric field gradients induce a dipole in the cell, dependent on its properties, and can be used to manipulate the cell through a specific balance of fluidic and electric forces [76]. The DEP force that is exerted on a cell depends on certain properties of that cell in an electric field, and can permit users to sort cells and small particles by features such as malignancy, size, viability, cell type, and other factors [77–81]. The application of microfluidic devices for dielectrophoresis allows this technology to be easily and efficiently transferred into low-cost medical devices [80].

13.4.1 Electrochemical Properties of Cells

The cellular environment has many different properties that can affect the polarizability of a cell, thus leading to a unique electromechanical behavior that can act as that cell's signature. Cellular properties create an intrinsic polarizability of the cell [73]. When the cell is placed under an electric field, free charges align to create a dipole within the cell. Some properties of the cells that can influence this cellular polarizability are amino acid content, interaction between charged areas of amino acids and the water molecules around the cells, structure and rigidity of the lipid bilayer membrane, as well as other factors [73, 82–84]. For more information on the biophysics of cells, please see the review by Pethig and Kell, "The passive electrical properties of biological systems: their significance in physiology, biophysics and biotechnology" [82].

13.4.2 DEP Theory

DEP relies on an important property of cells—their intrinsic polarizability, which allows them to induce an electric dipole in the presence of an electric field. When this dipole is induced within the cell, a force of attraction or repulsion can form between the cell and another cell, or between the cell and objects in the microfluidic channel. These forces induced by dielectrophoresis have been derived in other publications, such as in the review of dielectrophoretic theory by Pethig [73]. The dielectrophoretic force is usually written as

$$\overrightarrow{F_{\mathrm{DEP}}} = 2\pi\varepsilon_{\mathrm{m}} r^3 Re[K(\omega)]\nabla\left(|\vec{E}|^2\right) \tag{13.49}$$

In this equation, ε_{m} is the permittivity of the medium, r is the radius of the particle of interest, and $K(\omega)$ is the Clausius-Mossotti factor as a function of ω, the angular frequency of the electric field. The root mean square of the electric field is represented by $\vec{E}$. The Clausius-Mossotti factor is a constant throughout the electric field, and is dependent on cell polarizability, medium polarizability, conductivity of the medium, and frequency of the electric field. The gradient of the electric field is

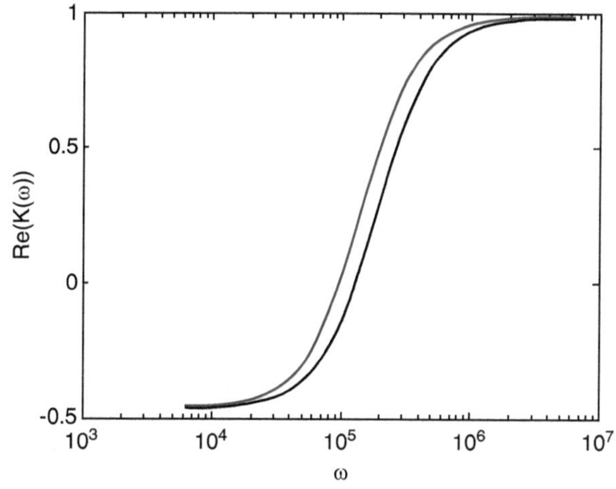

Fig. 13.8 Clausius-Mossotti factor plotted for different frequencies using the single-shell model [70]

spatially dependent, and can be determined computationally for complex geometries. The real part of the Clausius-Mossotti factor reduces to [73]

$$Re[K(\omega)] = Re\left[\frac{\epsilon_c^* - \epsilon_m^*}{\epsilon_c^* + 2\epsilon_m^*}\right] \tag{13.50}$$

In this equation, ϵ_c^* is the complex permittivity of the cell and ϵ_m^* is the complex permittivity of the medium. In both cases, the complex permittivity is described by $\epsilon^* = \epsilon + \frac{\sigma}{i\omega}$, with ω being the angular frequency of the applied electric field, and σ the conductivity of the medium [73].

Cells with different permittivities will have distinct values for Clausius-Mossotti factor, leading to a difference in the force each cell feels within the chip at a given frequency and local field gradient. This difference in forces is what permits separation. In a microfluidic chip with laminar flow, the trapping force can either act to attract or repel a cell or have very little influence on it. If the force on the cell is not great, the cell will not deviate much from its streamline and will continue to flow as if no electric field were applied. In the presence of a dielectric force, the cell may stick to a part of the device or deviate from its normal streamline.

Frequency-dependent curves for the Clausius-Mossotti factor can have a shape such as what is shown in Fig. 13.8. The differences between these curves allow for differences in DEP forces felt by cells and variation of the cell pathway in a device.

13.4.2.1 Negative vs. Positive Dielectrophoresis

Depending on the applied frequency and the electrical properties of the cell and the suspending medium, the DEP force can be either negative or positive. Negative dielectrophoresis (nDEP) is when the cell experiences a repelling force from regions of higher electric field gradient. The electrodes within the channel push away any cell with the appropriate polarizability given the frequency and voltage of current applied in the channel. Conversely, positive dielectrophoresis (pDEP) is a system in which the cells are attracted to regions with higher field gradients. While negative dielectrophoresis is good for redirecting flow of cells based on their polarizability, positive dielectrophoresis is good for cell trapping. However, cells can be trapped or redirected with either method, depending on the force balance between the drag on the particle and the dielectrophoresis forces.

In a proof of concept, researchers have designed negative dielectrophoresis traps to immobilize cells on a chip [85]. Another group reported using negative dielectrophoretic design to pattern liver cells on a chip, by repelling cells from certain areas of the chip and causing them to land in designated patterns [86]. Using a quadrupole device, another group was able to determine blood type by localizing the crossover frequency (where negative dielectrophoresis switches from positive dielectrophoresis) for a set of blood cells [87].

The magnitude of the dielectric force is a function of the medium permittivity, cell permittivity, magnitude of the applied electric field, and cell radius. The direction of the force exhibited on the cell, however, is a function only of the electric field in the chip and whether positive or negative DEP is being exerted, and thus can be determined independently of cellular properties, creating an effective field of possible cell-chip interactions. An effective magnitude of force can be developed by normalizing the cell-specific properties, such as the Clausius-Mossotti factor, to the cell-dependent properties. While the magnitude of the force depends on the properties of each individual cell type, the relative magnitude of force in one region vs. another is independent of cellular properties. This effective field of DEP forces is useful in determining chip design and predicting trapping regions. The effective field is

$$\vec{\Gamma} = \nabla\left(\vec{E} \cdot \vec{E}\right) \tag{13.51}$$

For a derivation of this field, please see the paper by Sano et al. *Multilayer contactless dielectrophoresis: theoretical considerations* [88].

13.4.2.2 Multi-Shell Model and Single-Shell Model for Measuring Cell Permittivity

For theoretical calculations of cell polarizability, the cell can be approximated as a set of concentric spherical shells as is shown in Fig. 13.9. These shells define the properties of each layer of the cell and the topology of the related regions. For a typical cell, one could estimate the outer membrane as one shell, the cytoplasm as another, and the nucleus/nuclear envelope as a third. By reducing shells of similar properties into effective shells, it is possible to condense the complex dielectric factor describing the set of concentric rings into its simplified form. Usually the cell

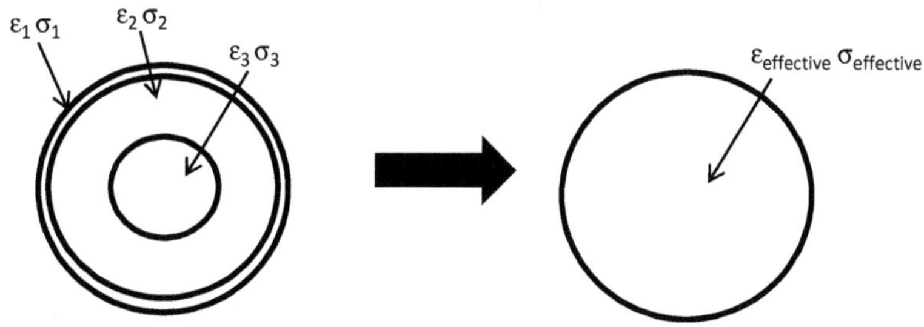

Fig. 13.9 Depiction of the multi-shell model

can be condensed to a single-shell model, taking an effective permittivity inside of the cell. For more information, see Pethig's review, *Dielectrophoresis: An assessment of its potential to aid the research and practice of drug discovery and delivery* [72].

13.4.2.3 Derivation of DEP Force

The DEP force is derived from the intrinsic polarizability of a cell, which can lead to an induced dipole when it flows through a chip. To derive this DEP force from the properties of a cell, we will first start with a dipole in an electric field as is shown in Fig. 13.10. For more detail and mathematical guidance, please see Chap. 2 of *Electromechanics of Particles* by Jones, Cambridge University Press [89].

The cell is not a perfect dipole, but the induced cellular dipole that forms in the presence of the electric field can be estimated as a dipole with a small distance between the two poles $\vec{d}$. Because the electric field is nonuniform in the chip, due to the presence of obstacles and other cells, we must consider the electric field at each point independently, rather than making an assumption about the form of the field. The force on a dipole is then

$$\vec{F} = q\vec{E}\left(\vec{r} + \vec{d}\right) - q\vec{E}(\vec{r}) \tag{13.52}$$

We will assume that we are measuring from a point $\vec{r}$, far away from the dipole. This allows us to say that the distance between the two points in the dipole, $\vec{d}$, is very small in comparison to the distance from which we are measuring. This assumption should be considered valid in a dielectrophoretic chip, as the internal cellular dipole is smaller than the relation between that dipole and other features of the chip. This assumption is made in order to simplify the nonuniform electric field. Otherwise, the electric field would need to be considered separately at $\vec{r}$ and $\vec{d} + \vec{r}$. We can make a Taylor expansion for $\vec{E}\left(\vec{r} + \vec{d}\right)$. This becomes

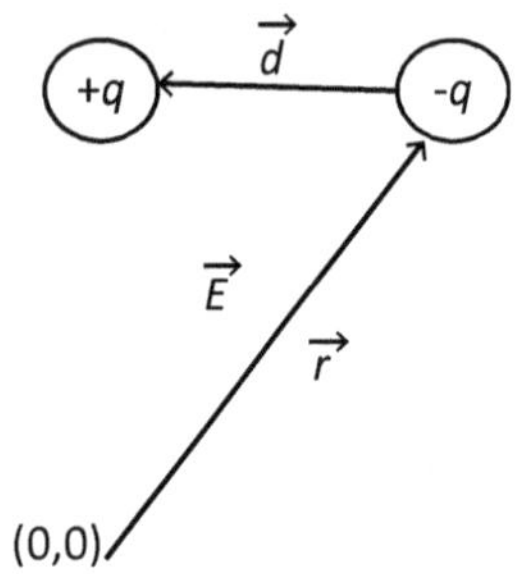

Fig. 13.10 Dipole in an electric field, analogous to a cell in an electric field

$$\vec{E}\left(\vec{r}+\vec{d}\right) = \vec{E}(\vec{r}) + \vec{d}\nabla\vec{E}(\vec{r}) + \cdots \tag{13.53}$$

Because we assumed that $\vec{d} \ll \vec{r}$, we can neglect higher terms than $\nabla E(r)$ as they will be very small. The force on the dipole then can be approximated as

$$\overrightarrow{F_{\text{DEP}}} \simeq q\,\vec{E}\left(\vec{r}+\vec{d}\right) - q\vec{E}(\vec{r}) = q\vec{d}\nabla\vec{E}(\vec{r}) \tag{13.54}$$

This is known as the dielectrophoretic approximation. By definition, the dipole moment, $\vec{p}$, is $\vec{p} = q\vec{d}$. Therefore,

$$\overrightarrow{F_{\text{DEP}}} = \vec{p}\nabla\vec{E}(\vec{r}) \tag{13.55}$$

gives the dielectrophoretic force in a chip [89].

13.4.2.4 Derivation of the Clausius-Mossotti Factor

This derivation also comes from *Electromechanics of Particles* by Jones [89]. Please refer to the book for further information. For a small dipole, the potential of the dipole is given by Φ:

$$\Phi = \frac{q}{4\pi\epsilon_1 r_+} - \frac{q}{4\pi\epsilon_1 r_-} = \frac{q}{4\pi\epsilon_1}\left(\frac{1}{r_+} - \frac{1}{r_-}\right) \tag{13.56}$$

This considers that the point charges are separated by a distance *2r*. In this case, the permittivity constant, ϵ_1, is the permittivity of the medium, as we are considering these point charges to be in the medium of the device. See *Electromechanics of Particles* for more details. This equation for electric potential can be rewritten as a Maclaurin expansion:

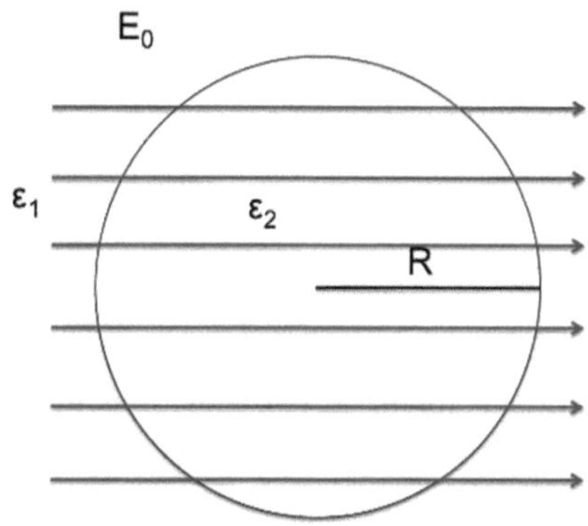

Fig. 13.11 Insulating sphere in uniform electric field

$$\Phi = \frac{qdP_1(\cos\theta)}{4\pi\epsilon_1 r^2} + \cdots \tag{13.57}$$

In this expansion, $P_1(\cos\theta)$ is the first-order Legendre polynomial, which is $P_1(\cos\theta) = \cos\theta$. Taking the first term of the Maclaurin expansion gives the first-order approximation for the potential between two point charges (in this case, the cellular dipole):

$$\Phi \simeq \frac{qd\ \cos\theta}{4\pi\epsilon_1 r^2} \tag{13.58}$$

As before, we know the dipole moment to be $p = qd$, which gives

$$\Phi \sim \frac{p\cos\theta}{4\pi\epsilon_1 r^2} \tag{13.59}$$

Again, here ϵ_1 would be the permittivity of the medium around the two point charges. We will keep this equation until a bit later. Now, we consider an insulating sphere of radius R in a uniform electric field as is shown in Fig. 13.11. Assuming a first order approximation, this can also be used to describe a lossless particle in a medium, as the cell is not particularly conductive. Because the cell is much smaller than the chip, we make the assumption that the electric field will be mostly uniform when passing through the cell. The permittivity of the medium is taken to be ϵ_1, and the permittivity of the cytoplasm is ϵ_2.

Using Gauss's law, we find the potential inside and outside the sphere. This gives

$$\Phi_1 = -E_0 r\ \cos\theta + \frac{A\cos\theta}{r^2} \text{ outside the sphere} \tag{13.60}$$

$$\Phi_2 = -Br\cos\theta \text{ inside the sphere} \tag{13.61}$$

The electric field is continuous at the boundary. The norm of the displacement flux is also continuous at the boundary. Therefore, we must consider the boundary conditions (at $r = R$) to be

$$\Phi_1 = \Phi_2 \tag{13.62}$$

$$-\epsilon_1 \frac{\partial \Phi_1}{\partial r}\bigg|_R = -\epsilon_2 \frac{\partial \Phi_2}{\partial r}\bigg|_R \tag{13.63}$$

Solving for these two equations given the boundary conditions leaves us with a system of equations that can be solved to get A and B. In this case, we find

$$A = R^3(E_0 - B) = R^3 E_0 \frac{\epsilon_2 - \epsilon_1}{\epsilon_2 + 2\epsilon_1} \tag{13.64}$$

$$B = \frac{3\epsilon_1 E_0}{\epsilon_2 + 2\epsilon_1} \tag{13.65}$$

If we plug these back into the equations for Φ_1, Φ_2, we get

$$\Phi_1 = -E_0 r \cos\theta + \frac{R^3}{r^2} E_0 \frac{\epsilon_2 - \epsilon_1}{\epsilon_2 + 2\epsilon_1} \cos\theta \text{ outside the sphere} \tag{13.66}$$

$$\Phi_2 = \frac{-3\epsilon_1}{\epsilon_2 + 2\epsilon_1} E_0 r \cos\theta \text{ inside the sphere} \tag{13.67}$$

Outside the sphere, we have Φ_1, but we also have the equation for two point charges that we derived earlier, $\Phi \simeq \frac{p\,\cos\theta}{4\pi\epsilon_1 r^2}$. We can set the $1/r^2$ terms equal to each other, as the linear term is referring to the uniform electric field rather than the induced dipole. Doing so, we have

$$\frac{p\cos\theta}{4\pi\epsilon_1 r^2} = \frac{R^3}{r^2} E_0 \frac{\epsilon_2 - \epsilon_1}{\epsilon_2 + 2\epsilon_1} \cos\theta \tag{13.68}$$

We know that for a homogeneous dielectric sphere, $p = 4\pi\epsilon_1 K R^3 E_0$ (see *Electromechanics of Particles*, Chap. 2 for details) [84]. Plugging this in gives

$$K = \frac{\epsilon_2 - \epsilon_1}{\epsilon_2 + 2\epsilon_1} \tag{13.69}$$

This K is the Clausius-Mossotti factor for a lossless dielectric sphere in a medium.

13.4.3 Design of Cell Separation Devices

There have been many studies to show the use of dielectrophoresis to separate cells in microfluidic channels. Remembering the DEP force, $\overrightarrow{F_{\text{DEP}}} = 2\pi\varepsilon_{\text{m}} r^3 Re[K(\omega)] \nabla \left(|\vec{E}|^2\right)$, we see that in order to have any force between the cell and an object in the channel it is necessary to have a gradient of the electric field. Cell separation can be done based on differences in cell radius and differences in Clausius-Mossotti factor between cells. The Clausius-Mossotti factor, $K(\omega) = \frac{\epsilon_c^* - \epsilon_m^*}{\epsilon_c^* + 2\epsilon_m^*}$, with $\epsilon^* = \epsilon + \sigma / i\omega$, is dependent on the permittivity of the medium, conductivity of the medium, and angular frequency of the electric field, which is the same for all cells in a device, and the permittivity of the cell, which can vary from one cell type to another. For this reason, factors such as the strength of the electric field, frequency of applied voltage, permittivity of the medium, and conductivity of the medium can all be modified to amplify the ratio of forces between two types of cells in suspension. However, as can be seen in Fig. 13.8, whether or not these two particles can be separated is dependent on the permittivity, conductivity, and radius of each cell type as well as the sensitivity of the device design.

Normally, a channel containing a design of electrodes and objects is fabricated, and cells are streamed through the channel at a constant rate. Objects in the channels create inhomogeneities in the electric field, which provide regions that can interact with the cells and elicit a force that will cause them to move according to their dielectric properties. Several of these types of dielectrophoretic mechanisms are discussed here, including contactless dielectrophoresis, insulator-based dielectrophoresis, and variations on these. Many different methods have been used to harness the polarizability of cells and other small molecules for separation, and this list is a small set of what has been done in this field.

13.4.3.1 Classical vs. Insulator-based Dielectrophoresis

Another method of defining dielectrophoresis is by contactless dielectrophoresis (cDEP). Classical dielectrophoresis requires an interdigitated electrode array to be in direct contact with the medium where the cells are, in order to affect the electric field and create a gradient/inhomogeneity for the dielectrophoresis force to be felt. In contrast, insulator-based dielectrophoresis (iDEP) uses insulating structures within the chip to create inhomogeneities in the electric field needed to drive DEP. iDEP has been widely used in a number of applications for cell separation. In addition, the technique has been employed as a method of trapping protein as well as DNA [90], as well as a method of separating membrane protein nanocrystals from solution [79].

13.4.3.2 Contactless Dielectrophoresis

As opposed to classical DEP or iDEP, contactless dielectrophoresis (cDEP) uses an insulating membrane to separate the cells/sample from the electrodes and deleterious electrochemical effects. Similar to iDEP, it utilizes inert insulating structures within the chip to create inhomogeneities in the electric field. This method improves cell viability by preventing direct contact with high-voltage sources [77, 88].

13.5 Conclusion

This chapter outlined the theory and practical implementation of methods of rare cell isolation using microfluidic and bioelectrical methodologies. The ability to capture and isolate rare cells is an important step in the process of being able to diagnose and treat cancer based on the presence of circulating tumor cells and other rare cells of interest to provide early and personalized diagnosis. The development of microfluidics and bioelectrical mechanics in recent years has provided a novel toolbox that can be utilized to improve our ability to obtain study and utilize these cells to improve patient outcomes.

References

1. Tjensvoll K, Nordgård O, Smaaland R (2014) Circulating tumor cells in pancreatic cancer patients: methods of detection and clinical implications. Int J Cancer 134:1–8. doi:10.1002/ijc.28134
2. Alix-Panabieres C, Pantel K (2014) Challenges in circulating tumour cell research. Nat Rev Cancer 14:623–631
3. Nagrath S, Sequist LV, Maheswaran S et al (2007) Isolation of rare circulating tumour cells in cancer patients by microchip technology. Nature 450:1235–1239. doi:10.1038/nature06385
4. Stott SL, Hsu C, Tsukrov DI et al (2010) Isolation of circulating tumor cells using a microvortex-generating herringbone-chip. Proc Natl Acad Sci U S A 107:18392–18397. doi:10.1073/pnas.1012539107
5. Wlodkowic D, Cooper JM (2010) Tumors on chips: oncology meets microfluidics. Curr Opin Chem Biol 14:556–567. doi:10.1016/j.cbpa.2010.08.016
6. Shelby JP, White J, Ganesan K et al (2003) A microfluidic model for single-cell capillary obstruction by Plasmodium falciparum-infected erythrocytes. Proc Natl Acad Sci U S A 100:14618–14622. doi:10.1073/pnas.2433968100
7. Gascoyne P, Satayavivad J, Ruchirawat M (2004) Microfluidic approaches to malaria detection. Acta Trop 89:357–369. doi:10.1016/j.actatropica.2003.11.009
8. Hou HW, Bhagat AAS, Chong AGL et al (2010) Deformability based cell margination—a simple microfluidic design for malaria-infected erythrocyte separation. Lab Chip 10:2605–2613. doi:10.1039/c003873c
9. Cheng X, Irimia D, Dixon M et al (2007) A microfluidic device for practical label-free CD4(+) T cell counting of HIV-infected subjects. Lab Chip 7:170–178. doi:10.1039/b612966h
10. Lee WG, Kim Y, Geun B et al (2010) Nano/microfluidics for diagnosis of infectious diseases in developing countries. Adv Drug Deliv Rev 62:449–457. doi:10.1016/j.addr.2009.11.016
11. Yang BD, Niu X, Liu Y et al (2008) Electrospun nanofibrous membranes: a novel solid substrate for microfluidic immunoassays for HIV. Adv Mater 20:4770–4775. doi:10.1002/adma.200801302
12. Baker CA, Duong CT, Grimley A, Roper MG (2009) Recent advances in microfluidic detection systems. Bioanalysis 1:967–975. doi:10.4155/bio.09.86
13. Chin CD, Laksanasopin T, Cheung YK et al (2011) Microfluidics-based diagnostics of infectious diseases in the developing world. Nat Med 17:1015–1019. doi:10.1038/nm.2408
14. Yager P, Domingo GJ, Gerdes J (2008) Point-of-care diagnostics for global health. Annu Rev Biomed Eng 10:107–144. doi:10.1146/annurev.bioeng.10.061807.160524
15. Chung TD, Kim HC (2007) Recent advances in miniaturized microfluidic flow cytometry for clinical use. Electrophoresis 28:4511–4520. doi:10.1002/elps.200700620

16. Dimov IK, Basabe-Desmonts L, Garcia-Cordero JL et al (2011) Stand-alone self-powered integrated microfluidic blood analysis system (SIMBAS). Lab Chip 11:845–850. doi:10.1039/c0lc00403k
17. Gad-el-Hak M (2001) Flow physics. MEMS Handb. doi:10.1201/9781420050905.ch4
18. Bruus H (2008) Theoretical microfluidics. Physics (College Park Md) 18:363. doi:10.1111/j.1574-6968.2009.01808.x
19. Lei H, Fedosov DA, Caswell B, Karniadakis GE (2013) Blood flow in small tubes: quantifying the transition to the non-continuum regime. J Fluid Mech 722:214–239. doi:10.1017/jfm.2013.91
20. Ali Beskok GEK (1999) report: a model for flows in channels, pipes, and ducts at micro and nano scales. Microscale Thermophys Eng 3:43–77. doi:10.1080/108939599199864
21. Thompson P, Troian SM (1997) A general boundary condition for liquid flow at solid surfaces. Nature 389:360–362. doi:10.1038/38686
22. Ghosh A (2011) Mechanics over micro and nano scales. In: Chakraborty S (ed). Springer, New York, pp 61–94
23. Pianka ER, Vitt LJ (2003) Lizards: windows to the evolution of diversity. University of California Press, Berkeley, CA
24. Trimmer W, Stroud R (2001) Scaling of micromechanical devices. MEMS Handb. doi:10.1201/9781420050905.ch2
25. Munson B, Young D, Okiishi T, Huebsch W (2009) Fundamentals of fluid mechanics, 6th edn. John Wiley & Sons Inc., New York
26. Sia SK, Whitesides GM (2003) Microfluidic devices fabricated in poly(dimethylsiloxane) for biological studies. Electrophoresis 24:3563–3576. doi:10.1002/elps.200305584
27. Kirby B (2010) Micro- and nanoscale fluid mechanics. Cambridge University Press, New York
28. Panton RL (2013) Flow at low Reynolds numbers. In: Incompressible flow. Wiley, New York, pp 607–649
29. Squires TM, Quake SRSR (2005) Microfluidics: fluid physics at the nanoliter scale. Rev Mod Phys 77:977–1026
30. Asmolov ES (1999) The inertial lift on a spherical particle in a plane Poiseuille flow at large channel Reynolds number. J Fluid Mech 381:63–87. doi:10.1017/S0022112098003474
31. Di Carlo D (2009) Inertial microfluidics. Lab Chip 9:3038–3046. doi:10.1039/b912547g
32. Dean WR (1928) Fluid motion in a curved channel. Proc R Soc A Math Phys Eng Sci 121:402–420. doi:10.1098/rspa.1928.0205
33. Berger SA, Talbot L, Yao LS (1983) Flow in curved pipes. Annu Rev Fluid Mech 15:461–512. doi:10.1146/annurev.fl.15.010183.002333
34. Bergman TL, Lavine AS, Incropera FP, Dewitt DP (2011) Fundamentals of heat and mass transfer, 7th edn. John Wiley & Sons, Inc., New York
35. Brody JP, Yager P (1997) Diffusion-based extraction in a microfabricated device. Sensors Actuators A Phys 58:13–18. doi:10.1016/S0924-4247(97)80219-1
36. Morgan H, Green NG (2003) Fluid dynamics. In: Pethig R (ed) AC electrokinetics: colloids nanoparticles. Institute of Physics Publishing, UK, pp 65–80
37. Lei U, Yang CY, Wu KC (2006) Viscous torque on a sphere under arbitrary rotation. Appl Phys Lett 89:20–23. doi:10.1063/1.2372704
38. Panton RL (2013) Ideal Flows in a plane. In: Incompressible flow. Wiley, New York, pp 438–501
39. Zeng L, Balachandar S, Fischer P (2005) Wall-induced forces on a rigid sphere at finite Reynolds number. J Fluid Mech 536:1–25. doi:10.1017/S0022112005004738
40. Rubinow SI, Keller JB (1961) The transverse force on a spinning sphere moving in a viscous fluid. J Fluid Mech 11:447. doi:10.1017/S0022112061000640
41. Segré G, Silberberg A (1961) Radial particle displacements in Poiseuille flow of suspensions. Nature 189:209–210. doi:10.1038/189209a0
42. Chun B, Ladd AJC (2006) Inertial migration of neutrally buoyant particles in a square duct: an investigation of multiple equilibrium positions. Phys Fluids 18:3–6. doi:10.1063/1.2176587

43. Ho BP, Leal LG (1976) Migration of rigid spheres in a two-dimensional unidirectional shear flow of a second-order fluid. J Fluid Mech 76:783. doi:10.1017/S002211207600089X
44. Cox RG, Brenner H (1968) The lateral migration of solid particles in Poiseuille flow—I: Theory. Chem Eng Sci 23:147–173. doi:10.1016/0301-9322(77)90001-5
45. Matas J-P, Morris JF, Guazzelli É (2004) Inertial migration of rigid spherical particles in Poiseuille flow. J Fluid Mech 515:171–195. doi:10.1017/S0022112004000254
46. Schonberg JA, Hinch EJ (1989) Inertial migration of a sphere in Poiseuille flow. J Fluid Mech 203:517–524
47. Di Carlo D, Edd JF, Humphry KJ et al (2009) Particle segregation and dynamics in confined flows. Phys Rev Lett 102:1–4. doi:10.1103/PhysRevLett.102.094503
48. Di Carlo D, Irimia D, Tompkins RG, Toner M (2007) Continuous inertial focusing, ordering, and separation of particles in microchannels. Proc Natl Acad Sci U S A 104:18892–18897. doi:10.1073/pnas.0704958104
49. Zhou J, Giridhar PV, Kasper S, Papautsky I (2013) Modulation of aspect ratio for complete separation in an inertial microfluidic channel. Lab Chip 13:1919–1929. doi:10.1039/c3lc50101a
50. Bhagat AAS, Kuntaegowdanahalli SS, Papautsky I (2008) Enhanced particle filtration in straight microchannels using shear-modulated inertial migration. Phys Fluids 20:101702. doi:10.1063/1.2998844
51. Bhagat AAS, Kuntaegowdanahalli SS, Papautsky I (2009) Inertial microfluidics for continuous particle filtration and extraction. Microfluid Nanofluid 7:217–226. doi:10.1007/s10404-008-0377-2
52. Hur SC, Brinckerhoff TZ, Walthers CM et al (2012) Label-free enrichment of adrenal cortical progenitor cells using inertial microfluidics. PLoS One 7:e46550. doi:10.1371/journal.pone.0046550
53. Wang X, Zandi M, Ho C-C et al (2015) Single stream inertial focusing in a straight microchannel. Lab Chip 15:1812–1821. doi:10.1039/C4LC01462F
54. Zhou J, Kasper S, Papautsky I (2013) Enhanced size-dependent trapping of particles using microvortices. Microfluid Nanofluid 15:611–623. doi:10.1007/s10404-013-1176-y
55. Hur SC, Mach AJ, Di Carlo D (2011) High-throughput size-based rare cell enrichment using microscale vortices. Biomicrofluidics 5:1–10. doi:10.1063/1.3576780
56. Sollier E, Go DE, Che J et al (2014) Size-selective collection of circulating tumor cells using Vortex technology. Lab Chip 14:63–77. doi:10.1039/c3lc50689d
57. Che J, Mach AJ, Go DE et al (2013) Microfluidic purification and concentration of malignant pleural effusions for improved molecular and cytomorphological diagnostics. PLoS One 8: e78194. doi:10.1371/journal.pone.0078194
58. Wang X, Zhou J, Papautsky I (2013) Vortex-aided inertial microfluidic device for continuous particle separation with high size-selectivity, efficiency, and purity. Biomicrofluidics 7:22–25. doi:10.1063/1.4818906
59. Wang X, Papautsky I (2015) Size-based microfluidic multimodal microparticle sorter. Lab Chip 15:1350–1359. doi:10.1039/c4lc00803k
60. Gossett DR, Di Carlo D (2009) Particle focusing mechanisms in curving confined flows. Anal Chem 81:8459–8465. doi:10.1021/ac901306y
61. Matas JP, Morris JF, Guazzelli E (2004) Lateral forces on a sphere. Oil Gas Sci Technol Rev IFP 59:59–70
62. Lee MG, Shin JH, Bae CY et al (2013) Label-free cancer cell separation from human whole blood using inertial micro fluidics at low shear stress. Anal Chem 85(13):6213–6218
63. Lee MG, Choi S, Park J-K (2009) Three-dimensional hydrodynamic focusing with a single sheath flow in a single-layer microfluidic device. Lab Chip 9:3155–3160. doi:10.1039/b910712f
64. Zhang J, Li M, Li WH, Alici G (2013) Inertial focusing in a straight channel with asymmetrical expansion–contraction cavity arrays using two secondary flows. J Micromech Microeng 23:085023. doi:10.1088/0960-1317/23/8/085023
65. Hou HW, Warkiani ME, Khoo BL et al (2013) Isolation and retrieval of circulating tumor cells using centrifugal forces. Sci Rep 3:1259. doi:10.1038/srep01259

66. Nivedita N, Papautsky I (2013) Continuous separation of blood cells in spiral microfluidic devices. Biomicrofluidics 7:1–14. doi:10.1063/1.4819275
67. Sun J, Liu C, Li M et al (2013) Size-based hydrodynamic rare tumor cell separation in curved microfluidic channels. Biomicrofluidics 7:1–11. doi:10.1063/1.4774311
68. Xiang N, Yi H, Chen K et al (2013) High-throughput inertial particle focusing in a curved microchannel: insights into the flow-rate regulation mechanism and process model. Biomicrofluidics 7:1–10. doi:10.1063/1.4818445
69. Tallapragada P, Hasabnis N, Katuri K et al (2015) Scale invariant hydrodynamic focusing and sorting of inertial particles by size in spiral micro channels. J Micromech Microeng 25:084013. doi:10.1088/0960-1317/25/8/084013
70. Xiang N, Chen K, Dai Q et al (2015) Inertia-induced focusing dynamics of microparticles throughout a curved microfluidic channel. Microfluid Nanofluid 18:29–39. doi:10.1007/s10404-014-1395-x
71. Reece AE, Kaastrup K, Sikes HD, Oakey J (2015) Staged inertial microfluidic focusing for complex fluid enrichment. RSC Adv 5:53857–53864. doi:10.1039/C5RA10634F
72. Pethig R (2013) Dielectrophoresis: an assessment of its potential to aid the research and practice of drug discovery and delivery. Adv Drug Deliv Rev 65:1589–1599. doi:10.1016/j.addr.2013.09.003
73. Pethig R (2010) Review article-dielectrophoresis: status of the theory, technology, and applications. Biomicrofluidics 4:1–35. doi:10.1063/1.3456626
74. Martinez-Duarte R, Camacho-Alanis F, Renaud P, Ros A (2013) Dielectrophoresis of lambda-DNA using 3D carbon electrodes. Electrophoresis 34:1113–1122. doi:10.1002/elps.201200447
75. Moncada-Hernandez H, Nagler E, Minerick AR (2014) Theoretical and experimental examination of particle-particle interaction effects on induced dipole moments and dielectrophoretic responses of multiple particle chains. Electrophoresis 35:1803–1813. doi:10.1002/elps.201300636
76. Gascoyne PRC, Vykoukal J (2002) Particle separation by dielectrophoresis. Electrophoresis 23:1973–1983. doi:10.1002/1522-2683(200207)23:13<1973::AID-ELPS1973>3.0.CO;2-1
77. Salmanzadeh A, Romero L, Shafiee H et al (2012) Isolation of prostate tumor initiating cells (TICs) through their dielectrophoretic signature. Lab Chip 12:182–189. doi:10.1039/c1lc20701f
78. Wang X, Becker FF, Gascoyne PRC (2002) Membrane dielectric changes indicate induced apoptosis in HL-60 cells more sensitively than surface phosphatidylserine expression or DNA fragmentation. Biochim Biophys Acta 1564:412–420
79. Abdallah BG, Chao T-C, Kupitz C et al (2013) Dielectrophoretic sorting of membrane protein nanocrystals. ACS Nano 7:9129–9137. doi:10.1021/nn403760q
80. Jones PV, Staton SJR, Hayes MA (2011) Blood cell capture in a sawtooth dielectrophoretic microchannel. Anal Bioanal Chem 401:2103–2111. doi:10.1007/s00216-011-5284-9
81. Shafiee H, Sano MB, Henslee EA et al (2010) Selective isolation of live/dead cells using contactless dielectrophoresis (cDEP). Lab Chip 10:438–445. doi:10.1039/b920590j
82. Pethig R, Kell DB (1987) The passive electrical properties of biological systems: their significance in physiology, biophysics and biotechnology. Phys Med Biol 32:933–970. doi:10.1088/0031-9155/32/8/001
83. Cemažar J, Kotnik T (2012) Dielectrophoretic field-flow fractionation of electroporated cells. Electrophoresis 33:2867–2874. doi:10.1002/elps.201200265
84. Chan KL, Gascoyne PR, Becker FF, Pethig R (1997) Electrorotation of liposomes: verification of dielectric multi-shell model for cells. Biochim Biophys Acta 1349:182–196
85. Nejad HR, Chowdhury OZ, Buat MD, Hoorfar M (2013) Characterization of the geometry of negative dielectrophoresis traps for particle immobilization in digital microfluidic platforms. Lab Chip 13:1823–1830. doi:10.1039/c3lc41292j

86. Puttaswamy SV, Sivashankar S, Chen R-J et al (2010) Enhanced cell viability and cell adhesion using low conductivity medium for negative dielectrophoretic cell patterning. Biotechnol J 5:1005–1015. doi:10.1002/biot.201000194
87. Leonard KM, Minerick AR (2011) Explorations of ABO-Rh antigen expressions on erythrocyte dielectrophoresis: changes in cross-over frequency. Electrophoresis 32:2512–2522. doi:10.1002/elps.201100077
88. Sano MB, Salmanzadeh A, Davalos RV (2012) Multilayer contactless dielectrophoresis: theoretical considerations. Electrophoresis 33:1938–1946. doi:10.1002/elps.201100677
89. Jones TB (1995) Electromechanics of particles. Cambridge University Press, Cambridge
90. Camacho-Alanis F, Gan L, Ros A (2012) Transitioning streaming to trapping in DC insulator-based dielectrophoresis for biomolecules. Sens Actuators B Chem 173:668–675. doi:10.1016/j.snb.2012.07.080

Chapter 14
Microfluidics for High-Throughput Cellular Isolation and Analysis in Biomedicine

Caroline N. Jones and Joseph M. Martel-Foley

Abstract Rare or low-abundance cells in a much larger population of background cells require extremely accurate as well as high-throughput selection and enumeration for a variety of biomedical applications. Conventional bench-top techniques have limited capabilities to isolate and analyze these rare cells because of their generally low selectivity, significant sample loss, and limited ability for bulk measurements in heterogeneous cell populations. Microfluidics has enabled facile handling of minute sample volumes and massively parallel multiplexing capabilities for high-throughput processing, making this platform excellent to deal with the transport, isolation, and analysis of rare cells. We classify the microfluidic rare cell isolation techniques based on the manner in which they achieve isolation through taking advantage of differences in cell properties such as size, surface marker expression, and behavior. In this chapter, we focus on recently published work utilizing microfluidic isolation techniques for circulating tumor cells (CTCs), immune cells, pathogens, and stem cells. We also cover methodologies for analyzing rare cell phenotypes, including migration patterns, using microfluidic platforms with integrated biosensors. Finally, we discuss future applications for microfluidic technology in advancing human health and basic biological understanding of rare cell types, such as CTCs.

Keywords Circulating tumor cells (CTCs) • Immune cell • Migration phenotype • Individualized medicine

C.N. Jones, Ph.D. (✉)
Department of Biological Sciences, Virginia Polytechnic Institute and State University, Life Sciences 1 Building, Room 221, 970 Washington St. SW, Blacksburg, VA 24061, USA
e-mail: jonescn@vt.edu

J.M. Martel-Foley, Ph.D.
Department of Biomedical Engineering, Wentworth Institute of Technology, 550 Huntington Avenue, Boston, MA 02115, USA
e-mail: martelj@wit.edu

© Springer International Publishing Switzerland 2016

C. Lu, S.S. Verbridge (eds.), *Microfluidic Methods for Molecular Biology*,
DOI 10.1007/978-3-319-30019-1_14

14.1 Introduction

Cell types with an abundance of less than 1000 in a milliliter sample are considered as rare [1]. The importance of these rare cells has been shown for the diagnosis and prognosis of many cancers, prenatal diagnosis, and diagnosis of viral infections as well as other various applications [2]. Typical rare cells in blood samples are circulating tumor cells (CTCs), subpopulations of immune cells, circulating fetal cells, stem cells [3], cells infected by a virus or parasite [4], or blood-borne pathogens [5]. Recent rapid advances in microfluidics provide robust solutions to the challenges associated with the isolation, enrichment, and analysis of these important clinical targets.

Rare cells can be isolated from a large background population of other cell types based on one or several unique properties including size, surface marker expression, and/or behavior. A typical microfluidic device consists of a microchannel network integrated with various sensors and actuators with typical channel dimensions on the order of hundreds of micrometers, while the size of cells ranges from several to tens of micrometers. This similar size scale allows for precise control of the interactions between cells, the fluid flow, and the microchannel geometry. The small size of microfluidic devices also allows for the implementation of new protocols such as single-cell analyses, which were not possible previously with macroscale techniques, shedding light on the importance of sample heterogeneity and rare cells. Another advantage of microfluidic devices is that they can provide highly precise processing of complex cellular fluids, with minimal damage to sensitive cell populations due to shear forces [6], and do not require cell fixation [7]. This allows for culture and analysis of live rare cells after isolation, which will be discussed in the second portion of this chapter.

Microfluidics for rare cell isolation is becoming more common in the clinic, including the enrichment of CTCs [8], hematopoietic stem cells (HSCs), and circulating fetal cells (CFCs) from blood [9]. Meanwhile, the growing interest in personalized medicine [10], in which treatments are tailored to the prognoses of patients, is further driving the demand for rapid and high-performance microfluidic platforms [11]. In the future, CTCs from a patient could be cultured and expanded for chemotherapy drug screening efficacy purposes. The up-front cost of such individualized methods utilizing microfluidics would be offset by elimination of months of costly, ineffective treatments.

Several parameters can be used to evaluate the performance of a microfluidic device for rare cell isolation, namely efficiency, purity, and viability. The efficiency is typically presented either as the recovery rate or yield, which is the ratio between the number of isolated cells at the exit of the device over the number of known targeted cells introduced at the inlet or enrichment, which is the ratio of the volumetric concentration of the isolated cells in the device with respect to the original supplied sample. Thus, the efficiency can only be determined in experiments with spiked cell samples where the input numbers are known quantities. The purity is defined as the ratio of isolated target cells with respect to the total number of collected cells in the output. Finally, viability is the percentage of cells surviving

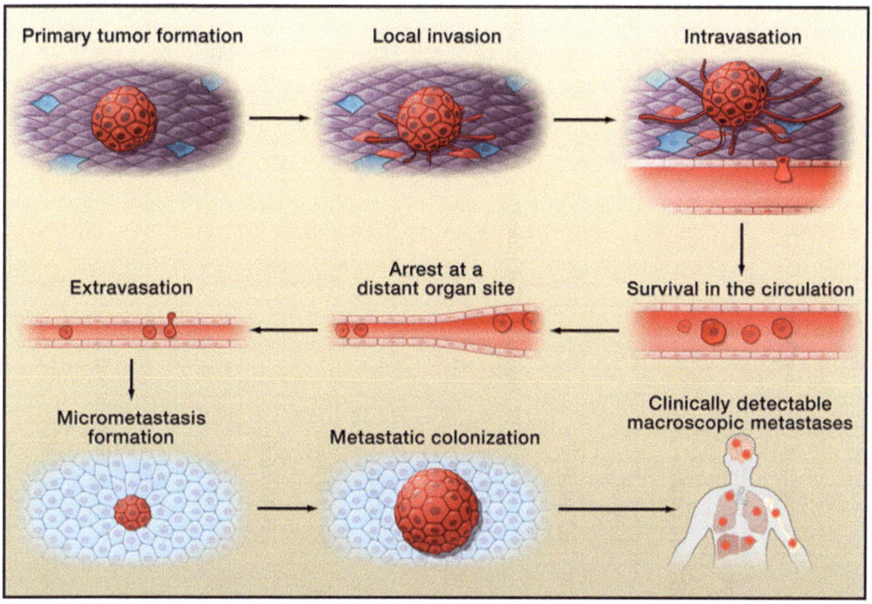

Fig. 14.1 Circulating tumor cells contribute to metastasis. Tumor cells that acquire invasive abilities may shed from the primary tumor. Some of them undergo an epithelial-to-mesenchymal transition (EMT), while others stay in an epithelial state. When they enter blood vessels and circulate with normal blood cells, they are considered circulating tumor cells (CTCs). CTCs can escape blood vessels through a process called extravasation and proliferate within distant tissues (like bone marrow) forming dormant microtumors or metastases. Cells in these dormant microtumors or metastases can also shed cancerous cells into blood vessels to form further metastases, and some cells in dormant microtumors may break away from dormancy and escape from immunosurveillance to form distant metastases. All these metastases may eventually grow into overt metastases and harm patient health. Reproduced with permission from [13]

(cell membrane intact or metabolically functioning) the isolation process with respect to the total number of target cells.

In this chapter, we first focus on current literature describing microfluidics engineered for the isolation of CTCs, as this has been a core motivation for much of the technology development for rare cell isolation. CTCs are shed by metastatic tumors into the blood and have generated much recent interest as the occurrence of metastasis is strongly associated with adverse patient outcomes. However, CTCs are extremely rare, intermixed with blood cells at concentrations of approximately one per billion in clinical samples [12] (Fig. 14.1). The ultimate goal is to efficiently isolate this rare population of cells in a viable and intact state and with high purity from the vast number of surrounding blood cells, a formidable technological challenge. This "liquid biopsy" will allow physicians to diagnose and monitor a patient's disease state without invasive surgery and enable personalized cancer treatments [12].

We describe several rare cell isolation microfluidic techniques based on cell properties—size, surface marker expression, and phenotype (Table 14.1). We then

Table 14.1 Rare cell isolation technology details

Isolation technique	Style	Target cell type	Yield	Purity	Throughput (WB equivalent U.O.S.)	Refs.
Affinity						
Surface capture	Post	CTCs (EpCAM)	60 %	100–200 WBCs/mL of input	1 mL/h	[14]
		CTCs (EpCAM)	80 %	330–850 WBCs/mL of input	10 mL/h	[15]
	GEDI	CTCs (PSMA)	85 %	68.0 %	1 mL/h	[16, 17]
	Flow patterning	CTCs (EpCAM)	92 %	~2600 WBCs per mL of input	1.2 mL/h	[18]
Fluorescence based		CTC (Multi)	70.9 %	72.9 %	150 cells/s (RBC lysed, WBC depleted)	[19, 20]
Magnetic bead capture		CTC (EpCAM)	93.2 %	78.4 %	14.4 mL/h	[21]
		HSCs (CD133)	96.4 %	78.4 %	7.2 mL/h	[21]
		EPCs (CD133)	82.3 %	78.4 %	7.2 mL/h	[21]
		CTC (EpCAM)	89.0 %	502 WBCs per mL of input	60–300 mL/h	[22]
		CTC (EpCAM)	88.0 %	N/A	2.5–10 mL/h	[23]
		CTC (EpCAM)	90.0 %	N/A	1.2 mL/h	[24]
		Bacteria	100.0 %	N/A	60 mL/h	[25]
Physical properties						
Size	Microfilters (engineered)	CTCs	85 %	N/A	60 mL/h (50 % blood in buffer)	[26]
	Microfilters (isotropic)	CTCs	85 %	N/A		[27]

		CTCs		~10–500 WBCs/mL of input		[28, 29]
	Inertial focusing	CTCs	98 %	~811 WBCs/mL of input	45 mL/h (WB equivalent)	[30]
		CTCs	85 %	~440 WBCs/mL of output	3 mL/h	[31]
		CTCs	80 %	~425 WBCs/mL of input	1.2 mL/h (WB equivalent)	[32]
		CTCs	~10–20 %	75.0 %		[33]
	Acoustophoresis	CTCs	~88 %	~99.8 %	6 mL/h (10× diluted RBC lysed blood)	[34]
Clustering		CTC clusters	100 %	N/A	2.5 mL/h	[35]
Deformability		Cancer cell lines	96–97 %	~4 %	0.167 mL/h (RBC lysed blood) or approximately 11,000 cells/s	[36]
Multimodal						
Size and affinity		CTCs from blood	95 %	1500 WBCs/mL input	8 mL/h or approximately 11,000 cells/s (WBCs and CTCs only)	[37]
Size and negative affinity		CTCs from blood	97 %	1200 WBCs/mL input	8 mL/h or approximately 11,000 cells/s (WBCs and CTCs only)	[38]
		CTCs from blood	93.3 %	300,000 WBCs/mL input	2 mL/h or approximately 2750 cells/s (WBCs and CTCs only)	[39]
		CTCs from blood	89.0 %	N/A	18.7–140 mL/h (WB equivalent)	[40]

provide examples of microfluidic devices for isolating other rare cell types such as subpopulations of immune cells, pathogens, and stem cells from a variety of bodily samples. Finally, we discuss how single-cell measurements of rare cell phenotypes post-isolation can enable better understanding of cancer metastases or dysregulation of immune cell function leading to sepsis or susceptibility to infectious diseases.

Microfluidic Isolation Techniques Based on Cell Properties

14.2 Immunochemical Signature: Specific Antibodies or Aptamers

One of the most popular means of differentiating CTCs from normal blood cells is through unique surface markers. These unique markers are used in three main modalities for isolating cells in microfluidic devices: fluorescence-activated cell sorting (FACS), magnetic activated cell sorting (MACS), and surface capture. FACS is similar to the well-known field of flow cytometry where microfluidics provides an improved sensitivity with fluorescence surface marker staining. MACS has also seen improvements with sensitivity as compared to bulk separations when translated to microfluidics. In MACS, magnetic beads are targeted to the cancer cells and then passed through a magnetic field where labeled and unlabeled cells can be separated. Surface capture technologies take advantage of well-controlled microflows (Fig. 14.2) in which cells interact with surfaces coated with a complementary antibody or aptamer that will cause targeted cells to arrest on a surface while nontarget cells continue to flow without sticking.

For tumors of epithelial origin, EpCAM is used as a targeting moiety. While this is useful, it has been shown that there are cases where EpCAM expression of CTCs can be low and even change dynamically. Other surface targets for isolation are epidermal growth factor receptor (EGFR) [42–44], human epidermal growth factor receptor 2 (HER2) [43], prostate-specific membrane antigen (PSMA) [16], and cell surface vimentin (CSV) [45, 46].

The adhesion force between a cell and a surface labeled with antibodies is proportional to the capturing surface area and the contact area with the cell. The proportionality factor is the bond strength between a single antigen–antibody complex [47]. The adhesion force needs to balance the drag force of the flow. Thus, an upper limit for the flow rate and the throughput can be estimated for isolation based on this affinity. The shear stress needed to overcome this cell adhesion is on the order of few millipascals. Microfluidic isolation methods based on the immunochemical signature are further subcategorized as either positive or negative isolation. In positive isolation, the target cells are captured and remain inside the microfluidic device. In negative isolation, all the other cell types are captured or filtered inside the device allowing only the intact target cells to proceed to the exit. The most widely used CTC isolation techniques rely on antibody-based

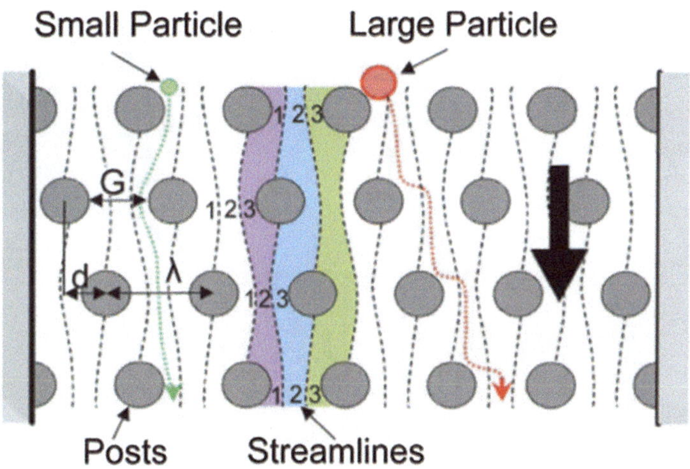

Fig. 14.2 Separation by deterministic lateral displacement. An array of microposts with a row shift fraction of one-third is used to sort small particles (*green dotted line*) from large ones (*red dotted line*). Depending on particle's size, it will follow either one of the three streamlines (*purple*, *blue*, or *green*). *Dashed lines* represent the critical streamline delimiting the particles' paths. The posts can be coated with capture moieties (e.g., antibodies, aptamers) specific to the rare cell type of interest. Reproduced with permission from Davis et al. [41]. Copyright © (2006) National Academy of Sciences, USA

capture of CTCs, which express epithelial cell surface markers that are absent from normal leukocytes. An established biomarker for CTCs is EpCAM, which is expressed by cells of epithelial origin but lacking in blood cells.

Mehmet Toner's group at the BioMEMS Resource Center at Massachusetts General Hospital developed several generations of microfluidic platforms for single-step isolation of CTCs from unprocessed blood specimens [14, 48, 49]. The first-generation CTC-chip is a silicon chamber etched with 78,000 microposts that are coated with an anti-EpCAM antibody. As 2–4 mL of whole blood flows through the chip, flow kinetics have been optimized for minimal shear stress on cells while enhancing contacts with the antibody-coated microposts. Captured CTCs attached to microposts are visualized by staining with antibodies against cytokeratin or tissue-specific markers. For CTC enumeration, the entire device is imaged at multiple planes using a semiautomated imaging system while on-chip lysis allows for DNA and RNA extraction and molecular analyses. The CTC-chip enables a high yield of capture (median, 50 CTCs per milliliter) and purity (ranging from 0.1 to 50 %). This microfluidic platform enables gentle one-step microfluidic processing, which is critical when purifying rare delicate cell populations [14]. Several other micropost-based capture chips have been developed including the geometrically enhanced differential immunocapture (GEDI) device developed at Cornell University (Fig. 14.3) [17] and a radially based flow design for improved throughput [15].

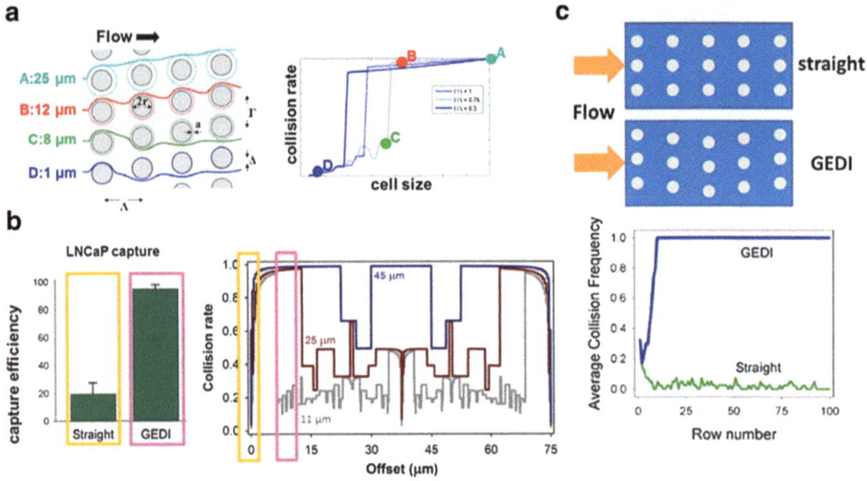

Fig. 14.3 GEDI microfluidic device—results in cell lines recapitulate simulations. (**a**) Microfluidic obstacle array with array geometric parameters. $\Delta =$ obstacle offset. Pathlines (*various colors*) denote trajectories of cells of different diameters. Obstacle array spacing and orientation parameters are also defined. *Right*: The rate of cell-wall collisions for cells traveling through the array is a strong function of the offset parameter of the array; the GEDI design methodology implies use of an offset parameter that leads to size-dependent collision rates. The results predicted for the flow through of the geometry at *left* are shown at *right* by the *solid line*; the four specific cell sizes lead to results denoted by the *four colored dots* on this graph. Other geometric arrangements lead to different results, shown at *right* in the *dotted and dashed lines*. (**b**) Straight arrays or arrays with small offsets (*gold boxes*) lead to lower capture efficiency (*left*) and size independence (*right*). Carefully chosen offsets (*magenta boxes*) lead to high capture rates (*left*) and size dependence (*right*). Capture rates at *left* compare GEDI (7 μm offset) and straight (no offset) geometry performance as measured by LNCaP capture efficiency on J591-functionalized devices. Rates at *right* describe simulated collision rates in these geometries. Both experimental results have the same surface-area-to-volume ratio. (**c**) Devices with the same surface area-to-volume ratio give vastly different results: straight arrays lead to collisions that decrease as the blood travels through the device; GEDI arrays lead to collisions that increase with travel through the device. Figure reproduced with permission from Kirby et al. [17]

A second-generation CTC-chip, the "herringbone chip," was engineered to further increase the number of cell-surface interactions by producing micro-vortices in the anti-EpCAM antibody-coated channel. The herringbone chip is fabricated by two-layer photolithography so that a herringbone pattern is present on the upper surface of the chip. In comparison with a conventional flat-walled microfluidic device, the herringbone-induced micro-vortices disrupt the laminar flow streamline in which cells travel, causing them to shift. The herringbone mixer generates transverse flows by a steady axial pressure gradient induced by diagonal ridges along one of the channel walls. This results in a steady chaotic flow where stretching and folding proceed exponentially with travelled distance; hence effective mixing is achieved. Using this device, prostate cancer cells were isolated with the high recovery rate of 91.8 ± 5.2 %, and 386 ± 238 CTCs/mL were separated from 93 % of metastatic prostate cancer patients (Fig. 14.4).

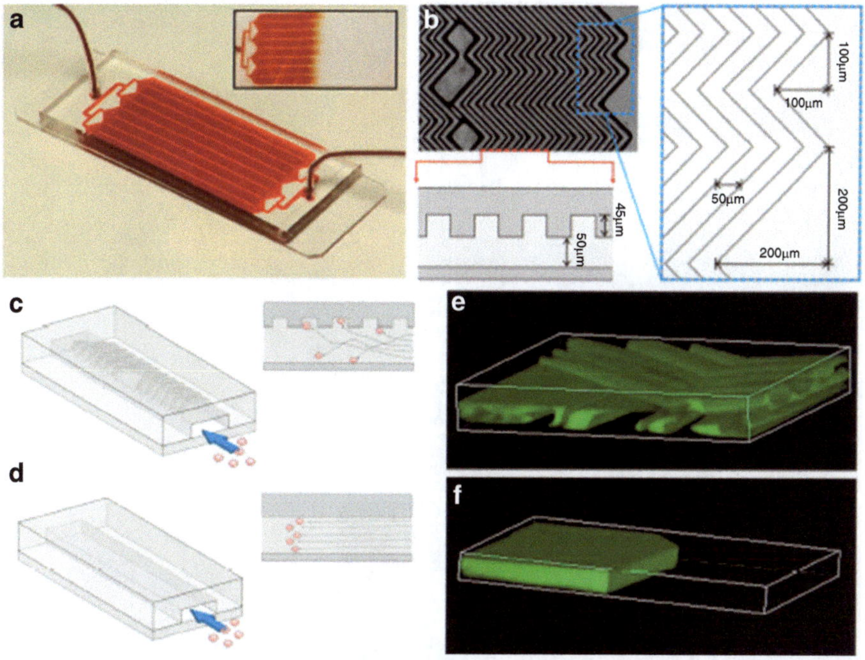

Fig. 14.4 The herringbone chip. (**a**) The HB-chip consists of a microfluidic array of channels with a single inlet and exit. *Inset* illustrates the uniform blood flow through the device. (**b**) A micrograph of the grooved surface illustrates the asymmetry and periodicity of the herringbone grooves. Cartoon illustrating the cell-surface interactions in (**c**) the HB-chip, and (**d**) a traditional flat-walled microfluidic device. Flow visualization studies using two paired streams of the same viscosity (one stream is *green*, the other clear) demonstrate (**e**) the chaotic micro-vortices generated by the herringbone grooves, and the lack of mixing in (**f**) traditional flat-walled devices. Figure reproduced with permission from Stott et al. [18]

Another advantage of the herringbone chip is that it preserves multicellular aggregates unlike the micropost chip, in which tight distances between the microposts presumably prevent the passage of large cellular clusters. Interestingly, when stained with hematoxylin and eosin, most CTC clusters have an increased staining intensity relative to individual CTCs, suggesting less autolysis and protection from the environment within the CTC clusters (Fig. 14.5).

The clinical relevance of CTC clusters versus single CTCs within the circulation and their potential implications for metastatic spread warrant further investigation and will be discussed in future sections of this chapter.

Releasing captured tumor cells from the device surface for subsequent molecular analysis or cell culture is also an important and challenging step. The strong adhesion forces between cells and antibody-functionalized surfaces need to be overcome to detach cells. Methods like thermodynamic release, electrochemical desorption, and proteolytic enzyme degradation have been used to achieve this. However, these require either elaborate design of sensors or specific enzymes that target cell

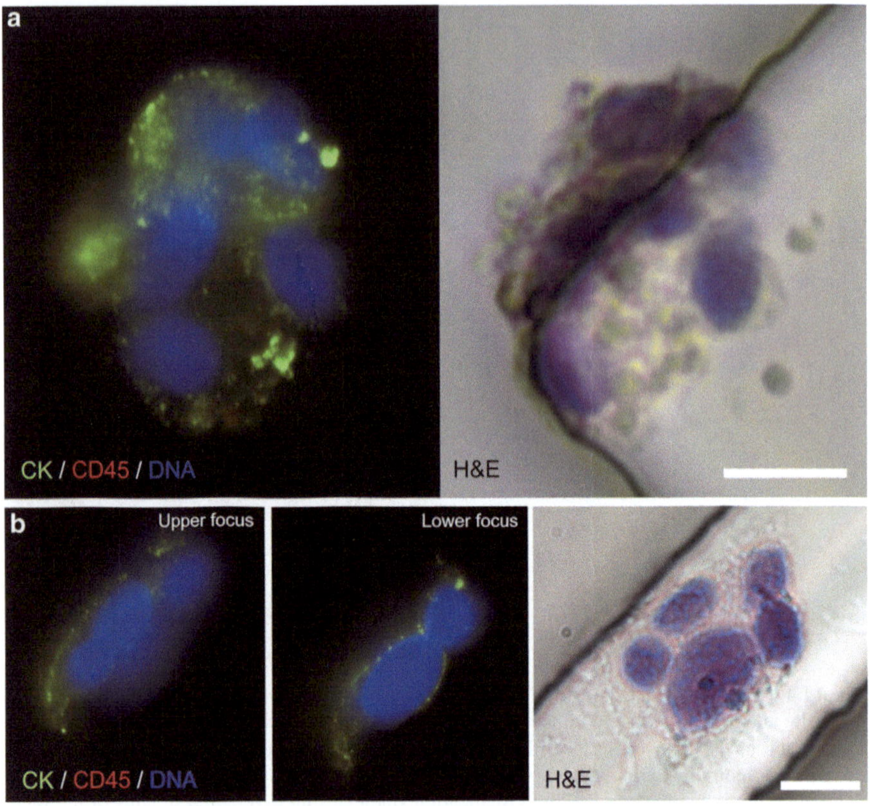

Fig. 14.5 CTC clusters captured from lung cancer patients' blood using an anti-EpCAM-coated HB-chip. (**a**, **b**) Fluorescent micrographs of cytokeratin 7/8 (CK; *green*)-, CD45 (*red*)-, and DAPI (*blue*)-stained CTC clusters taken at different focal planes and corresponding hematoxylin and eosin (H&E) stains are shown. Bars, 10 μm. Figure reproduced with permission from Stott et al. [18]

receptors and/or antibodies [50]. In addition, the spreading and flatness of captured cells on the surface lead to an extremely high threshold to detachment force, which must be exceeded in order to completely elute captured cells from the surfaces.

Wan et al. have shown that anti-EGFR aptamer-grafted glass bead arrays can recognize and capture hGBM cells that overexpress EGFR, which is an important biomarker. The device flow through velocity of 1 mL/h gives a single-pass capture of tumor cells. The cells are detached from the glass beads for further analysis using an antisense release oligonucleotide which increases the recovery of tumor cells to 92 % [51]. Microfluidic FACS is another more traditional manner in which to take advantage of unique surface markers on CTCs and tagging them with fluorescent probes in order to distinguish CTCs from normal cells.

CellSearch, the only US Food and Drug Administration (FDA)-cleared CTC isolation method for prognostic use to date, uses an epithelial cell adhesion molecule (EpCAM)-dependent immunomagnetic enrichment method commercialized by Veridex [52, 53]. This successful bulk method relies on the attachment of magnetic microbeads to the target cells and then using an external magnetic field to wash away nontarget cells. Using microfluidics can assist in the more uniform treatment of individual cells as well as improved sensitivity for cells with fewer attached magnetic beads.

14.3 Isolation Based on Hydrodynamic Signatures

Isolation of rare cells can take advantage of the difference in their physical properties (size, deformability, density, etc.), which are independent of their biochemical surface markers. Cell separation techniques based on physical properties share some common advantages. First of all, the approach is label free; these separation methods only rely on the physical properties of the cells, requiring no immunostaining or antibody labeling. Expensive chemical reagents are not needed, decreasing the sample preparation time as well as the risk of cell damage. Second, a high sorting throughput is possible for cell separation based on physical characteristics. With a proper design, this separation method is able to handle a large amount of cellular samples in a short time. Although not absolute, several physical properties distinguish CTCs from most normal blood cells. These include the larger size of most epithelial cells and differences in density, charge, and migratory properties [12].

One of the older techniques used for CTC isolation uses microporous filters to trap larger cells. This procedure known as isolation by size of epithelial tumor cells or ISET is widely used for the isolation of CTCs with 8 μm pores. Several other groups of researchers have used more engineered systems of micropores as well [54, 55].

Inertial forces can result in the induced migration of cells or particles across streamlines in laminar flow streams. Typically, inertial forces emanate from boundary effects of fluid flow adjacent to the walls of a microfluidic channel, causing lift. Inertial focusing in curved channels refers to a subset of distinct phenomenological techniques for cell fractionation, which includes the use of serpentine or Archimedean spiral patterns for cell ordering and sorting [56] (Fig. 14.6). Di Carlo et al. demonstrated that cells could be differentially focused and sorted based on size under laminar flow using a serpentine pattern [57]. A major benefit of this system is its high throughput (e.g., 1.5 mL/min) without sheath flow or sequential cell manipulation, which is useful for processing native biological fluids and flow cytometry [58, 59].

The Toner research group has also developed a new technology called "cluster-chip" to analyze CTC clusters of 2–19 cells in a group of patients with metastatic breast, prostate, and melanoma cancers (Fig. 14.7). The cluster-chip captures CTC clusters by relying on the strength of cell-cell junctions as clusters flow at physiological speed through a set of triangular pillars without relying on labeling or surface markers of the cells. Three pillars make up the basic unit of the chip; two

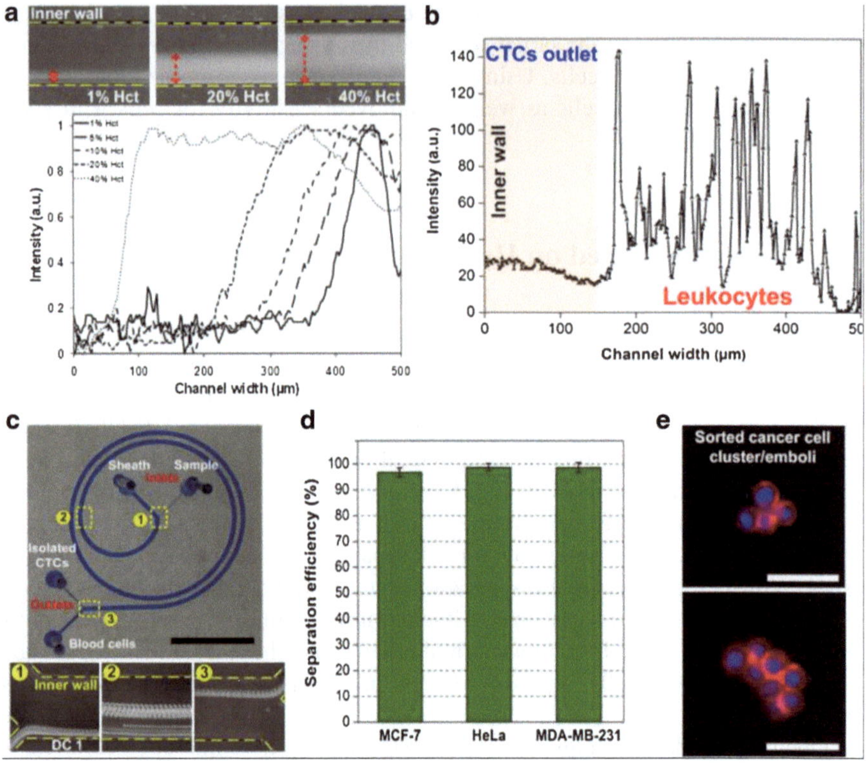

Fig. 14.6 Device characterization using hematological cells and cancer cell lines at Dean cycle 1 (DC 1). (**a**) Averaged composite images and intensity plot illustrate broadening RBC-occupied regions (*red dashed line*) for increasing hematocrit prior to outlet bifurcation. *Yellow dotted lines* indicate position of channel walls. (**b**) Line scan indicating lateral positions of leukocytes near the outlet at DC 1. The larger leukocytes behaved similar to RBCs and were transposed to outer half of the channel under the influence of Dean drag forces, resulting in minimal leukocytes entering the CTC outlet (150 μm width). (**c**) Average composite images representing the focusing position of MCF-7 cells at corresponding locations of the spiral channel. Scale bar is 1 cm. (**d**) Histogram plot indicating a high separation efficiency of >85 % for different cancer cell lines tested. (**e**) Images of sorted MCF-7 clusters stained with anti-EpCAM antibodies (*red*) and Hoechst dye (*blue*) to identify the cell membrane and nucleus, respectively. Scale bar is 100 μm. Reproduced with permission from Hou et al. [31]

form a narrowing channel that funnels the cells into an opening, where the edge of the third pillar is positioned to bifurcate the laminar flow. As blood flows, single blood and tumor cells divert to one of the two streamlines at the bifurcation and pass through the 12 μm × 100 μm opening. In contrast, CTC clusters are held at the edge of the bifurcating pillar, even if they are deformable enough to squeeze through either one of the openings. This is because the bifurcating edge retains the captured cluster in both streamlines simultaneously under a dynamic force balance—cell-cell junctions support a stable equilibrium (not possible for a single cell) across the bifurcating pillar, which serves as a fulcrum. The three-pillar unit is repeated in multiple rows for

Fig. 14.7 Capture of CTC clusters from blood samples of patients with metastatic cancer. (**a**) Image of a working cluster-chip. Blood from a single inlet is uniformly distributed over 4096 parallel trapping paths and then collected at a single outlet. Close up (*inset*) shows a CTC cluster-trapping region with part of the microfluidic distribution and collection networks. The size of the glass slide is 3 in. × 1.5 in. (**b**) Representative images of three CTC clusters isolated from patients with metastatic breast cancer. *Left*, bright-field and fluorescence images of a live CTC cluster stained for common breast cancer surface markers. *Center*, SEM micrograph of a fixed CTC cluster. *Inset* is ~2.1× magnified. *Right*, fluorescence image of a highly deformable CTC cluster stained for cytokeratin. Note that this CTC cluster is not split but is highly strained even under slow flow in the cluster-chip [35]

redundancy (Fig. 14.7b and c). To ensure that CTC clusters do not dissociate, the chip is optimized to handle cellular aggregates with flow speeds (and therefore shear forces) well below those found in human capillaries. The peak flow speed of ~70 μm/s at the bifurcation is much lower than that of existing microfluidic and filter-based CTC isolation technologies. Yet the chip can interrogate clinical blood specimens at a rate of 2.5 mL/h owing to its highly parallel architecture.

In addition to cell size and clustering, it is also possible to isolate cancer cells based upon their mechanical properties, exploiting a difference in stiffness between cancer and blood cells. Using inertial focusing, researchers were able to classify and separate several cell lines from RBC lysed blood based upon a difference in equilibrium position caused by the different stiffnesses of cells. This label-free technique achieved a 96–97 % yield for modMCF7s and SAOS-2 from RBC lysed blood samples but had a purity of about 4 %. The researchers also noted a difference in equilibrium position between benign MCF7s and malignant modMCF7s [36].

14.4 Multi-Method Integration for Enrichment of Rare Cells

Both immunochemical and physical property techniques have been proven to accomplish rare cell isolation, however, it is important to note that each makes assumptions about the target population. In the case of immunochemical isolation the assumption is that we know a priori what targets are on CTCs and that all CTCs express a similar level of that surface marker. Due to the rareness, heterogeneity, and vulnerability of CTCs, it is still challenging for any of these methods alone to achieve all the goals at the same time with satisfactory results. For example, the EpCAM antibody-based methods may miss EpCAM-negative CTCs due to the varied expression levels of ligands on CTCs [42]. For physical properties we assume that all CTCs are larger or more deformable than other cells. As such, combinations of technologies both for positive selection and negative selection have been developed.

One means of circumventing the low yield of either a surface marker-based or size-based selection method is to use both sequentially on the same sample. This addition of technologies (EpCAM-based positive selection, then size-based positive selection microfilter) has been shown to mitigate some of the losses associated with variable size and EpCAM expression. So while for high-expressing cells the increase in overall yield of spiked cell lines was minimal (from 87 to 89 %), the increase in yield for a low-expressing cell line T24 was greater (2 to 61 %) [40]. A similar approach was taken using magnetic particles coated with anti-CD45 for depleting leukocyte prior to size-based filtration. This part negative, part positive depletion technique yielded ~91 % of spiked MCF7s and H1975 cells with an average white blood cell depletion of 2.3 log at a rate of 60 mL of whole blood per hour [39].

The CTC-iChip uses an integrated set of devices including deterministic lateral displacement, inertial focusing, and magnetophoresis to sort up to 10^7 cells/s. The CTC-iChip is able to operate in both positive and negative selection modalities as a function of the antibodies used on the magnetic beads (positive: anti-EpCAM, negative: anti-CD45). In positive selection mode the yield is dependent upon the EpCAM expression levels of the individual cancer cells leading to a yield of 97 % for highly EpCAM-expressing SKBR3 cells and a depletion of white blood cells of 3.8-log [37]. By using two-stage magnetophoresis and depletion antibodies against leukocytes, the same performance was achieved (3.8-log depletion of white blood

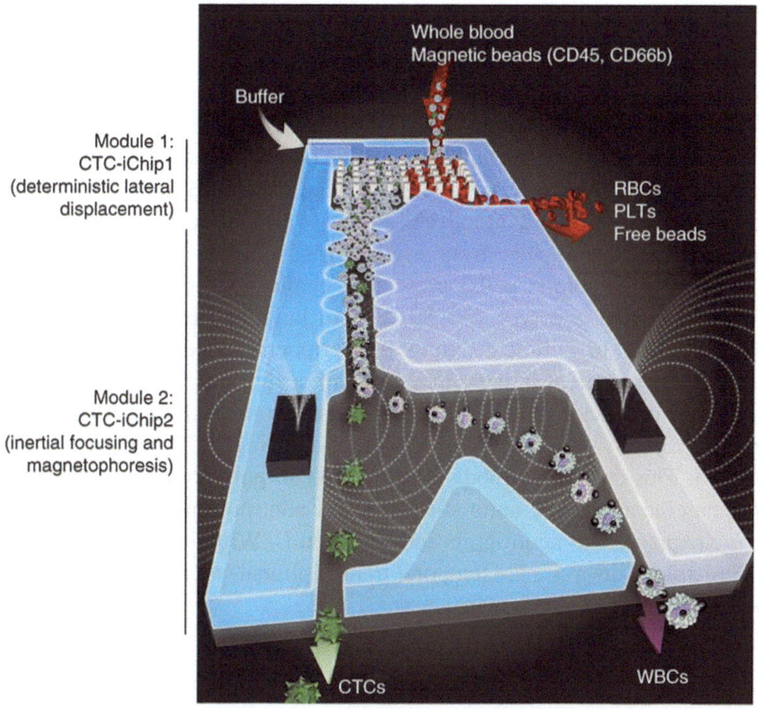

Fig. 14.8 The CTC-iChip: The CTC-iChip is composed of two separate microfluidic devices that house three different microfluidic components engineered for inline operation: DLD to remove nucleated cells from whole blood by size-based deflection by using a specially designed array of posts performed in CTC-iChip1, inertial focusing to line up cells to prepare for precise magnetic separation and magnetophoresis for sensitive separation of bead-labeled WBCs and unlabeled CTCs, which are performed in CTC-iChip2. *PLTs* platelets. Reproduced with permission from Karabacak et al. [38]

cells and a 97 % yield of rare cells) in negative selection mode with a sample processing rate of 8 mL of whole blood/h (Fig. 14.8) [38].

EpCAM-expressing circulating tumor cells, detected by CellSearch, are predictive of short survival in several cancers and may serve as a liquid biopsy to guide therapy. De Witt et al. have also investigated the presence of EpCAM+ CTC detected by CellSearch and EpCAM− CTC discarded by CellSearch, after EpCAM-based enrichment. EpCAM− CTCs were identified by filtration and fluorescent labeling. The majority of spiked EpCAM+ cells could be detected with CellSearch, whereas most spiked cells with EpCAMlow or EpCAM− expression were detected using filtration. Five or more CTCs were detected in 15 % of the patient samples; this increased to 41 % when adding the CTC detected in the discarded blood. In that pilot study, the presence of EpCAM+ CTC was associated with poor outcome, whereas the EpCAM− CTCs were not [60].

Other Applications of Rare Cell Isolation and Analysis

14.5 Immune Cells or Infected Cells

In clinical diagnostics, analysis of blood cellular components can yield information about disease state and progression, as well as patient response to treatment [61]. Microfluidic platforms have been developed for the isolation of platelets, erythrocytes, leukocytes [62, 63], lymphocytes [64], and B-cells [65]. Leukocytes make up 0.1–0.2 % of all blood cells ($4.1–10.9 \times 10^6$ cells/mL) and therefore rapidly isolating these cells from erythrocytes (93–96 % of blood cells) and platelets (4–7 % of blood cells) presents a challenge. Neutrophils are the most abundant leukocyte cell type (30–80 %) and have key roles in modulating the immune response. In a clinical setting, the peripheral blood is an easily accessible tissue, and there is great interest in using leukocyte transcript profiling to understand disease processes [66]. Kotz et al. designed a microfluidic device to capture neutrophils directly from 150 μl of whole blood using a CD66b-specific antibody (Fig. 14.9). This capture antibody is specific to all polymorphonuclear leukocytes bearing carcinoembryonic antigen-related cell adhesion molecule-8, an adhesion molecule expressed on neutrophils and eosinophils. With this robust methodology, they were able to discriminate temporal transcriptional events of neutrophils within a clinical setting.

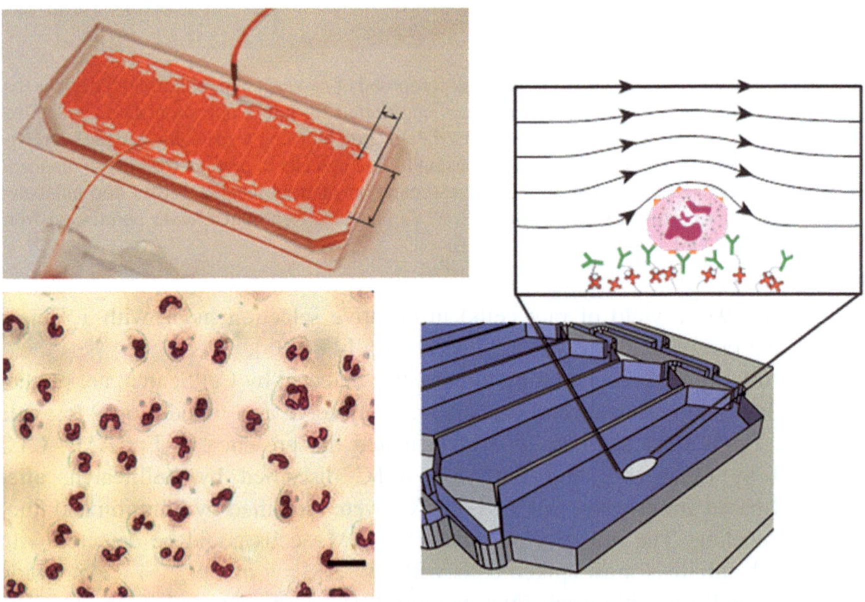

Fig. 14.9 Rapid purification microfluidic systems to isolate neutrophils from whole blood. Microfluidic device to purify neutrophils within minutes using antibody-based capture for subsequent diagnostic or research analysis. The microfluidic device is shown at the upper left with stained neutrophils that have been sorted from whole blood below (scale bar 520 mm); an illustration of the neutrophils captured within the microchannels by antibodies (zoomed view in *inset*) is also shown [67]

Although bulk populations of immune cells are not considered "rare" there is an increased interest in the ability to isolate and profile phenotypes of rare subpopulations of immune cells from whole blood to reveal the hidden variation in these cells. Individual cells behave differently from the average of many cells types, so the isolation and profiling of individual cells using microfluidics is advantageous in clinical and basic scientific research [68]. Research utilizing isolated single cells in sealed microfluidic chambers shows that less than 1 % of the population of cells, termed precocious cells, initiated the entire population response of the immune system. In the future, microfluidics can be used to isolate and detect this rare population of cells.

14.6 Pathogens

Standard diagnostic methods for typical bacterial infections involve several sequential growth steps followed by biochemical assays to identify the species and antibiotic susceptibility patterns [69], requiring 48–72 h. This lag time leads to excessive up-front use of broad-spectrum antibiotics or incorrect antibiotic administration due to resistant bacteria. Microfluidics has the advantages of reduced sample consumption, high analysis speed, and improved sensitivity compared with conventional analytical techniques, so this technology was suitable for rapid detection of pathogens.

Rapid sepsis diagnosis requires development of methods to identify rare pathogen cells in small samples of human blood. Magnetic beads functionalized with pathogen-binding ligands have been used to rapidly isolate microbes from blood; however, it is commonly difficult to optically detect the captured species because the excess numbers of beads required for pathogen binding physically interfere with light transmission after they have been concentrated. Inger's group has recently published on a microfluidic device to carry out magnetic isolation and optical detection of rare blood pathogens is to add ligand-coated magnetic beads to blood that bind specifically to pathogens, and to isolate the bound beads from the blood when flowing through a microfluidic channel using an external magnet to pull the beads up to the ceiling of the channel (Fig. 14.10).

Hou et al. utilize inertial microfluidics to directly determine pathogen identity and antibiotic susceptibility with hybridization-based RNA detection [71]. Using the same principle of Dean flow fractionation that was previously used to isolate CTCs, bacteria are separated from host blood cells. This label-free separation method has an efficient recovery of even low-abundance bacteria (<1 μm), which are significantly smaller than CTCs (~10–20 μm). A sheath flow at the inlet to "pinch" the bacteria-containing sample demonstrated well-controlled Dean migration of bacteria towards the outer wall to achieve efficient bacteria recovery, even over a large background of other cells. This enabled continuous, species-independent isolation of bacteria from whole blood without laborious affinity-based target labeling. Ribosomal RNA detection can then be applied for direct identification of low-abundance pathogens (~100 per mL) from blood without

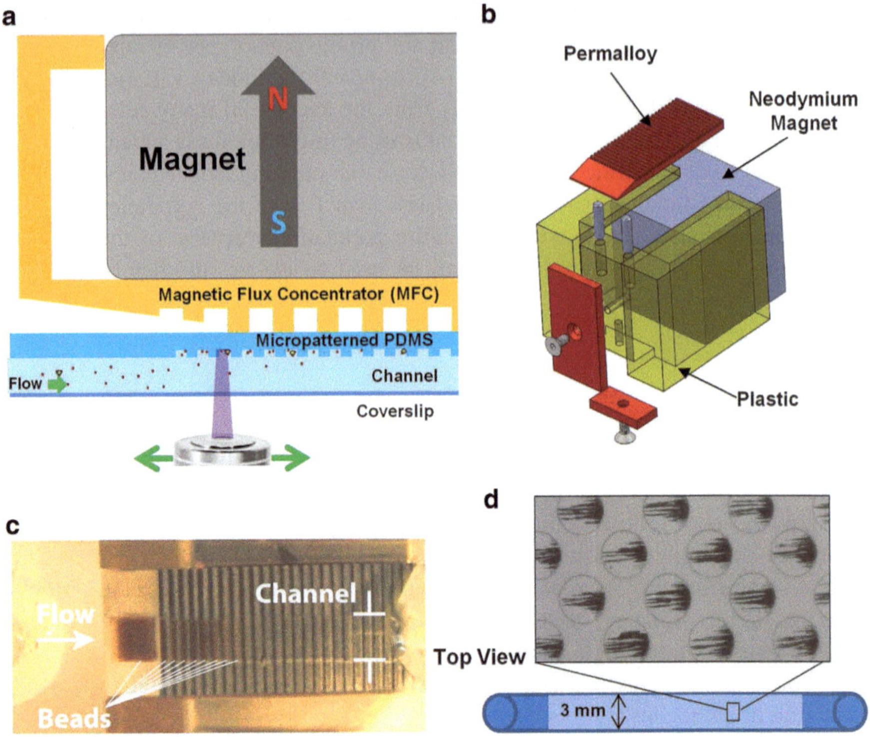

Fig. 14.10 Design of the magnetic capture microfluidic device. (**a**) Cross section of the entire device, showing orientation of the magnet, magnetic flux concentrator (MFC), and microfluidic channel (not to scale). (**b**) MFC made out of permalloy (*red*), neodymium magnet (*grey*), and plastic casing (*yellow*), forming a single, simple-to-use unit for capturing and displaying magnetic beads and tagged pathogens. (**c**) Photograph of beads captured in the device and spread into a thin layer by the presence of the MFC. (**d**) Higher magnification view of the microfluidic channel showing aligned chains of magnetic beads that were magnetically captured in the microwells molded into the ceiling of the PDMS microchannel. Figure reproduced with permission from Cooper et al. [70]

culturing or enzymatic amplification. Messenger RNA detection of antibiotic-responsive transcripts after brief drug exposure permits rapid susceptibility determination from bacteria with minimal culturing (~105 per mL). This unique coupling of microfluidic cell separation with RNA-based molecular detection techniques represents significant progress towards faster diagnostics (~8 h) to guide antibiotic therapy. By coupling microfluidic cell separation techniques with validated RNA detection methods to improve the interface between clinical samples and downstream molecular diagnostics, the potential exists for further improvements to increase the sensitivity and decrease the time required for diagnosing bacteremias in patients.

Microfluidics have recently been applied to isolate rare bacterium from air samples [72]. A staggered herringbone microfluidic mixer ensures cell contact with the protein A-coated microspheres for immunodetection (Fig. 14.11).

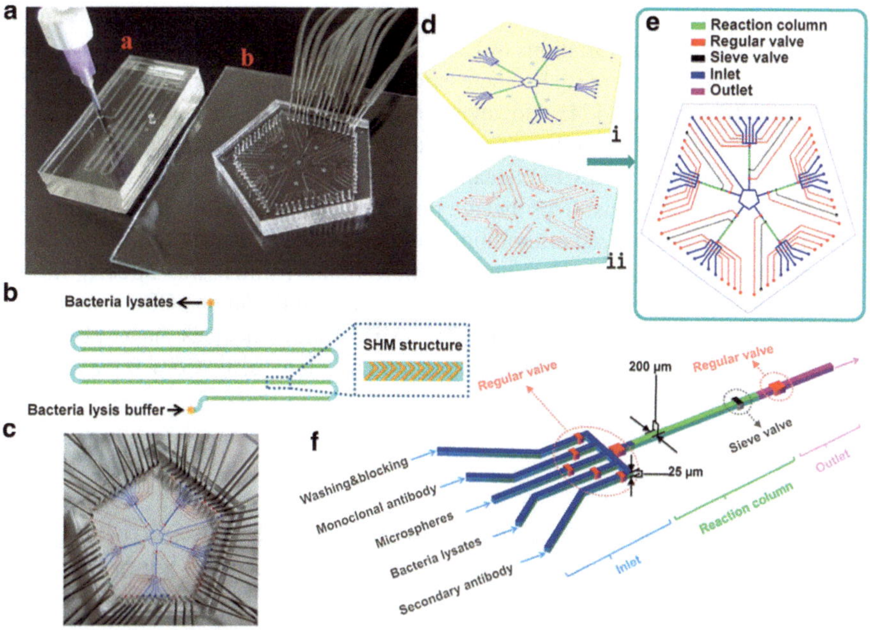

Fig. 14.11 Microfluidic platform for direct capture and analysis of airborne *Mycobacterium tuberculosis*. (**A**) Image of the system for airborne bacteria rapid enrichment and bacteriological diagnosis: (**a**) airborne bacteria enrichment microfluidic chip, (**b**) immunoassay microfluidic chip. (**B**) Schematic illustration and detailed structure of enrichment microfluidic chip. (**C**) Image of the microfluidic immunoassay chip. The various channels had been loaded with food dyes to visualize the structure of the microfluidic chip. (**D**, **E**) Schematic illustrations of designed immunoassay microfluidic chip. Valves and columns were illustrated by *different colors*: (*i*) fluidic layer, (*ii*) control layer. (**F**) Enlarged diagram showing detailed structure of microfluidic immunoassay chip. Figure reproduced with permission from Jing et al. [73]

Compared with the conventional analysis method, the whole analysis time is about 50 min including bacteria capture and enrichment, bacteria lysis, immunoreaction, and result readout, suggesting that this integrated system is suitable for rapid detection of airborne bacteria. Above all, this direct analysis system has advantages including simple operation, short test time, and accurate and reliable analysis, showing the potential applications in rapid detection of airborne infectious disease.

14.7 Stem Cells

Endothelial colony-forming cells (ECFCs) are endothelial progenitors that circulate in peripheral blood and are currently the subject of intensive investigation due to their therapeutic potential. The identification of ECFCs in human blood created a promising opportunity to noninvasively derive large quantities of autologous endothelial cells for therapeutic use. Besides their therapeutic potential, mounting evidence indicates that variations of ECFC levels are likely associated with various pathologies and thus there is also increasing interest in accounting levels of ECFCs for diagnostic purposes [74]. However, in adults, ECFCs comprise a very small subset among circulating cells, which makes their isolation a challenge. Currently, the standard method for ECFC isolation relies on the separation of mononuclear cells and erythrocyte lysis, steps that are time consuming and known to increase cell loss. Lin et al. have utilized microfluidics with antibody posts to capture these cells efficiently from whole blood samples. Whole blood (6 mL) was directly flowed through 20 parallel microfluidic devices with anti-human CD34 antibody-functionalized posts at 5 mL/min [75]. Microfluidic devices have also been engineered to capture and enrich hematopoietic stem cells from bone marrow samples up to 98 % [76].

14.8 Analysis of Rare Cell Phenotype Using Microfluidics: New Tools for Studying Cancer and Cancer Stem Cell Migration

Downstream of rare cell capture, microfluidics offers the unique capability to precisely quantify cell phenotype, including cell migration patterns. Microfluidic devices allow for a lab-on-a-chip array to simplify single-cell analysis by providing a microenvironment that is of micrometer dimension and containing nanomoles of reagent/media. Furthermore, microfluidic devices allow for controlled placement of cells and precise delivery of factors. Therefore, microfluidic technology has features of reliable, efficient, and cost-effective single-cell selection and navigation [77]. Precisely defined geometries have also been shown to influence cell migration patterns. Studies utilizing microfluidics have also provided a framework on the impact of external constraints on collective cell migration [78]. During cancer progression, malignant cells in the tumor invade surrounding tissues. This transformation of adherent cells to a motile phenotype has been associated with the epithelial-mesenchymal transition (EMT). One advantage of in vitro microfluidic assays such as this one is the capability for sampling rare events while interrogating an extraordinary range of experimental conditions, including combinatorial screening of cancer cell lines and patient samples against potential therapeutic compounds (Fig. 14.12).

Microfluidic devices provide both mechanical and 3D environments, as well as complicated chemical environment where chemotaxis is guided by concentration gradients of growth factors, chemokines, and surface ligands. These studies have enhanced our understanding of cell deformation and focal adhesion dynamics as well as their responses to subtle variations in their microenvironment [77].

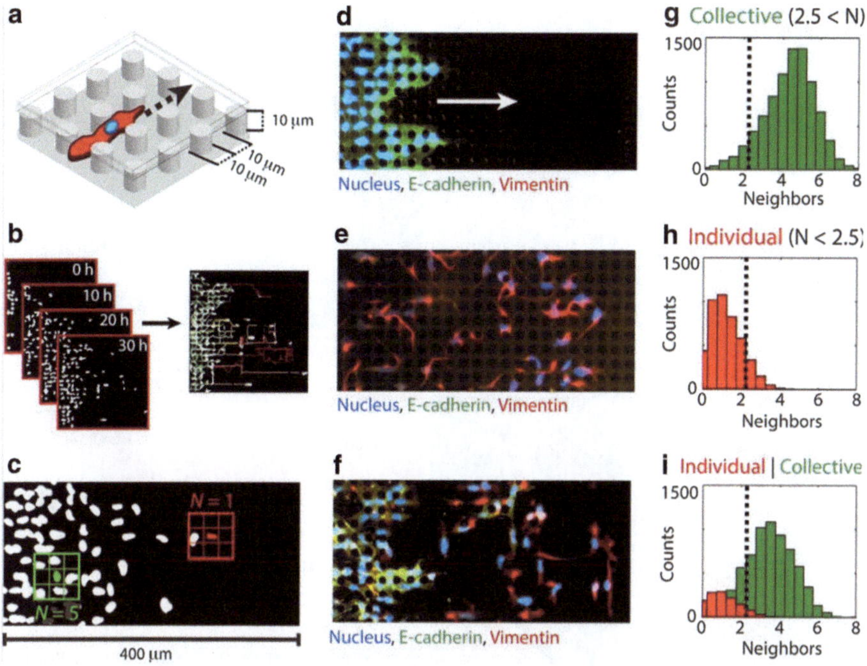

Fig. 14.12 Collective and individual migration following the epithelial-mesenchymal transition. (**a**) Cells invaded an enclosed array of fibronectin-coated PDMS micropillars with height, diameter, and spacing of 10 μm. (**b**) Cell migration was automatically tracked from time-lapse microscopy by segmenting fluorescently labeled nuclei. (**c**) A representative metric for collective or individual migration is based on the lifetime-average number of nearest neighbors within one pillar spacing. Immunofluorescent staining reveals biomarker expression associated with (**d**) epithelial phenotype (E-cadherin, *green*) in MCF-10A, (**e**) mesenchymal phenotype (vimentin, *red*) in MDA-MB-231, (**f**) both epithelial (E-cadherin, *green*) and mesenchymal phenotypes (vimentin, *red*) in MCF-10A Snail. Histograms of the number of lifetime-averaged nearest neighbors per cell indicate (**g**) collective migration in MCF-10A, (**h**) individual migration in MDA-MB-231, and (**i**) collective and individual migration in MCF-10A Snail [79]

14.9 Measuring Immune Cell Trafficking and Phenotype

Microfluidics has offered many solutions for next-generation chemotaxis assays [80, 81]; however, none of these methods have seen widespread adoption at the level of traditional assays, such as the Boyden chamber. A wide variety of analytical measurements can be used to characterize the state and capability of the immune system. The resulting data help reveal the fundamental biology of immunity, provide insight into the evolution of disease, aid the design of clinical diagnostics or interventions, and establish distinct signatures for effective immune responses. Improving the resolution of our measurements to capture the full complexity encompassing time-varying states and interconnectedness of cell subsets presents a substantial challenge [82].

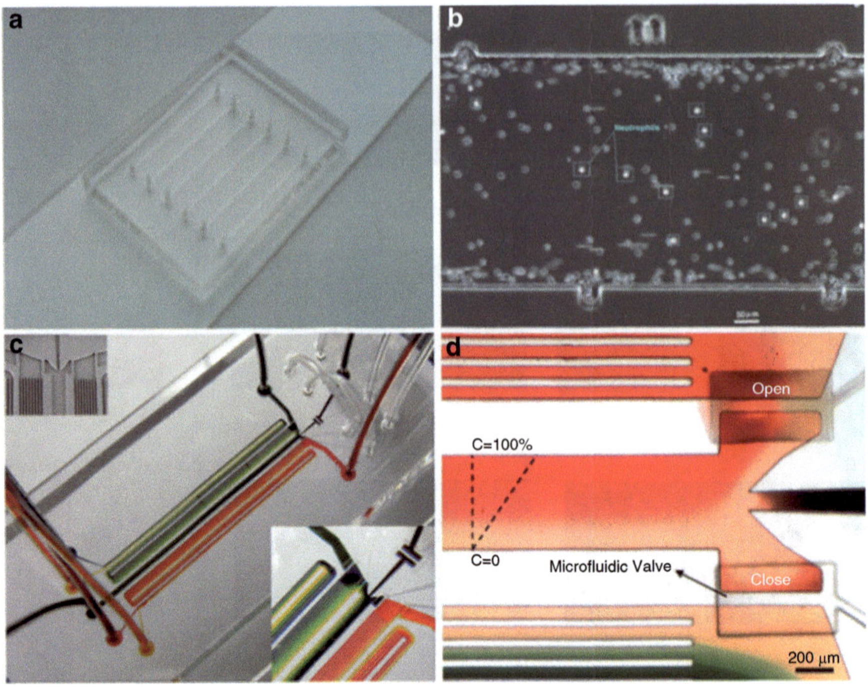

Fig. 14.13 Overall design of the microfluidic devices for neutrophil separation from whole blood and chemotaxis assay. (**a**) Microfluidic device with six parallel channels for neutrophil capture characterization. (**b**) Magnified image of the cell capture chamber with ruler markers on the top and bottom. Dimensions for each section between the two marks are 750 μm × 450 μm. (C) Microfluidic device for combined cell capture and chemotaxis assay. Two gradient generators are connected to the cell capture and migration chamber through microfluidic valves. Inset figures present details of the microstructured membrane valves employed in the device. (**d**) Close-up picture of the gradient flow chamber. The valve connecting the top gradient network is open allowing fluids to flow through the cell capture and migration chamber while the bottom valve is closed diverting fluids from the second gradient network to the waste reservoir. Dimensions of the separation and chemotaxis chamber are 450 μm wide × 80 μm tall [83]

With recent developments in microfluidic technologies, cell migration research has gained significant attention in recent years and a variety of in vitro migration assays have been developed. Agrawal et al. designed a microfluidic chip for chemotaxis studies which can use neutrophils isolated on the chip, directly from whole blood (Fig. 14.13). Subsequent analysis of neutrophil migration in chemoattractant gradients of N-formyl-methyl-leucyl-phenylalanine (fMLP) or interleukin-8 (IL-8) shows higher average velocities over E-selectin as compared to the P-selectin. These assays took less than 10 min and neutrophil chemotaxis at individual cell level was observed in the early cell migration events following exposure to different chemoattractant gradients.

Microfluidic devices can also be used to measure immune cell phenotypes as a biomarker for diseases, such as sepsis [84, 85]. Immune cell chemotaxis chambers

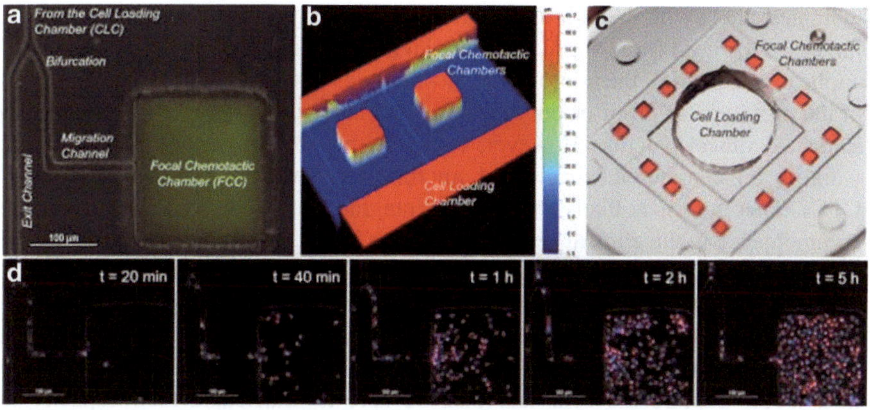

Fig. 14.14 Characterization of microfluidic postdiapedesis inflammation model. (**a**) Chemoattractant (*green*) is primed into the device and a gradient is formed along the migration channel towards the FCC. (**b**) Profilometer image of two adjacent FCCs illustrates distinct heights of FCCs and migration channels. (**c**) Large-scale model of device (magnification, 5×). Sixteen FCCs (*red*) surround each CLC. After washing, the chemoattractant only remains in the FCC, and a linear gradient is formed along the migration channel. (**d**) Double-stained (*blue*, nucleus; *red*, membrane) neutrophils begin migrating along the LTB4 gradient after 20 min and fill the chamber by 5 h [86]

can also be used to monitor leukocyte migration patterns and phenotype after treatment with inflammatory mediators. The microfluidic chambers for neutrophil and monocyte trafficking that we describe can be used to study the phlogistic and nonphlogistic recruitment and interactions between the two cell populations responding to various chemoattractants and modulators of inflammation. A key feature of the assay is the ability to probe the pro- or anti-inflammatory status of the cells trafficking towards the FCCs in the presence of lipid mediators and NPRMs (Fig. 14.14).

Finally, microfluidic devices have the potential to facilitate quantification of host immune cell-pathogen interactions at a single-cell level [87]. The microfluidic platform developed in this study may eventually become a useful tool for measuring the ability of neutrophils from patients to mount effective immune responses against fungi in vitro. This tool may help better estimate the risk for fungal infections for each patient and have implications for the faster diagnosis and earlier start of antifungal treatment during infections.

14.10 Conclusions

Microfluidic technologies have demonstrated great promise for the capture of rare cells with high throughput, yield, purity, specificity, and viability for clinical applications. Advances in cell biology, cell-based diagnostics, and cellular therapies, along with the promise of point-of-care and personalized medicine, have increased the need for rare cell isolation devices in both basic research and clinical applications. Most of the microfluidic technologies described in this review are still in the prototype or proof-of-concept stage and have yet to become integrated into standard clinical practice. However, currently there are more than 30 companies working on commercializing novel microfluidic platforms for the isolation of CTCs. Therefore it is likely that a fully integrated and validated system will be commonplace in the clinical setting in the near future. The substantial commercial investment for standardization, manufacturability, and repeatable performance will likely position microfluidic technologies at the forefront of rare cell isolation. The current market for CTC diagnostics is approximately $250 million annually, with most of the testing being performed in the USA. The next challenge is to go to the next level—capture the cells and then be able to understand their capacity to metastasize as well as to understand the most efficient drug regimens for a specific cancer in an individual patient. Microfluidics will likely also enable these high-level, post-sorting analyses such as biochemical, secretion, and cell culture assays for true lab-on-a-chip devices. Microfluidics will also enable us to get higher quality information from blood and other bodily fluids for other diseases, such as sepsis. Due to the superior benefits of microfluidics, this promising technology will be critical in advancing applications in biology, medicine, and industry.

Acknowledgements We would like to thank Mark Lenzi for his assistance with figure editing.

References

1. Dharmasiri U, Witek MA, Adams AA, Soper SA (2010) Microsystems for the capture of low-abundance cells. Annu Rev Anal Chem (Palo Alto Calif) 3:409–431. doi:10.1146/annurev.anchem.111808.073610
2. Hyun KA, Jung HI (2013) Microfluidic devices for the isolation of circulating rare cells: a focus on affinity-based, dielectrophoresis, and hydrophoresis. Electrophoresis 34:1028–1041. doi:10.1002/elps.201200417
3. Faley SL et al (2009) Microfluidic single cell arrays to interrogate signalling dynamics of individual, patient-derived hematopoietic stem cells. Lab Chip 9:2659–2664. doi:10.1039/b902083g
4. Chen Y et al (2014) Rare cell isolation and analysis in microfluidics. Lab Chip 14:626–645. doi:10.1039/c3lc90136j
5. Patterson AS et al (2013) Microfluidic chip-based detection and intraspecies strain discrimination of Salmonella serovars derived from whole blood of septic mice. Appl Environ Microbiol 79:2302–2311. doi:10.1128/AEM.03882-12
6. Lee MG, Shin JH, Bae CY, Choi S, Park JK (2013) Label-free cancer cell separation from human whole blood using inertial microfluidics at low shear stress. Anal Chem 85:6213–6218. doi:10.1021/ac4006149
7. Xu L et al (2015) Optimization and evaluation of a novel size based circulating tumor cell isolation system. PLoS One 10:e0138032. doi:10.1371/journal.pone.0138032
8. Huang X et al (2015) Meta-analysis of the prognostic value of circulating tumor cells detected with the Cell Search System in colorectal cancer. BMC Cancer 15:202. doi:10.1186/s12885-015-1218-9

9. Fiddler M (2014) Fetal cell based prenatal diagnosis: perspectives on the present and future. J Clin Med 3:972–985. doi:10.3390/jcm3030972
10. Wang CH, Weng CH, Che YJ, Wang K, Lee GB (2015) Cancer cell-specific oligopeptides selected by an integrated microfluidic system from a phage display library for ovarian cancer diagnosis. Theranostics 5:431–442. doi:10.7150/thno.10891
11. Shields CW, Reyes CD, Lopez GP (2015) Microfluidic cell sorting: a review of the advances in the separation of cells from debulking to rare cell isolation. Lab Chip 15:1230–1249. doi:10.1039/c4lc01246a
12. Yu M, Stott S, Toner M, Maheswaran S, Haber DA (2011) Circulating tumor cells: approaches to isolation and characterization. J Cell Biol 192:373–382. doi:10.1083/jcb.201010021
13. Faltas B (2012) Cornering metastases: therapeutic targeting of circulating tumor cells and stem cells. Front Oncol 2:68. doi:10.3389/fonc.2012.00068
14. Nagrath S et al (2007) Isolation of rare circulating tumour cells in cancer patients by microchip technology. Nature 450:1235–1239. doi:10.1038/nature06385
15. Murlidhar V et al (2014) A radial flow microfluidic device for ultra-high-throughput affinity-based isolation of circulating tumor cells. Small 10:4895–4904. doi:10.1002/smll.201400719
16. Gleghorn JP et al (2010) Capture of circulating tumor cells from whole blood of prostate cancer patients using geometrically enhanced differential immunocapture (GEDI) and a prostate-specific antibody. Lab Chip 10:27–29. doi:10.1039/b917959c
17. Kirby BJ et al (2012) Functional characterization of circulating tumor cells with a prostate-cancer-specific microfluidic device. PLoS One 7:e35976. doi:10.1371/journal.pone.0035976
18. Stott SL et al (2010) Isolation of circulating tumor cells using a microvortex-generating herringbone-chip. Proc Natl Acad Sci U S A 107:18392–18397. doi:10.1073/pnas.1012539107
19. Watanabe M, Serizawa M (2014) A novel flow cytometry-based cell capture platform for the detection, capture and molecular characterization of rare tumor cells in blood. J Transl Med 12:143. doi:10.1186/1479-5876-12-143
20. Watanabe M et al (2013) Multicolor detection of rare tumor cells in blood using a novel flow cytometry-based system. Cytometry A 85:206–213. doi:10.1002/cyto.a.22422
21. Plouffe BD, Mahalanabis M, Lewis LH, Klapperich CM, Murthy SK (2012) Clinically relevant microfluidic magnetophoretic isolation of rare-cell populations for diagnostic and therapeutic monitoring applications. Anal Chem 84:1336–1344. doi:10.1021/ac2022844
22. Chang C-L et al (2015) Circulating tumor cell detection using a parallel flow micro-aperture chip system. Lab Chip 15:1677–1688. doi:10.1039/C5LC00100E
23. Hoshino K et al (2011) Microchip-based immunomagnetic detection of circulating tumor cells. Lab Chip 11:3449–3457. doi:10.1039/c1lc20270g
24. Kang JH et al (2012) A combined micromagnetic-microfluidic device for rapid capture and culture of rare circulating tumor cells. Lab Chip 12:2175. doi:10.1039/c2lc40072c
25. Lee JJ et al (2014) Synthetic ligand-coated magnetic nanoparticles for microfluidic bacterial separation from blood. Nano Lett 14:1–5. doi:10.1021/nl3047305
26. Riahi R, Gogoi P, Sepehri S (2014) A novel microchannel-based device to capture and analyze circulating tumor cells (CTCs) of breast cancer. Int J Oncol 44:1870–1878. doi:10.3892/ijo.2014.2353
27. Desitter I, Guerrouahen BS, Benali-Furet N (2011) A new device for rapid isolation by size and characterization of rare circulating tumor cells. Anticancer Res 31(2):427–441
28. Vona G et al (2000) Isolation by size of epithelial tumor cells: a new method for the immunomorphological and molecular characterization of circulating tumor cells. Am J Pathol 156:57–63
29. Zabaglo L et al (2003) Cell filtration-laser scanning cytometry for the characterisation of circulating breast cancer cells. Cytometry A 55A:102–108. doi:10.1002/cyto.a.10071
30. Warkiani M et al (2014) An ultra-high-throughput spiral microfluidic biochip for the enrichment of circulating tumor cells. Analyst 139:3245–3255
31. Hou HW et al (2013) Isolation and retrieval of circulating tumor cells using centrifugal forces. Sci Rep 3:1259. doi:10.1038/srep01259

32. Bhagat A, Hou H, Li L, Lim C, Han J (2011) Pinched flow coupled shear-modulated inertial microfluidics for high-throughput rare blood cell separation. Lab Chip 11:1870–1878
33. Sollier E et al (2014) Size-selective collection of circulating tumor cells using Vortex technology. Lab Chip 14:63–77. doi:10.1039/c3lc50689d
34. Antfolk M, Magnusson C, Augustsson P, Lilja H, Laurell T (2015) Acoustofluidic, label-free separation and simultaneous concentration of rare tumor cells from white blood cells. Anal Chem 87:9322–9328. doi:10.1021/acs.analchem.5b02023
35. Sarioglu AF et al (2015) A microfluidic device for label-free, physical capture of circulating tumor cell clusters. Nat Methods 12:685–691. doi:10.1038/nmeth.3404
36. Hur SC, Henderson-Maclennan NK, Mccabe ERB, Di Carlo D (2011) Deformability-based cell classification and enrichment using inertial microfluidics. Lab Chip 11:912–920. doi:10.1039/c0lc00595a
37. Ozkumur E et al (2013) Inertial focusing for tumor antigen-dependent and -independent sorting of rare circulating tumor cells. Sci Transl Med 5:179ra147. doi:10.1126/scitranslmed.3005616
38. Karabacak NM et al (2014) Microfluidic, marker-free isolation of circulating tumor cells from blood samples. Nat Protoc 9:694–710. doi:10.1038/nprot.2014.044
39. Sajay BNG et al (2014) Microfluidic platform for negative enrichment of circulating tumor cells. Biomed Microdevices 16:537–548. doi:10.1007/s10544-014-9856-2
40. de Wit S et al (2015) The detection of EpCAM+ and EpCAM− circulating tumor cells. Sci Rep: 1–10. doi:10.1038/srep12270.
41. Davis JA et al (2006) Deterministic hydrodynamics: taking blood apart. Proc Natl Acad Sci U S A 103:14779–14784. doi:10.1073/pnas.0605967103
42. Aceto N, Toner M, Maheswaran S, Haber DA (2015) En route to metastasis: circulating tumor cell clusters and epithelial-to-mesenchymal transition. Trends Cancer 1:44–52. doi:10.1016/j.trecan.2015.07.006
43. Javaid S et al (2015) MAPK7 regulates EMT features and modulates the Generation of CTCs. Mol Cancer Res 13:934–943. doi:10.1158/1541-7786.MCR-14-0604
44. Sundaresan TK et al (2015) Detection of T790M, the acquired resistance EGFR mutation, by tumor biopsy versus noninvasive blood-based analyses. Clin Cancer Res. doi:10.1158/1078-0432.CCR-15-1031.
45. Satelli A, Brownlee Z, Mitra A, Meng QH, Li S (2014) Circulating tumor cell enumeration with a combination of epithelial cell adhesion molecule- and cell-surface vimentin-based methods for monitoring breast cancer therapeutic response. Clin Chem 61:259–266. doi:10.1373/clinchem.2014.228122
46. Satelli A et al (2014) Universal marker and detection tool for human sarcoma circulating tumor cells. Cancer Res 74:1645–1650. doi:10.1158/0008-5472.CAN-13-1739
47. Schwesinger F et al (2000) Unbinding forces of single antibody-antigen complexes correlate with their thermal dissociation rates. Proc Natl Acad Sci U S A 97:9972–9977
48. Maheswaran S et al (2008) Detection of mutations in EGFR in circulating lung-cancer cells. N Engl J Med 359:366–377. doi:10.1056/NEJMoa0800668
49. Stott SL et al (2010) Isolation and characterization of circulating tumor cells from patients with localized and metastatic prostate cancer. Sci Transl Med 2:25ra23. doi:10.1126/scitranslmed.3000403
50. Zhu H, Yan J, Revzin A (2008) Catch and release cell sorting: electrochemical desorption of T-cells from antibody-modified microelectrodes. Colloids Surf B Biointerfaces 64:260–268. doi:10.1016/j.colsurfb.2008.02.010
51. Wan Y et al (2012) Capture, isolation and release of cancer cells with aptamer-functionalized glass bead array. Lab Chip 12:4693–4701. doi:10.1039/c2lc21251j
52. Allard WJ et al (2004) Tumor cells circulate in the peripheral blood of all major carcinomas but not in healthy subjects or patients with nonmalignant diseases. Clin Cancer Res 10:6897–6904. doi:10.1158/1078-0432.CCR-04-0378

53. de Bono JS et al (2008) Circulating tumor cells predict survival benefit from treatment in metastatic castration-resistant prostate cancer. Clin Cancer Res 14:6302–6309. doi:10.1158/1078-0432.CCR-08-0872
54. Ilyas A, Asghar W, Kim YT, Iqbal SM (2014) Parallel recognition of cancer cells using an addressable array of solid-state micropores. Biosens Bioelectron 62:343–349. doi:10.1016/j.bios.2014.06.048
55. Asghar W et al (2012) Electrical fingerprinting, 3D profiling and detection of tumor cells with solid-state micropores. Lab Chip 12:2345–2352. doi:10.1039/c2lc21012f
56. Di Carlo D (2009) Inertial microfluidics. Lab Chip 9:3038–3046. doi:10.1039/b912547g
57. Di Carlo D, Irimia D, Tompkins RG, Toner M (2007) Continuous inertial focusing, ordering, and separation of particles in microchannels. Proc Natl Acad Sci U S A 104:18892–18897. doi:10.1073/pnas.0704958104
58. Di Carlo D, Edd JF, Irimia D, Tompkins RG, Toner M (2008) Equilibrium separation and filtration of particles using differential inertial focusing. Anal Chem 80:2204–2211. doi:10.1021/ac702283m
59. Oakey J et al (2010) Particle focusing in staged inertial microfluidic devices for flow cytometry. Anal Chem 82:3862–3867. doi:10.1021/ac100387b
60. de Wit S et al (2015) The detection of EpCAM(+) and EpCAM(−) circulating tumor cells. Sci Rep 5:12270. doi:10.1038/srep12270
61. Jackson EL, Lu H (2013) Advances in microfluidic cell separation and manipulation. Curr Opin Chem Eng 2:398–404. doi:10.1016/j.coche.2013.10.001
62. Geislinger TM, Eggart B, Ller SB, Schmid L, Franke T (2012) Separation of blood cells using hydrodynamic lift. Appl Phys Lett 100:183701. doi:10.1063/1.4709614
63. Zheng SY, Liu JQ, Tai YC (2008) Streamline-based microfluidic devices for erythrocytes and leukocytes separation. J Microelectromech Syst 17:1029–1038. doi:10.1109/Jmems.2008.924274
64. Rosenbach AE et al (2011) Microfluidics for T-lymphocyte cell separation and inflammation monitoring in burn patients. Clin Transl Sci 4:63–68. doi:10.1111/j.1752-8062.2010.00255.x
65. Murthy SK, Sin A, Tompkins RG, Toner M (2004) Effect of flow and surface conditions on human lymphocyte isolation using microfluidic chambers. Langmuir 20:11649–11655. doi:10.1021/la048047b
66. Calvano SE et al (2005) A network-based analysis of systemic inflammation in humans. Nature 437:1032–1037. doi:10.1038/nature03985
67. Kotz KT et al (2010) Clinical microfluidics for neutrophil genomics and proteomics. Nat Med 16:1042–1047. doi:10.1038/nm.2205
68. Shalek AK et al (2014) Single-cell RNA-seq reveals dynamic paracrine control of cellular variation. Nature 510:363–369. doi:10.1038/nature13437
69. Weile J, Knabbe C (2009) Current applications and future trends of molecular diagnostics in clinical bacteriology. Anal Bioanal Chem 394:731–742. doi:10.1007/s00216-009-2779-8
70. Cooper RM et al (2014) A microdevice for rapid optical detection of magnetically captured rare blood pathogens. Lab Chip 14:182–188. doi:10.1039/c3lc50935d
71. Hou HW, Bhattacharyya RP, Hung DT, Han J (2015) Direct detection and drug-resistance profiling of bacteremias using inertial microfluidics. Lab Chip 15:2297–2307. doi:10.1039/c5lc00311c
72. Jing W et al (2013) Microfluidic device for efficient airborne bacteria capture and enrichment. Anal Chem 85:5255–5262. doi:10.1021/ac400590c
73. Jing W et al (2014) Microfluidic platform for direct capture and analysis of airborne Mycobacterium tuberculosis. Anal Chem 86:5815–5821. doi:10.1021/ac500578h
74. Munoz-Hernandez R et al (2014) Decreased level of cord blood circulating endothelial colony-forming cells in preeclampsia. Hypertension 64:165–171. doi:10.1161/HYPERTENSIONAHA.113.03058
75. Lin RZ, Hatch A, Antontsev VG, Murthy SK, Melero-Martin JM (2015) Microfluidic capture of endothelial colony-forming cells from human adult peripheral blood: phenotypic and

functional validation in vivo. Tissue Eng Part C Methods 21:274–283. doi:10.1089/ten.TEC.2014.0323

76. Schirhagl R, Fuereder I, Hall EW, Medeiros BC, Zare RN (2011) Microfluidic purification and analysis of hematopoietic stem cells from bone marrow. Lab Chip 11:3130–3135. doi:10.1039/c1lc20353c
77. Huang Y, Agrawal B, Sun D, Kuo JS, Williams JC (2011) Microfluidics-based devices: New tools for studying cancer and cancer stem cell migration. Biomicrofluidics 5:13412. doi:10.1063/1.3555195
78. Vedula SR et al (2012) Emerging modes of collective cell migration induced by geometrical constraints. Proc Natl Acad Sci U S A 109:12974–12979. doi:10.1073/pnas.1119313109
79. Wong IY et al (2014) Collective and individual migration following the epithelial-mesenchymal transition. Nat Mater 13:1063–1071. doi:10.1038/nmat4062
80. Keenan TM, Folch A (2008) Biomolecular gradients in cell culture systems. Lab Chip 8:34–57. doi:10.1039/b711887b
81. Irimia D (2010) Microfluidic technologies for temporal perturbations of chemotaxis. Annu Rev Biomed Eng 12:259–284. doi:10.1146/annurev-bioeng-070909-105241
82. Chattopadhyay PK, Gierahn TM, Roederer M, Love JC (2014) Single-cell technologies for monitoring immune systems. Nat Immunol 15:128–135. doi:10.1038/ni.2796
83. Agrawal N, Toner M, Irimia D (2008) Neutrophil migration assay from a drop of blood. Lab Chip 8:2054–2061. doi:10.1039/b813588f
84. Jones CN et al (2014) Spontaneous neutrophil migration patterns during sepsis after major burns. PLoS One 9:e114509. doi:10.1371/journal.pone.0114509
85. Kurihara T et al (2013) Resolvin D2 restores neutrophil directionality and improves survival after burns. FASEB J 27:2270–2281. doi:10.1096/fj.12-219519
86. Jones CN et al (2012) Microfluidic chambers for monitoring leukocyte trafficking and humanized nano-proresolving medicines interactions. Proc Natl Acad Sci U S A 109:20560–20565. doi:10.1073/pnas.1210269109
87. Jones CN et al (2016) Human neutrophils are primed by chemoattractant gradients for blocking the growth of Aspergillus fumigatus. J Infect Dis 213:465. doi:10.1093/infdis/jiv419

Chapter 15
Microfluidics for Cell Culture

Deepika Devadas and Edmond W.K. Young

Abstract Recent developments in biomicrofluidics research have led to myriad device technologies that enable cell culture in miniaturized geometries, often with additional functionalities that allow biological analyses at the molecular and cellular levels. This chapter describes key aspects of microscale cell culture enabled by microfluidics, includes illustrative examples from some of the latest advances in the field, and discusses important molecular biology techniques that can be performed on cells cultured in microscale environments, both in on-chip and off-chip formats. The aim of this chapter is to provide a basic guide and a list of key examples of reliable and robust microfluidic systems that can be used as reference for designing other systems.

Keywords Biomicrofluidics • Cell culture • Organ-on-a-chip • Cell-based assay • Microscopy • Cell function

15.1 Introduction

Laboratory cell culture has been around for more than 100 years, dating back to Ross Harrison's first-ever culture of nerve fibers from a frog embryo [1]. In the past 50 years, it has enabled significant advances in our understanding of the biology of living cells, the basic building blocks of life. While existing cell culture techniques have served the community with incredible reliability and facilitated tremendous research advancements, they also have various limitations that temper their utility, especially for more advanced and complex research problems.

D. Devadas, Ph.D. • E.W.K. Young, Ph.D. (✉)
Department of Mechanical and Industrial Engineering, University of Toronto,
5 King's College Road, Toronto, ON M5S 3G8, Canada
e-mail: deepikad@mie.utoronto.ca; eyoung@mie.utoronto.ca

© Springer International Publishing Switzerland 2016

C. Lu, S.S. Verbridge (eds.), *Microfluidic Methods for Molecular Biology*,
DOI 10.1007/978-3-319-30019-1_15

Recent advances in *biomicrofluidics* have led to a major shift in our ability to carry out laboratory experimentation on cells in culture. Biomicrofluidics, which is the study and integration of biology, microscale engineering, and fluid flow, offers several advantages to the conventional cell culture done on plastic dishes, flasks, and well plates. This includes reduced consumption of expensive reagents, increased throughput and parallelization, enhanced spatiotemporal control, and improved physiologic and pathophysiologic relevance of cultured cells in micro-environments. More recently, microfluidic cell cultures have led a major revolution toward highly complex models with organ-level function, i.e., "organs-on-chips" [2, 3]. Thus, the area is rapidly expanding, and technical knowledge on the methods and approaches for creating these advanced systems is desirable.

The focus of this chapter is on the basic principles and key considerations for culturing living cells in microfluidic devices. The aim is to provide guidelines for practitioners interested in pursuing microfluidic cell culture, and to use illustrative examples of state-of-the-art technology to offer the most reliable and robust techniques available for applying microfluidics to the biological question of interest. Since the focus of this chapter is on cell culture, other cell-based microfluidic applications such as cell handling, cell manipulation, counting, sorting, and encapsulation are beyond the scope of this discussion. First, general cell culture principles are discussed both in terms of macroscale and microscale factors. Examples of microfluidic cell culture systems are then examined in sequence, from 2D to 3D and from one cell type to multiple cell types, to illustrate key aspects of designing and developing the proper culture conditions. Biological assays incorporated with cell culture (either on-chip or off-chip) are described and discussed. Finally, technical challenges and future directions of the field are presented, with a focus on single-cell analyses and clinical applications.

15.2 Key Factors of Cell Culture

Living cells extracted from their native "in vivo" environments can be adapted to grow on laboratory substrates such as glass or plastic by providing the cells with liquid medium rich in nutrients and growth factors. This culturing of living cells "in vitro" enables controlled experimentation and examination of cellular responses and functions to assist our advancement of biology and medicine. In this section, key factors of cell culture that are critical to proper cell maintenance and viability at both macro- and microscales are introduced, including cell type, cell culture confluence, media volume, and nutrient and waste balance. Factors that are specific to microscale cultures are further examined, with the aim of providing procedural guidance to assist in achieving advanced microfluidic cell cultures, such as those described in subsequent sections.

15.2.1 Cell Type

The choice of cell type is clearly dictated by the biological question at hand, but is also influenced by the availability of sample material, and whether a model cell type can provide sufficient biological relevance to be useful in the proposed experiment. There are two main categories of cell sources: (1) cell lines, and (2) primary cells. Cell lines, or continuous cell lines, are often the first cells tested in a specific line of inquiry. Cell lines are derived from a single historic clone that has the ability to divide and be subcultured in vitro indefinitely, if handled properly. They are widely available, either commercially or from neighboring laboratories, and numerous established protocols exist to ensure high cell viability when regular maintenance is performed. The large majority of microfluidic cell-based systems have been characterized with cell line experiments as proof of concept (see Sect. 15.3). However, because cell lines have been "immortalized," they possess mutated DNA that often impacts their true biological behavior, which can in turn mask the underlying mechanisms that are to be investigated.

In contrast, primary cells are freshly derived from humans or other animals without immortalization, and therefore better mimic the true physiology of cells. However, primary cells cannot be propagated indefinitely like cell lines, and must be derived from raw samples for every experiment. They have a limited (and typically small) number of cell divisions (i.e., population doublings) before entering cell senescence, and have more fastidious culture requirements than cell lines. Thus, they are much more challenging to maintain and study in vitro. The challenge of managing primary cells extends to the microscale, and this has partially hindered progress in the field. Recent developments, however, show that the use of primary cells in microfluidic devices is on the rise.

15.2.2 Confluence

Once seeded onto a 2D substrate, cells rapidly attach and proliferate to cover the available surface. Some cell types (e.g., HS-5 human bone marrow stromal cell; L929 mouse fibroblast) can grow to confluence (100 % area coverage) or become over-confluent, i.e., overlapping each other in multiple layers when the substrate is already fully covered. Other cell types such as endothelial cells or epithelial cells establish cell-cell contacts with adjacent cells via cell surface proteins such as cadherins and cell-adhesion molecules (CAMs), and stop growing further due to contact inhibition. The result is a 2D confluent "monolayer" of cells with a thickness of one cell layer. This is often a sufficient model for simulating certain cellular behaviors, such as the endothelial barrier function of the inner lining of blood vessels [4, 5]. Confluence is typically reached after several days of growth, depending on the initial cell seeding density and volume of media (see Sect. 15.2.3). While reaching a desired confluence at the macroscale may be routine based on well-established protocols, reaching confluence at the microscale requires more experimentation, and consideration of other key factors such as device geometry and critical perfusion rates [6].

15.2.3 Media Volume

Cells cultured in conventional labware (e.g., Petri dishes, wells, and flasks) are supplied with enough media volume—and thus nutrients—to support multiple population doublings, and to maintain a manageable number of media exchanges before expansion (or subculturing) is needed. Selecting the proper media volume supplied in macroscale culture is also based on the need to maintain a proper fluid depth such that a controlled distance between cells and the gas-liquid interface is provided for gas exchange. Importantly, the volume of media also serves to control the concentration of cellular waste products. Metabolism of sugars in the medium results in acidic by-products that can significantly alter the pH of the medium. Low pH levels can irreversibly inhibit enzyme activities, and the synthesis of DNA, RNA, and protein, thus decreasing cell viability. Therefore, careful control of media pH is crucial to cell culture. This can often be addressed by adding chemical buffers to media that can control pH changes (e.g., HEPES). This method, combined with frequent media changes, can help extend culture times without loss in viability or quality.

Choosing the proper media volume becomes especially important at the microscale. Previous work has shown that pH and osmolality changes can become significant in microfluidic devices [7] and frequency of media changes increases due to rapid waste accumulation. These issues can often be addressed by incorporating perfusion, or by designing devices to account for their effective culture volumes (ECVs) [8] and effective culture times (ECTs) [6]. The key factor at the microscale is the significantly higher surface area-to-volume (SAV) ratio, which dictates that cultures have much less volume of media available per cell. While managing media volumes in microdevices can be challenging, especially when dealing with other factors such as evaporation, microscale culture also has its benefits in cases where signaling mechanisms are dictated by concentrated soluble factors that would otherwise be diluted in bulk volumes at the macroscale [9].

15.2.4 Specific Considerations for Microfluidic Cell Culture

While macro- and microscale cell cultures share certain similarities in several key factors, microscale cell culture also involves unique aspects that distinguishes it from traditional macroscale cultures, including the choice of device material and the geometric complexity of the design. These aspects have previously been discussed [6], and a brief overview is presented for convenience.

1. *Device material.* While the vast majority of traditional culture labware is made of either glass or polystyrene (PS), microfluidic devices have been fabricated with a wider range of materials, most notably poly(dimethylsiloxane) (PDMS), a soft elastomeric material that can be bonded to glass, plastic, and other layers of PDMS to create multilayer devices. Owing to a plethora of microfabrication methods and growing interest in the field, more options for device material are becoming available for fabricating microfluidic cell culture platforms. A comparison between PDMS and PS has been discussed in detail [10]. Furthermore, cyclo-olefin polymers and copolymers (COP and COC) are increasingly being studied and applied for microfluidic cell culture because of their lower autofluorescence properties that help enhance the quality of fluorescence images [11–13]. The key consideration is that device materials can differ in their bio-inertness, gas permeability, evaporation rate, autofluorescence and optical properties, and cell adhesion capability. PDMS and PS are hydrophobic by nature, and typically require oxygen plasma or UV treatment to facilitate cell attachment. Additionally, optical features and multiplexing abilities of materials may have to be considered depending on the experimental objectives. These considerations are not usually necessary at the macroscale because of the more limited number of available options, but are critical in the design of microfluidic devices.
2. *Geometries.* Lithography techniques offer great flexibility in the design of potentially complex microfluidic geometries and configurations. Cell culture microchannels are typically wider (~10^2–10^3 μm) compared to other microfluidic applications, mostly to allow growth of sufficient cells necessary for biological assays and for creating specific biological structures. In some instances, however, such as the culture of neurons, a combination of channel sizes is required to accommodate both neuron cell bodies (~10^2–10^3 μm wide channels) and axon processes (~10 μm wide channels) [14]. Depending on the experimental objectives and the readouts required, design geometries should account for soluble factor secretion, molecular diffusion, chemokine gradients, cell migration distances, and other functions that are dependent on the control and measurement of spatiotemporal factors.
3. *Effective culture volumes and times.* In contrast to conventional cultures, cells in microfluidic devices have much smaller volumes of media for the same surface area (i.e., a higher surface area-to-volume (SAV) ratio). This translates to more frequent media changes to maintain pH and therefore cell viability, if perfusion is not used. As discussed above, ECVs and ECTs have been derived to assist in the design of microfluidic cell culture systems [6, 8]. Briefly, a rough estimate of media change frequency can be calculated based on comparing the height of the microchannel to the height of media in a conventional culture system. For a typical ~1.2 mm height of media (common in macroscale cultures), media should be changed once every 48 h. If the height is halved (~600 μm), frequency should be doubled (i.e., once every 24 h). Alternatively, a perfusion setup can be used to convert a non-static (diffusion-based only) culture into a dynamic (or convective) culture as a means to replenish consumed nutrients and reduce

the frequency of media exchanges. In either case, the consideration of culture volumes and frequency of media changes at the microscale is more prominent than at the macroscale, as discussed previously in multiple reviews [6, 15].

The key principles of general and microscale cell culture identified above have become increasingly important for maintaining viability and function of cells cultured in microfluidic devices, especially as devices have become more biologically sophisticated. Many of these factors are known not to pose significant challenges for macroscale cell cultures when they deviate slightly from the norm. For example, pH fluctuations that lie within the range of pH ~7.4 ± 0.3 are commonly observed, and have relatively moderate effects only on cell culture viability, especially if the normal pH is restored on the order of hours. However, at the microscale, diffusional processes occur at much faster time scales, and issues such as pH imbalance become more pronounced, compelling the experimentalist to monitor them more closely to ensure proper maintenance [16]. Thus, many current microfluidic cell-based systems are accompanied by specific culture protocols to ensure that the procedures are aligned specifically for the particular microfluidic design.

15.3 Microfluidic Cell Culture for Biological Models: Recent Advances

Integrating microfluidics and biology has led to many innovative technologies, and in particular a significant advance in the complexity of biological models. Over the years, as more devices have been developed, and more protocols have become established using the abovementioned design guidelines, certain methods and techniques have emerged that have shown good reliability and repeatability. In this section, we review specific examples and recent advances, classified by model complexity (i.e., from 2D monocultures and cocultures to 3D monocultures and cocultures), in order to provide strong supporting evidence of the most reliable methods developed to date.

15.3.1 2D Monoculture Models

The simplest cell culture system (at both macro- and microscales) is a single cell type on a basic 2D substrate. There are two key advantages of microscale over macroscale cultures that are apparent even at this basic 2D level: (1) increased complexity of the channel geometries, and (2) increased functionality and control of spatiotemporal factors. Many of the first-generation microfluidic culture systems began as 2D monocultures, and several interesting advances in this category have developed in the past few years.

Neurons are one cell type in particular that has garnered significant interest since the beginning of microfluidic cell cultures. This is mostly due to the importance of controllable chemical gradients and the need for physical confinement and directional guidance of neurons, all of which can be conveniently achieved within microfluidic channels. Various devices, most often PDMS-based molds bonded to glass, have been developed to guide axon navigation and neuron development [17–19]. Typically, microchannel surfaces are first coated with extracellular matrix (ECM) proteins to facilitate cell adhesion, spreading, and mechanosensitive cell functions such as migration. In the case of neurons, laminin is most often used due to its abundance in neural tissues [20]. Other microfluidic devices have also employed poly-D-lysine (PDL) [21], a positively charged lysine homopolymer that facilitates cell attachment onto negatively charged surfaces like plastics. Both forms of polylysine (D and L forms) function well, but since the naturally available poly-L-lysine is susceptible to enzymatic degradation, the artificial poly-D-lysine is often preferred to prevent its digestion. As mentioned, the key to culturing neurons is the judicious combination of different-sized microchannels to accommodate the larger neuronal cell bodies while enabling extension of much narrower axons into microgrooves [14] (Fig. 15.1).

Another popular cell type studied in microfluidic channels is the endothelial cell (EC). Microfluidic channels are highly appropriate for studies of ECs because they can serve as microscale versions of traditional parallel plate flow chambers used to examine the effects of shear stress on endothelial function [22]. Culture devices have been developed to study EC adhesion [23], endothelial permeability [5], migration for wound healing, and others [24–27]. Numerous other examples of 2D microfluidic cultures for studying ECs and vascular functions are available in the literature, and have been critically reviewed [28]. Depending on the specific type of EC (e.g., aortic EC vs. aortic valve EC), the preferred ECM protein for coating the microchannels may be different (e.g., fibronectin vs. fibronectin and/or type I collagen) [23]. Furthermore, while geometries for neurons are dictated by the need to separate and confine different cell entities like axons and neuron cell bodies, geometries for EC culture are typically dictated by the desire to control shear stress to model the mechanobiological forces induced on an endothelium.

The simplicity of 2D microfluidic monocultures, through the examples of neuron and EC culture above, allows us to examine the most basic components and steps needed to achieve viable microfluidic cultures. Indeed, the key factors of geometry, media volume, device material, confluence, and cell type have already been covered. The one other factor that is clear from this section is the need to choose the proper ECM protein for cell adhesion. In the case of ECs, varying the ECM protein type and concentration can have significant effects on cell attachment and cell spread [23]. Furthermore, as more advanced functions are studied (e.g., stem cell differentiation), the choice of ECM will also lead to mechanical stiffness effects that may drive cell fate [29]. These mechanobiological factors should be accounted for when developing microfluidic cultures, not only in 2D monoculture models but more advanced cultures as well.

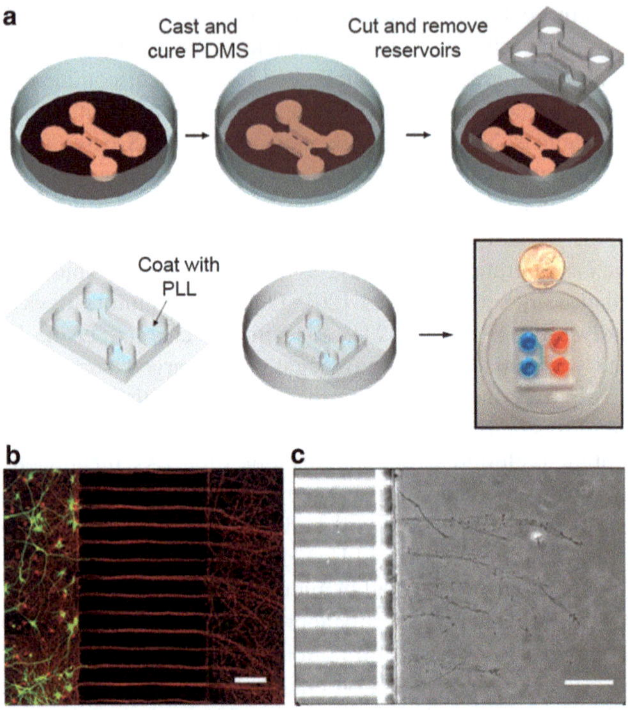

Fig. 15.1 Microfluidic 2D monoculture of neurons in PDMS-glass device. (**a**) The commonly used design involves media reservoirs at the microchannel inlets and outlets that can be easily cut out. The PDMS mold is then bonded to a glass cover slip (or other substrate), and coated with PLL (or PDL). (**b**) Immunostain of cultured neurons (*green* = Tau; *red* = MAP5) showing cell bodies in left chamber and axons extending along straight microgrooves to the right chamber. (**c**) Phase-contrast image of axonal processes in right chamber. Scale bars = 100 μm. Adapted from Park et al. [14] with permission

15.3.2 2D Coculture Models

To improve mimicry of in vivo tissue microenvironments, coculture models are employed that incorporate multiple relevant cell types in the same system. Cells influence the biological processes of their neighbors through various cell-cell interactions [30]. These interactions facilitate cell division, growth, movement, migration, survival, apoptosis, as well as signaling in disease pathways including cancer and tumor microenvironments. Cell-cell communication is mediated through various mechanisms, including via autocrine [31, 32], paracrine, and juxtacrine (or attachment-mediated) signaling [33]. In some instances, cocultures have merely consisted of mixing two different cell types in a culture, but in an effort to spatially separate cells without allowing physical contact, biologists have resorted to Transwell inserts that employ membrane filters, typically made of polyester or polycarbonate with fixed pore sizes, to support the culture of one cell type on the membrane, and segregate it from the culture of a second cell type below the membrane. Transwell inserts, however, are limited to two compartmentalized cell types, and tend to have long diffusion times due to the large distance from Transwell membrane to the base of the well [16].

Microfluidics offer important advantages for cocultures because of the precise spatial and temporal control it provides [9, 34]. This includes achieving physiologically relevant distances and concentrations, precise control over dynamic diffusion rather than uncontrolled convective mixing [35], and spatial arrangement of multiple cell types [36]. Indeed, many microfluidic systems are currently employed to study the interactions between different cell types in the same culture environment. On the topic of angiogenesis and endothelial morphogenesis, ECs have been cocultured with fibroblasts [37–39], or with both fibroblasts and macrophages [40]. In the latter case, Theberge and co-workers used a micropipette-based method to load one microfluidic compartment with macrophages while sequentially loading a neighboring compartment with fibroblasts (FBs) and ECs. The ECs relied on the FBs as a "feeder" layer, and enabled the tubule formation in the presence or absence of macrophages in close proximity (Fig. 15.2). In cancer research, various tumor cells have been studied in microfluidic coculture with fibroblasts for breast cancer [41], stromal cells for multiple myeloma [42], and ECs to understand the process of tumor extravasation [43]. Cells separated by ECM-coated, thin PDMS membranes have also been used to set up cocultures to mimic the lung microenvironment of a vascularized pulmonary edema [44]. Neto and co-workers adapted the microfluidic neuron culture system by Park et al. (see Fig. 15.1) for the coculture of sensory neurons and osteoblasts to model the in vivo interactions between nerve fibers and the bone microenvironment [18]. The design had previously demonstrated robustness in achieving axonal and neurite growth within straight microgrooves [14], and was extended to allow cell-cell communication between neurons and osteoblasts, with successful demonstration of axons extending from the neuron compartment, through the microgrooves, and into the osteoblast compartment.

Finally, coculture techniques have also extended to "multi-kingdom" studies, with the best example to date involving the coculture of intestinal epithelial cells with gut bacteria, specifically *Lactobacillus rhamnosus* [45]. This "gut-on-a-chip" has been useful for mimicking the interactions in the gut microenvironment during an infection, and the symbiotic culture of bacteria and mammalian cells under appropriate mechanobiological conditions revealed that organ-level functions can be recapitulated in vitro using microfluidics. This, and the other examples discussed herein, shows that microfluidic coculture techniques have greatly improved the mimicry of biological tissues, and that the microscale allows physiological concentrations of chemokines and appropriate cell compartmentalization that both play a role in cell growth, proliferation, metabolism, and migration that is better mimicked in microfluidic geometries.

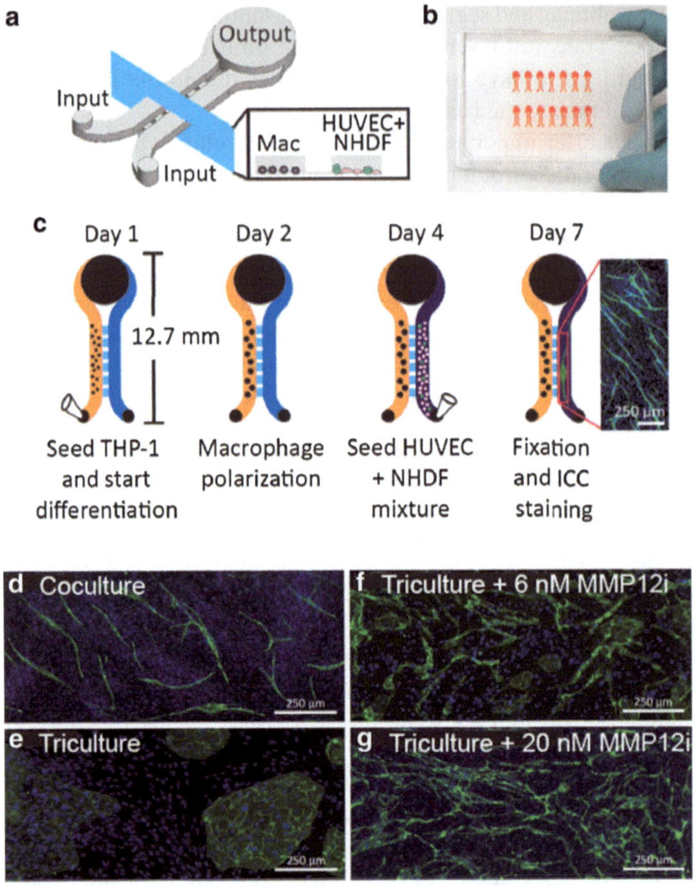

Fig. 15.2 Microfluidic triculture device for studying angiogenesis in the presence of fibroblasts and macrophages. (**a**) Two straight microchannels are interconnected by diffusion ports, and share a large output reservoir. *Mac* = macrophages, *HUVEC* = human umbilical vein endothelial cells, *NHDF* = neonatal human dermal fibroblasts. (**b**) The small footprint and open-port design enables fabrication of arrays of triculture systems on a single plate. (**c**) Different cell types are seeded sequentially via the input ports; HUVECs and NHDFs are seeded as a mixture at the same time, and lead to self-assembly into feeder layers and endothelial islands or tubules (**d**–**g**). Adapted from Theberge et al. [40] with permission

15.3.3 3D Culture Models

While 2D models are convenient and allow high-quality imaging due to proximity to the glass coverslip surface, they do not properly represent the three-dimensionality of cells in their native microenvironments. The need for 3D cultures to better recapitulate in vivo cellular behavior and function is well established [46, 47], and serves as strong motivation to pursue 3D cell culture platforms. The most popular 3D cellular model is the multicellular tumor spheroid (MCTS).

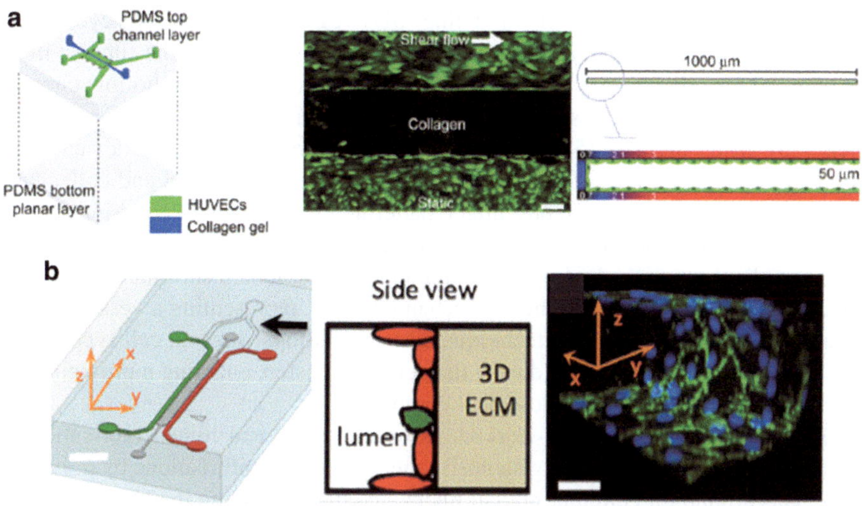

Fig. 15.3 Two different 3D microfluidic culture devices that culture endothelial cells (ECs) on three sides to model three-dimensionality of the endothelium. (**a**) Microdevice for studying shear force effects on angiogenesis. *HUVEC* = human umbilical vein endothelial cells. Endothelial monolayers are cultured on the ceiling, floor, and side wall of the channel. Adapted from Song et al. [52] with permission. (**b**) Microdevice used for studying extravasation of tumor cells through endothelium. Side view schematic shows ECs (*red*) cultured on the side of a 3D hydrogel layer, enabling a tumor cell (*green*) to migrate through the layer. Confocal imaging shows the C-shaped endothelium. Adapted from Zervantonakis et al. [50] with permission

Spheroids are comprised of large aggregates of cancer cells with a proliferating outer core and a quiescent, often necrotic inner core devoid of oxygen and nutrients. This results in a decreased availability of nutrients and oxygen as one moves away from the core, typical of a tumor microenvironment. 3D spheroid cultures have become the standard in vitro model in drug discovery and high-throughput screening applications. It is therefore not surprising that researchers have pursued the development of spheroids in microfluidic devices [48, 49]. The use of microfluidics, however, goes well beyond spheroids, and extends to more complicated 3D microenvironments that have not been achieved before with traditional methods.

To create models of endothelialized vessels, for example, researchers have fabricated rectangular cross-sectional microchannels, and simply seeded ECs on three (or four) sides of the microchannel [50, 51] (Fig. 15.3). Such systems have allowed the study of angiogenic sprouting in the presence of shear, as well as the dynamic monitoring of extravasation of tumor cells for elucidating mechanisms of metastasis. In both cases, gravity is initially used to allow the ECs to settle and attach on one of the surfaces, and the device is rotated to allow settling on all three (or four) surfaces, thus creating the full rectangular cross-sectional lumen of ECs. In particular, the presence of the vertical side wall of ECs enables planar visualization of certain biological processes, such as the sprouting of neovessels or the migration of other cell types.

An alternative method of modeling 3D vessels is to create cylindrical (as opposed to rectangular) lumens that more closely represent the tubular morphology of vessels. Several methods have been employed to create hollow circular lumens for tube-shaped tissues, including a method for embedding ECs in fibrin gels [53], using alginate [54], or using cellulose-based tubes [55] as a sacrificial template for the culture of ECs in a circular microchannel. Bischel and colleagues used a technique called viscous finger patterning (VFP) where they introduced liquid media into square microchannels containing pre-polymerized hydrogel. Flow of the media through the hydrogel left behind a circular lumen surrounded by a thin layer of hydrogel attached to the microchannel walls, representing a vessel wall or mammary duct [56, 57]. This technique allows additional stromal cell types to be incorporated into the gel layer during the gel-loading step, enabling multi-cell-type cocultures for additional physiological complexity.

Recently, Mannino and co-workers extended cylindrical lumen fabrication to include geometries for studying stenosis, aneurysms, and bifurcations by polymerizing gels around thin circular optical fibers, and then removing the fibers to create hollow structures for EC culture [58] (Fig. 15.4). Similar to the method by Bischel, Mannino et al. also coated the inner surface by steady mechanical rotation, creating what they referred to as "do-it-yourself" vasculature. Yeon et al. [59] developed a PDMS-glass microfluidic device that comprised a narrow ladder structure filled

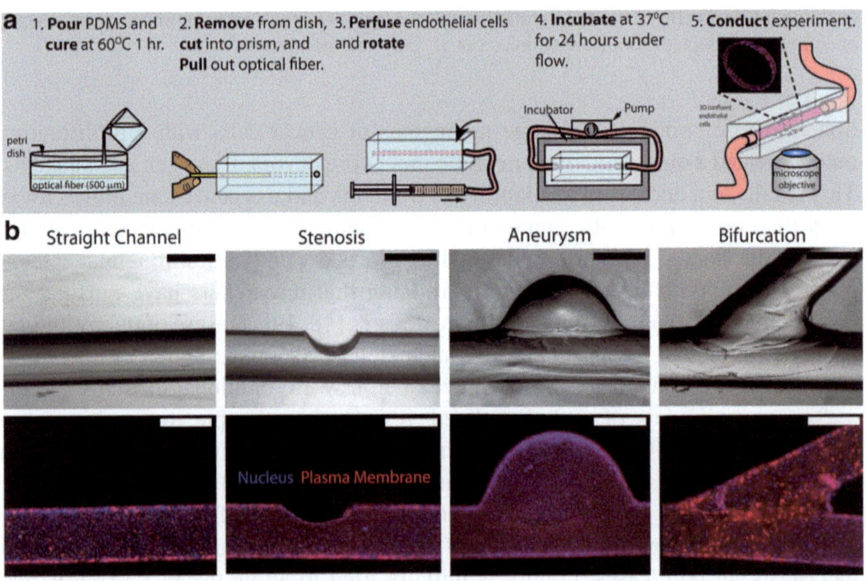

Fig. 15.4 Low-cost "do-it-yourself" vascular models created by (**a**) casting PDMS molds over cylindrical optical fibers, removing the fiber to create an empty core, and then perfusing and rotating the core with endothelial cells (ECs) so that the ECs coat the inner surface. (**b**) The shape of the optical fiber determines the shape of the vessel, and is sufficiently flexible to enable models for normal physiology, stenosis, aneurysm, and regions with bifurcations. Scale bar = 500 μm. Adapted from Mannino et al. [58] with permission

with polymerized fibrin. HUVECs were grown on the gel surfaces on both sides of the ladder structure, and in the presence of lung fibroblasts sprouted toward each other until they fused into a single, perfusable capillary tube.

Cells can often be incorporated within the bulk of a hydrogel layer that when polymerized creates a 3D scaffold that surrounds the cells and mediates 3D cell attachment [60]. This technique significantly improves the physiologic relevance of 3D cell-matrix interactions, and supports better growth and maturation of cells in vitro to develop relevant 3D tissues, like blood vessels [61]. Additionally, cells that do not naturally form a 2D monolayer are not forced to conform to a 2D surface. Such studies have contributed to our understanding of the influence of ECM stress forces on the direction of tumor cell migration [62]. The key design feature in these cases, where cells are embedded in hydrogels, is to account for cell contractility that leads to gel contraction, and to ensure that sufficient media is supplied, especially since these cell constructs tend to have significantly lower media volume available than their 2D counterparts.

One particularly unique method was developed by Zheng and colleagues, who cultured ECs with neighboring smooth muscle cells and pericytes, which are the cells that wrap around the outer adventitial layer of blood vessels and are known to be implicated in angiogenic processes [63]. The design consisted of a collagen bed with microchannels cast into the gel, and accommodated physiologic geometries including bifurcations and junctions, allowing studies of both angiogenesis and thrombosis. The most unique result from this study was that after long-term culture, the ECs that were originally cultured into a rectangular cross section remodeled into cylindrical vessels.

Indeed, there are numerous 3D coculture systems currently under development, and the ones described represent only the tip of the iceberg. Some other notable systems include: (1) a 3D breast cancer metastasis model to study tumor cell extravasation into a fibrin gel containing ECs, mesenchymal stem cells, and osteoblast-differentiated cells [64]; (2) a PDMS-based microfluidic device developed to mimic brain vasculature-ECM microenvironment of the neural stem cell niche of mice [65]; and (3) "organ-on-chip" models that recreate organ-level function with 3D microfluidic technology, reviewed elsewhere [3]. Between these highly complex models and low-cost "do-it-yourself" methods, the examples and options for achieving successful culture of cells in microfluidic systems are extensive.

15.4 Controlling the Microenvironment In Vitro: Revisited

The plethora of microfluidic cell culture systems developed thus far serves as an indication of the significant advancement of microfluidics for culturing cells. The ability to achieve such rapid development in the field has been aided by improved techniques that have circumvented various challenges associated with controlling the microenvironment, as previously reviewed [6]. In the early years of microfluidic cell culture, issues revolved around mitigating bubbles, dealing with evaporation, devising a cell seeding approach, and choosing the appropriate device material (see Sect. 15.2.4). These challenges have largely been met by technical solutions, but have also been addressed by increasing awareness and bringing the issue to the forefront [10].

Aside from these previously examined practical considerations, two recent challenges associated with controlling the microenvironment have emerged as systems have become increasingly complex: (1) control of conditions during long-term live-cell microscopy, and (2) control and reproducibility of hydrogels within microchannels. These issues are borne out of the desire to extract increasingly more spatial and temporal information from living cells, and doing so within microenvironments that better mimic the in vivo condition.

In terms of live-cell microscopy, the issue is associated with maintaining control of the temperature, humidity, and CO_2 concentration while the microfluidic culture is atop the microscope stage. The two most common apparatuses used with common microscopes to control environmental conditions are the on-stage incubation chamber and the large environmental enclosure or "cage" that envelops the stage, the optical components, and parts of the microscope frame. In either case, controlling environmental conditions are different in this setting compared to the normal cell culture incubator, where temperature fluctuations are minimal and methods to manage evaporation are already established [66]. One exception is the work by Chen et al. [67] who were able to perform long-term continuous imaging (every 10 min, for 9 days) by perfusing culture media containing 30 % FBS at a flow rate of 0.4 μl/min through microwells containing single cells until they clonally expand in 9–14 days. A second example is the work by Lecault [68], where they created a separate PDMS layer for the device intended to provide a media bath that can help temper temperature fluctuations and bubble nucleation. Nevertheless, integrating live-cell microscopy with microfluidic cell culture will likely be a major ongoing development in the field, and careful attention to account for this is highly recommended.

In terms of managing gel polymerization to create 3D scaffolds for cell culture, hydrogels are clearly gaining prominence in microfluidic cell culture, especially in enabling the development of more realistic 3D cell-matrix interactions. Commonly used hydrogels, in particular Matrigel, are naturally derived and provide cells with a suitable physiological environment for attachment due to the availability of binding sites. However, naturally derived hydrogels have some limitations that affect reproducibility and experimental control. These materials are derived from live tissue and can contain up to ~1800 proteins that include various growth factors with cell signaling capacity. Growth factor-reduced versions of these materials have largely overcome this particular obstacle, but the material remains poorly defined and suffers from high batch-to-batch variation. These limitations have led to the increased use of synthetic gels like polyacrylamide, whose composition is known and well characterized, and its stiffness can be tuned to mimic different tissues of interest. Polyacrylamide may suffer from a lack of biocompatibility that can be exacerbated in 3D culture [69], but this can often be circumvented. Alternate synthetic materials that hold promise for 3D-based hydrogel culture include hyaluronic acid, polyethylene glycol, synthetic peptides, and polyesters like polylactic acid [69, 70]. These materials offer defined compositions, controlled mechanical properties, and better stability. Effective control on microenvironmental conditions could thus result in high cell viability and increased reproducibility of these hydrogels in microfluidic cell culture.

15.5 Integrating Bioassays with Microfluidic Cell Culture

A major objective of preparing cell culture to model various healthy and diseased microenvironments is to enable cellular analysis that can reveal changes in the measurable quantities of biomolecules like enzymes, proteins, and nucleic acids. The detection of these changes may be crucial in improving our understanding of disease, studying the efficacy and toxicity of a drug, or advancing our basic knowledge of cells and tissues within those environments. There are a wide variety of conventional bioassays to detect and measure these biomolecules, and the field of microfluidics has already adapted many of them for the microscale, including immunostaining, polymerase chain reaction (PCR), fluorescent in situ hybridization (FISH), enzymatic-linked immunosorbent assays (ELISA), Western blotting, and many others. While an in-depth discussion of these assays is beyond the scope of this specific chapter, we briefly summarize some examples of the application of these assays to cells that have been cultured in microfluidic systems. The main distinction in these examples is whether the analysis was performed "on-chip" or "off-chip."

One of the most commonly used biological assays for characterizing culture cells is immunostaining because it helps identify and confirm the presence (or absence) of cellular proteins and their localizations within cells. Immunostaining has been the most widely adopted "on-chip" biological assay for microfluidic cell cultures to date because of its inherent simplicity and natural integration with common microfluidic steps: immunostaining procedures rely heavily on the ability to sequentially introduce and remove liquid reagents, and this is aligned with basic microfluidic control. Once stained, microscopy and image analysis can be performed, and because microfluidic devices can be constructed with PDMS and coverslip glass, the quality and resolution of images can be comparable to traditional immunostained cultures [37, 53, 71]. Furthermore, by incorporating image analysis software, immunostaining can provide valuable quantitative information, such as the level of nuclear translocation of transcription factors as an alternative to electrophoretic mobility shift assays (EMSAs) [42]. Besides glass, an increasing number of microfluidic systems are being fabricated in thermoplastics as a bridge toward commercialization [10]. These devices have also been used in conjunction with immunostaining applications. For example, Mottet et al. developed a cyclo-olefin copolymer (COC)-based chip to culture cancer cells within a 3D chamber and analyzed the cells by immunofluorescence

[72]. COC is preferred over polystyrene (PS) for immunostaining applications given its lower autofluorescence [13].

The integration of imaging with microfluidic cell culture systems goes well beyond basic immunofluorescence, and has excellent utility in biomedical research from the lab bench to the field [73]. Other imaging methods employed with microfluidic cell culture systems include the use of plate readers [74], and infrared imaging for protein quantification [11]. In addition, the applications have extended beyond mammalian cells toward other biological entities such as viruses for infection assays. Viral plaque assays for vesicular stomatitis virus were performed on cells cultured on microfluidic devices saving valuable reagents as well as flow-enhanced assay sensitivity [75]. These applications can greatly improve the availability and functionality of point-of-care diagnostics in biomedical research.

Aside from fluorescence and imaging-based assays, many other biological assays have been conducted on microfluidic cell cultures, ranging from "off-chip" to "on-chip" methods. This compromise between off-chip and on-chip strategies depends mostly on the researchers' needs to balance convenience and availability of infrastructure with the importance of full system integration. For example, Nevill et al. [76] established an integrated microfluidic platform to perform not only successful cell culture of a variety of cell types but also electrochemical cell lysis to generate lysate samples for off-chip analysis like ELISA or PCR. One particular "hybrid" microfluidic device allowed on-chip extraction of mRNA from cocultured cells, followed by subsequent off-chip RT (reverse transcriptase)-PCR to detect RNA expression [77]. The chip itself used magnetic capture beads within a lysis buffer to capture the mRNA and performed this on the different cell types within the coculture with minimal cross contamination.

Another key biological assay is the ELISA, which allows antibody-based detection of proteins in solution. As an example of an on-chip demonstration, Huang and colleagues combined the steps of cell loading, stimulation, and chemokine detection within a single microfluidic device using a fully customized chemiluminescent signal detection setup [78]. As an extension of traditional ELISA techniques that detect only a single biomarker, multiplex assays have been developed that can simultaneously detect a panel of biomolecules, allowing for increased data output while reducing sample and reagent consumption. The Luminex® platform is one such example, and has already been used "off-chip" to detect angiogenesis markers from media collected from the outlet of the device [40].

Western blotting, which separates proteins based on structure and polypeptide length via gel electrophoresis, has been used to analyze protein expression of cells cultured in microfluidic systems. For example, in a study that employed a microfluidic system for testing cancer cell adhesion on endothelium [79], Western blots were performed on microfluidic cultures to reveal levels of expression of cell surface adhesion molecules. Another microfluidic system studying coculture of neurons and Schwann cells and their sensitivity to botulinum neurotoxin used off-chip Western blots to characterize and compare sensitivity between macroscale and microscale techniques [80].

FISH, which is a quantitative method for gene expression within cells as well as a detection method for chromosomal abnormalities, was successfully miniaturized on a COC-based chip to enable all steps including cell adhesion, washing steps, enzymatic digestions, drying, probe injection, hybridization, and staining [81]. Researchers demonstrated HER-2 expression, a breast cancer marker, in cell lines as well as in patient samples.

These recent developments in integrating conventional bioassays with microfluidic cell cultures have great potential for advancing biology research, especially since microfluidic devices are increasingly demonstrating greater sensitivity and lower costs than conventional assays. However, the field continues to tackle challenges of balancing off-chip and on-chip analyses, as technical hurdles for full on-chip integration remain and the availability of current infrastructure supports the use of off-chip alternatives.

15.6 Recent Developments and Future Outlook

With rapid technical advances in microfluidic culture, the field is progressing toward increasingly complex applications. Organ-on-a-chip technologies have developed rapidly in the past few years, and there is growing interest in achieving the vision of human-on-chip technologies [82]. Amidst this interest in pushing the boundaries of microfluidic cell culture toward tissue- and organ-level models, technical challenges remain in realizing the often promised but unfulfilled goal of widespread adoption, automation, and massive throughput capabilities of microfluidic devices. The issue of automation and throughput has been discussed at length in other reviews [6], and there is little debate that there has been incredible progress on these fronts in the last few years. Researchers have developed high-throughput automated systems for the multistep process of staining various samples cultured within PDMS microchannels [83]. Acquired confocal images of cells grown on PDMS in this device were comparable to those on conventional glass slides. Several high-throughput microfluidic systems have also been developed for the growth of tumor spheroids [84]. The hanging drop technique for spheroid culture has also been automated using digital microfluidics [85]. Others have generated 3D masses of bone marrow-derived mesenchymal stem cells in a high-throughput system that interrogates the type, concentration, and temporal sequences of regulatory factors that drive tissue regeneration [86]. Advances in 3D culture have also led to automation in these systems, including 3D culture of mammary fibroblasts within collagen-containing hydrogels integrated to subsequent staining steps in an effort to automate the entire process [87]. Apart from these demonstrations, however, it is becoming evident that automation and throughput are merely a means to an end in achieving new insight into biological research. The most recent developments have appeared to tackle two other major directions in research that has potential for significant impact, of which microfluidic cell culture will likely have a major role: (1) single-cell analysis, and (2) clinical applications.

15.6.1 Single-Cell Analysis

Conventional biological assays provide readouts from a cell population that produces total or averaged values across the entire population. With an increasing appreciation for the heterogeneity within a population and its effect on understanding variability and functionality [88, 89] there is now significant demand for single-cell data and analyses. Microfluidics applied within complex microscale geometries is well suited for capturing and probing single cells, as has been demonstrated in applications toward microfluidic flow cytometry [90, 91], cell fusion studies [92], and single-cell real-time PCR [93]. Thus, the ability to handle and manipulate individual cells in microfluidic channels has advanced rapidly and proven to be quite powerful, particularly when the cells are probed without culture.

Recent advances, however, have also enabled single-cell analyses of cultured cells in microfluidic systems, which allows monitoring and analyses of cell behavior and function in controlled microenvironments. Microfluidic devices have recently been combined with effective image analyses to generate data for single cells from high-quality immunofluorescence images, and then recombined to study population distributions of cell function. Cao et al. coupled a microfluidic platform with quantum dot-based immunofluorescence to measure glycan expression on the cell surface of single cells [94], while Young et al. studied the single-cell nuclear translocation of the NF-κB transcription factor in multiple myeloma cells with and without drug treatment to understand the effects of stromal cells on drug resistance [42]. Such quantitative data not only reflects single-cell heterogeneity, but also sheds light on population dynamics when entire population distributions are reconstructed from the data.

The above examples demonstrated single-cell data acquired via imaging of cultured populations, but did not require a priori capture of cells in separate chambers or compartments. Kobel et al. have recently reported using hydrodynamic trapping methods in microfluidics, followed by perfusion, to capture and culture single cells [95]. Another approach developed by Lin et al. involved capturing single cells in small "capture" microwells with a high rate of efficiency (80–99 %), and then flipping the chip over so the captured cells fell into wider "culture" wells that supported cell culture for up to 1 week without the need for perfusion [96]. Hughes and colleagues also employed a massive array of microwells together with a 30 μm thick polyacrylamide gel to develop a novel single-cell Western blot [97]. Finally, one of the most elaborate microfluidic systems was developed by Lecault et al., which consisted of a massive array of thousands of nanoliter-scale cell culture chambers, assisted by microfluidic valving technology, that could image and track the clonal proliferation of a single captured hematopoietic stem cell [68].

Overall, these examples indicate a growing trend (and need) toward microfluidic systems that can capture, culture, and study (via imaging or other means) the cellular behavior and function with single-cell resolution. This data can then be reconstructed into population-based information, with significantly more detail and potential for insight.

15.6.2 Clinical Applications

Aside from pushing the technical boundaries of microfluidic cell culture systems, recent advances have also pushed the applications toward more clinical relevance [98]. Increasingly, microfluidic systems are being employed to study and examine primary clinical samples acquired directly from patients "at the bedside" (or in the field).

For example, the microfluidic chip designed by Young et al. [42] for multiple myeloma (MM) was originally characterized with MM and stromal cell lines, but was recently applied to clinical samples extracted from raw bone marrow to identify patient-specific therapy effectiveness [99]. The patient's own malignant bone marrow plasma cells were cocultured with the patient's mononuclear cells, a significant part of the bone marrow microenvironment, and tested for responsiveness to bortezomib-containing therapeutics. The results correlated with clinical outcomes for all 17 patient samples tested, demonstrating its potential for predicting clinical outcomes in response to drugs.

Another major clinical application of microfluidics for cancer research has been the capture, enumeration, and analysis of circulating tumor cells (CTCs). The ability to detect and culture CTCs in cancer patients can greatly improve disease outcome by providing disease status and drug sensitivity information, since these cells serve as indicators for metastasis [100]. Various microfluidic devices have reported the successful micromagnetic or antibody-based capture of CTCs and their subsequent culture [101–103]. The latest advance in this area involves the capture and characterization of CTC clusters, or aggregates of multiple CTCs with occasional macrophages attached [104].

Microfluidics research is poised for major developments in preclinical and clinical applications in the near future, from diagnostics to therapeutics, and from point-of-care instruments to implantable systems [105]. A unique example of how far technology has reached involves a microfluidic device used as an effective blood-cleansing device for sepsis [106]. The device, which was effectively a microfluidic "spleen," was tested with rats and shown to clear the blood of bacteria, fungi, endotoxins at a flow rate of 1.25 L/h, and then deliver clean blood back to the animal, improving their survival rates. While this example does not involve the culture of cells, it provides a glimpse of the potential for microfluidics in the clinic.

The future will likely involve advancing capture and manipulation to the culture and functional analyses of clinical samples, building more and more toward personalized medicine.

15.6.2.1 Outlook

The significant role of microfluidics in biomedical research has been established and well documented [107]. The latest advances point toward increased attention to adding further functionality (*throughput and automation*), extracting more sample information (*single-cell analyses*), and applying microfluidic cultures to more clinical applications. While progress is evident by the rising applications of microfluidic devices to cell culture, this may not be as advanced as initially promised, although this is not uncommon during technological development and adoption. Technical challenges remain, including the ability to expand our microfabrication toolkit, improving control for long-term live-cell imaging, and adding more on-chip functionality toward effective biomolecule detection that is integrated with the cell culture system. These anticipated developments will have potential to significantly increase the applicability of microfluidic devices in biology labs, both for cell culture and beyond.

15.7 Summary

In summary, key principles of cell culture at both the macro- and microscales were discussed and examined with the aim of providing guidelines for culturing cells in microfluidic devices. Specific examples of successful cultures leading to advanced biological models were used to illustrate some of the most reliable methods and techniques for incorporating multiple cell types, 3D scaffolds and gels, and on-chip functionality onto the microfluidic cell-based systems. The rapidly growing field has moved from cell- to tissue- to organ-level functionality, and while technical challenges remain, there is potential growth in the areas of increasing throughput, automation, single-cell analysis, and clinical applications to further advance the technical capabilities of microfluidic cell culture.

References

1. Harrison RG (1911) On the stereotropism of embryonic cells. Science 34:279–281
2. Young EWK (2013) Cells, tissues, and organs on chips: challenges and opportunities for the cancer tumor microenvironment. Integr Biol 5:1096–1109
3. Huh D, Hamilton GA, Ingber DE (2011) From 3D cell culture to organs-on-chips. Trends Cell Biol 21:745–754

4. Mehta D, Malik AB (2006) Signaling mechanisms regulating endothelial permeability. Physiol Rev 86:279–367
5. Young EWK, Watson MWL, Srigunapalan S, Wheeler AR, Simmons CA (2010) Technique for real-time measurements of endothelial permeability in a microfluidic membrane chip using laser-induced fluorescence detection. Anal Chem 82:808–816
6. Young EWK, Beebe DJ (2010) Fundamentals of microfluidic cell culture in controlled microenvironments. Chem Soc Rev 39:1036–1048
7. Heo YS, Cabrera LM, Song JW, Futai N, Tung Y-C, Smith GD, Takayama S (2007) Characterization and resolution of evaporation-mediated osmolality shifts that constrain microfluidic cell culture in poly(dimethylsiloxane) devices. Anal Chem 79:1126–1134
8. Walker GM, Zeringue HC, Beebe DJ (2004) Microenvironment design considerations for cellular scale studies. Lab Chip 4:91–97
9. Domenech M, Yu H, Warrick J, Badders NM, Meyvantsson I, Alexander CM, Beebe DJ (2009) Cellular observations enabled by microculture: paracrine signaling and population demographics. Integr Biol (Camb) 1:267–274
10. Berthier E, Young EWK, Beebe D (2012) Engineers are from PDMS-land, biologists are from polystyrenia. Lab Chip 12:1224
11. Su X, Young EWK, Underkofler HAS, Kamp TJ, January CT, Beebe DJ (2011) Microfluidic cell culture and its application in high-throughput drug screening: cardiotoxicity assay for hERG channels. J Biomol Screen 16:101–111
12. Van Midwoud P, Janse A, van Midwoud PM, Merema MT, Groothuis GMM, Verpoorte E (2012) Comparison of biocompatibility and adsorption properties of different plastics for advanced microfluidic cell and tissue culture models. Anal Chem 84:3938–3944
13. Young EWK, Berthier E, Beebe DJ (2013) Assessment of enhanced autofluorescence and impact on cell microscopy for microfabricated thermoplastic devices. Anal Chem 85:44–49
14. Park JW, Vahidi B, Taylor AM, Rhee SW, Jeon NL (2006) Microfluidic culture platform for neuroscience research. Nat Protoc 1:2128–2136
15. Kim L, Toh Y-C, Voldman J, Yu H (2007) A practical guide to microfluidic perfusion culture of adherent mammalian cells. Lab Chip 7:681–694
16. Paguirigan AL, Beebe DJ (2008) Microfluidics meet cell biology: bridging the gap by validation and application of microscale techniques for cell biological assays. Bioessays 30:811–821
17. Millet LJ, Stewart ME, Nuzzo RG, Gillette MU (2010) Guiding neuron development with planar surface gradients of substrate cues deposited using microfluidic devices. Lab Chip 10:1525–1535
18. Neto E, Alves CJ, Sousa DM et al (2014) Sensory neurons and osteoblasts: close partners in a microfluidic platform. Integr Biol (Camb) 6:586–595
19. Shi P, Nedelec S, Wichterle H, Kam LC (2010) Combined microfluidics/protein patterning platform for pharmacological interrogation of axon pathfinding. Lab Chip 10:1005–1010
20. Powell SK, Kleinman HK (1997) Neuronal laminins and their cellular receptors. Int J Biochem Cell Biol 29:401–414
21. Harris J, Lee H, Vahidi B, Tu C, Cribbs D, Jeon NL, Cotman C (2007) Fabrication of a microfluidic device for the compartmentalization of neuron soma and axons. J Vis Exp (7):261
22. Young EWK, Simmons CA (2010) Macro- and microscale fluid flow systems for endothelial cell biology. Lab Chip 10:143–160
23. Young EWK, Wheeler AR, Simmons CA (2007) Matrix-dependent adhesion of vascular and valvular endothelial cells in microfluidic channels. Lab Chip 7:1759–1766
24. van der Meer AD, Vermeul K, Poot AA, Feijen J, Vermes I (2010) A microfluidic wound-healing assay for quantifying endothelial cell migration. Am J Physiol Heart Circ Physiol 298:H719–H725
25. Wallace C, Jacob JT, Stoltz A, Bi J, Bundy K (2005) Corneal epithelial adhesion strength to tethered-protein/peptide modified hydrogel surfaces. J Biomed Mater Res A 72:19–24

26. Lovchik RD, Bianco F, Matteoli M, Delamarche E (2009) Controlled deposition of cells in sealed microfluidics using flow velocity boundaries. Lab Chip 9:1395–1402
27. Zhang W, Choi DS, Nguyen YH, Chang J, Qin L (2013) Studying cancer stem cell dynamics on PDMS surfaces for microfluidics device design. Sci Rep 3:2332
28. Wong KHK, Chan JM, Kamm RD, Tien J (2012) Microfluidic models of vascular functions. Annu Rev Biomed Eng 14:205–230
29. Engler AJ, Sen S, Sweeney HL, Discher DE (2006) Matrix elasticity directs stem cell lineage specification. Cell 126:677–689
30. Marimuthu M, Kim S (2011) Microfluidic cell coculture methods for understanding cell biology, analyzing bio/pharmaceuticals, and developing tissue constructs. Anal Biochem 413:81–89
31. Cooper GM (2000) Signaling molecules and their receptors. In: The cell: a molecular approach, 2nd edn. Sinauer Associates, Sunderland
32. Gao SP, Mark KG, Leslie K et al (2007) Mutations in the EGFR kinase domain mediate STAT3 activation via IL-6 production in human lung adenocarcinomas. J Clin Invest 117:3846–3856
33. Braga VMM (2002) Cell-cell adhesion and signalling. Curr Opin Cell Biol 14:546–556
34. Guo F, French JB, Li P, Zhao H, Chan CY, Fick JR, Benkovic SJ, Huang TJ (2013) Probing cell-cell communication with microfluidic devices. Lab Chip 13:3152–3162
35. Przybyla L, Voldman J (2012) Probing embryonic stem cell autocrine and paracrine signaling using microfluidics. Annu Rev Anal Chem (Palo Alto Calif) 5:293–315
36. Nilsson J, Evander M, Hammarström B, Laurell T (2009) Review of cell and particle trapping in microfluidic systems. Anal Chim Acta 649:141–157
37. Chung S, Sudo R, Mack PJ, Wan C-R, Vickerman V, Kamm RD (2009) Cell migration into scaffolds under co-culture conditions in a microfluidic platform. Lab Chip 9:269–275
38. Jeong GS, Han S, Shin Y, Kwon GH, Kamm RD, Lee S-H, Chung S (2011) Sprouting angiogenesis under a chemical gradient regulated by interactions with an endothelial monolayer in a microfluidic platform. Anal Chem 83:8454–8459
39. Vickerman V, Kamm RD (2012) Mechanism of a flow-gated angiogenesis switch: early signaling events at cell-matrix and cell-cell junctions. Integr Biol (Camb) 4:863–874
40. Theberge AB, Yu J, Young EWK, William A, Bushman W, Beebe DJ (2015) Microfluidic multiculture assay to analyze biomolecular signaling in angiogenesis. Anal Chem 87:3239–3246
41. Lang JD, Berry SM, Powers GL, Beebe DJ, Alarid ET (2013) Hormonally responsive breast cancer cells in a microfluidic co-culture model as a sensor of microenvironmental activity. Integr Biol (Camb) 5:807–816
42. Young EWK, Pak C, Kahl BS, Yang DT, Callander NS, Miyamoto S, Beebe DJ (2012) Microscale functional cytomics for studying hematologic cancers. Blood 119:e76–e85
43. Chen MB, Whisler JA, Jeon JS, Kamm RD (2013) Mechanisms of tumor cell extravasation in an in vitro microvascular network platform. Integr Biol (Camb) 5:1262–1271
44. Huh D, Leslie DC, Matthews BD, Fraser JP, Jurek S, Hamilton GA, Thorneloe KS, McAlexander MA, Ingber DE (2012) A human disease model of drug toxicity-induced pulmonary edema in a lung-on-a-chip microdevice. Sci Transl Med 4:159ra147
45. Kim HJ, Huh D, Hamilton G, Ingber DE (2012) Human gut-on-a-chip inhabited by microbial flora that experiences intestinal peristalsis-like motions and flow. Lab Chip 12:2165
46. Schmeichel KL, Bissell MJ (2003) Modeling tissue-specific signaling and organ function in three dimensions. J Cell Sci 116:2377–2388
47. Schwartz MA, Chen CS (2013) Cell biology. Deconstructing dimensionality. Science 339:402–404
48. Hsiao AY, Torisawa Y, Tung Y-C, Sud S, Taichman RS, Pienta KJ, Takayama S (2009) Microfluidic system for formation of PC-3 prostate cancer co-culture spheroids. Biomaterials 30:3020–3027

49. Tung Y-C, Hsiao AY, Allen SG, Torisawa Y, Ho M, Takayama S (2011) High-throughput 3D spheroid culture and drug testing using a 384 hanging drop array. Analyst 136:473–478
50. Zervantonakis IK, Hughes-Alford SK, Charest JL, Condeelis JS, Gertler FB, Kamm RD (2012) Three-dimensional microfluidic model for tumor cell intravasation and endothelial barrier function. Proc Natl Acad Sci U S A 109:13515–13520
51. Song JW, Munn LL (2011) Fluid forces control endothelial sprouting. Proc Natl Acad Sci U S A 108:15342–15347
52. Song JW, Bazou D, Munn LL (2015) In: Ribatti D (ed) Vascular morphogenesis: methods and protocols. Springer, New York, pp 243–254
53. Kim S, Lee H, Chung M, Jeon NL (2013) Engineering of functional, perfusable 3D microvascular networks on a chip. Lab Chip 13:1489–1500
54. Wang X-Y, Jin Z-H, Gan B-W, Lv S-W, Xie M, Huang W-H (2014) Engineering interconnected 3D vascular networks in hydrogels using molded sodium alginate lattice as the sacrificial template. Lab Chip 14:2709–2716
55. Wang X-Y, Pei Y, Xie M, Jin Z-H, Xiao Y-S, Wang Y, Zhang L-N, Li Y, Huang W-H (2015) An artificial blood vessel implanted three-dimensional microsystem for modeling transvascular migration of tumor cells. Lab Chip 15:1178–1187
56. Bischel LL, Young EWK, Mader BR, Beebe DJ (2013) Tubeless microfluidic angiogenesis assay with three-dimensional endothelial-lined microvessels. Biomaterials 34:1471–1477
57. Bischel LL, Beebe DJ, Sung KE (2015) Microfluidic model of ductal carcinoma in situ with 3D, organotypic structure. BMC Cancer 15:12
58. Mannino RG, Myers DR, Ahn B et al (2015) Do-it-yourself in vitro vasculature that recapitulates in vivo geometries for investigating endothelial-blood cell interactions. Sci Rep 5:12401
59. Yeon JH, Ryu HR, Chung M, Hu QP, Jeon NL (2012) In vitro formation and characterization of a perfusable three-dimensional tubular capillary network in microfluidic devices. Lab Chip 12:2815–2822
60. Shin Y, Han S, Jeon JS, Yamamoto K, Zervantonakis IK, Sudo R, Kamm RD, Chung S (2012) Microfluidic assay for simultaneous culture of multiple cell types on surfaces or within hydrogels. Nat Protoc 7:1247–1259
61. Park YK, Tu T-Y, Lim SH, Clement IJM, Yang SY, Kamm RD (2014) In vitro microvessel growth and remodeling within a three-dimensional microfluidic environment. Cell Mol Bioeng 7:15–25
62. Polacheck WJ, German AE, Mammoto A, Ingber DE, Kamm RD (2014) Mechanotransduction of fluid stresses governs 3D cell migration. Proc Natl Acad Sci U S A 111:2447–2452
63. Zheng Y, Chen J, Craven M et al (2012) From the cover: in vitro microvessels for the study of angiogenesis and thrombosis. Proc Natl Acad Sci U S A 109:9342–9347
64. Jeon JS, Bersini S, Gilardi M, Dubini G, Charest JL, Moretti M, Kamm RD (2015) Human 3D vascularized organotypic microfluidic assays to study breast cancer cell extravasation. Proc Natl Acad Sci U S A 112:214–219
65. Shin Y, Yang K, Han S, Park H-J, Seok Heo Y, Cho S-W, Chung S (2014) Reconstituting vascular microenvironment of neural stem cell niche in three-dimensional extracellular matrix. Adv Healthc Mater 3:1457–1464
66. Berthier E, Warrick J, Yu H, Beebe DJ (2008) Managing evaporation for more robust microscale assays. Part 1. Volume loss in high throughput assays. Lab Chip 8:852–859
67. Chen H, Sun J, Wolvetang E, Cooper-White J (2015) High-throughput, deterministic single cell trapping and long-term clonal cell culture in microfluidic devices. Lab Chip 15:1072–1083
68. Lecault V, Vaninsberghe M, Sekulovic S et al (2011) High-throughput analysis of single hematopoietic stem cell proliferation in microfluidic cell culture arrays. Nat Methods 8:581–586

69. Gill BJ, West JL (2014) Modeling the tumor extracellular matrix: tissue engineering tools repurposed towards new frontiers in cancer biology. J Biomech 47:1969–1978
70. Worthington P, Pochan DJ, Langhans SA (2015) Peptide hydrogels—versatile matrices for 3D cell culture in cancer medicine. Front Oncol 5:92
71. Ehsan SM, Welch-Reardon KM, Waterman ML, Hughes CCW, George SC (2014) A three-dimensional in vitro model of tumor cell intravasation. Integr Biol 6:603
72. Mottet G, Perez-Toralla K, Tulukcuoglu E, Bidard F-C, Pierga J-Y, Draskovic I, Londono-Vallejo A, Descroix S, Malaquin L, Louis Viovy J (2014) A three dimensional thermoplastic microfluidic chip for robust cell capture and high resolution imaging. Biomicrofluidics 8:024109
73. Wu J, Zheng G, Lee LM (2012) Optical imaging techniques in microfluidics and their applications. Lab Chip 12:3566–3575
74. Wen Y, Zhang X, Yang S-T (2010) Microplate-reader compatible perfusion microbioreactor array for modular tissue culture and cytotoxicity assays. Biotechnol Prog 26:1135–1144
75. Zhu Y, Warrick JW, Haubert K, Beebe DJ, Yin J (2009) Infection on a chip: a microscale platform for simple and sensitive cell-based virus assays. Biomed Microdevices 11:565–570
76. Nevill JT, Cooper R, Dueck M, Breslauer DN, Lee LP (2007) Integrated microfluidic cell culture and lysis on a chip. Lab Chip 7:1689–1695
77. Berry SM, Singh C, Lang JD, Strotman LN, Alarid ET, Beebe DJ (2014) Streamlining gene expression analysis: integration of co-culture and mRNA purification. Integr Biol (Camb) 6:224–231
78. Huang N-T, Chen W, Oh B-R, Cornell TT, Shanley TP, Fu J, Kurabayashi K (2012) An integrated microfluidic platform for in situ cellular cytokine secretion immunophenotyping. Lab Chip 12:4093–4101
79. Song JW, Cavnar SP, Walker AC, Luker KE, Gupta M, Tung Y-C, Luker GD, Takayama S (2009) Microfluidic endothelium for studying the intravascular adhesion of metastatic breast cancer cells. PLoS One 4:e5756
80. Hong WS, Young EWK, Tepp WH, Johnson EA, Beebe DJ (2013) A microscale neuron and Schwann cell coculture model for increasing detection sensitivity of botulinum neurotoxin type A. Toxicol Sci 134:64–72
81. Perez-Toralla K, Mottet G, Guneri ET et al (2015) FISH in chips: turning microfluidic fluorescence in situ hybridization into a quantitative and clinically reliable molecular diagnosis tool. Lab Chip 15:811–822
82. Marx U, Walles H, Hoffmann S, Lindner G, Horland R, Sonntag F, Klotzbach U, Sakharov D, Tonevitsky A, Lauster R (2012) "Human-on-a-chip" developments: a translational cutting-edge alternative to systemic safety assessment and efficiency evaluation of substances in laboratory animals and man? Altern Lab Anim 40:235–257
83. Shen J, Zhou Y, Lu T, Peng J, Lin Z, Huang L, Pang Y, Yu L, Huang Y (2012) An integrated chip for immunofluorescence and its application to analyze lysosomal storage disorders. Lab Chip 12:317–324
84. Liu W, Wang J-C, Wang J (2015) Controllable organization and high throughput production of recoverable 3D tumors using pneumatic microfluidics. Lab Chip 15:1195–1204
85. Aijian AP, Garrell RL (2015) Digital microfluidics for automated hanging drop cell spheroid culture. J Lab Autom 20:283–295
86. Occhetta P, Centola M, Tonnarelli B, Redaelli A, Martin I, Rasponi M (2015) High-throughput microfluidic platform for 3D cultures of mesenchymal stem cells, towards engineering developmental processes. Sci Rep 5:10288
87. Montanez-Sauri SI, Sung KE, Puccinelli JP, Pehlke C, Beebe DJ (2011) Automation of three-dimensional cell culture in arrayed microfluidic devices. J Lab Autom 16:171–185
88. Huang S (2009) Non-genetic heterogeneity of cells in development: more than just noise. Development 136:3853–3862
89. Lecault V, White AK, Singhal A, Hansen CL (2012) Microfluidic single cell analysis: from promise to practice. Curr Opin Chem Biol 16:381–390

90. Lim SW, Abate AR (2013) Ultrahigh-throughput sorting of microfluidic drops with flow cytometry. Lab Chip 13:4563–4572
91. Piyasena ME, Graves SW (2014) The intersection of flow cytometry with microfluidics and microfabrication. Lab Chip 14:1044–1059
92. Skelley AM, Kirak O, Suh H, Jaenisch R, Voldman J (2009) Microfluidic control of cell pairing and fusion. Nat Methods 6:147–152
93. White AK, VanInsberghe M, Petriv OI, Hamidi M, Sikorski D, Marra MA, Piret J, Aparicio S, Hansen CL (2011) High-throughput microfluidic single-cell RT-qPCR. Proc Natl Acad Sci U S A 108:13999–14004
94. Cao J-T, Chen Z-X, Hao X-Y, Zhang P-H, Zhu J-J (2012) Quantum dots-based immunofluorescent microfluidic chip for the analysis of glycan expression at single-cells. Anal Chem 84:10097–10104
95. Kobel S, Valero A, Latt J, Renaud P, Lutolf M (2010) Optimization of microfluidic single cell trapping for long-term on-chip culture. Lab Chip 10:857–863
96. Lin C-H, Hsiao Y-H, Chang H-C, Yeh C-F, He C-K, Salm EM, Chen C, Chiu I-M, Hsu C-H (2015) A microfluidic dual-well device for high-throughput single-cell capture and culture. Lab Chip 15:2928–2938
97. Hughes AJ, Spelke DP, Xu Z, Kang C-C, Schaffer DV, Herr AE (2014) Single-cell western blotting. Nat Methods 11:749–755
98. Young EWK, Moraes C (2015) Patients are a virtue: advances in microengineered systems for clinical applications. Integr Biol 7:962. doi:10.1039/C5IB90031J
99. Pak C, Callander NS, Young EWK, Titz B, Kim K, Saha S, Chng K, Asimakopoulos F, Beebe DJ, Miyamoto S (2015) MicroC(3): an ex vivo microfluidic cis-coculture assay to test chemosensitivity and resistance of patient multiple myeloma cells. Integr Biol (Camb) 7:643–654
100. Cristofanilli M, Budd GT, Ellis MJ et al (2004) Circulating tumor cells, disease progression, and survival in metastatic breast cancer. N Engl J Med 351:781–791
101. Kang JH, Krause S, Tobin H, Mammoto A, Kanapathipillai M, Ingber DE (2012) A combined micromagnetic-microfluidic device for rapid capture and culture of rare circulating tumor cells. Lab Chip 12:2175–2181
102. Zheng X, Jiang L, Schroeder J, Stopeck A, Zohar Y (2014) Isolation of viable cancer cells in antibody-functionalized microfluidic devices. Biomicrofluidics 8:024119
103. Sheng W, Ogunwobi OO, Chen T, Zhang J, George TJ, Liu C, Fan ZH (2014) Capture, release and culture of circulating tumor cells from pancreatic cancer patients using an enhanced mixing chip. Lab Chip 14:89–98
104. Sarioglu AF, Aceto N, Kojic N et al (2015) A microfluidic device for label-free, physical capture of circulating tumor cell clusters. Nat Methods 12:685–691
105. Kong JE, Koh J, Lin J, Di Carlo D (2015) Research highlights: translating chips. Lab Chip 15:1984–1988
106. Kang JH, Super M, Yung CW et al (2014) An extracorporeal blood-cleansing device for sepsis therapy. Nat Med 20:1211–1216
107. Sackmann EK, Fulton AL, Beebe DJ (2014) The present and future role of microfluidics in biomedical research. Nature 507:181–189

Chapter 16
Microfluidic Chromatin Immunoprecipitation for Analysis of Epigenomic Regulations

Yan Zhu and Chang Lu

Abstract Epigenetic modifications, such as DNA methylation and histone modifications, play important roles in gene expression and regulation, and are highly involved in cellular processes such as stem cell pluripotency/differentiation and tumorigenesis. Chromatin immunoprecipitation (ChIP) is the technique of choice for examining in vivo DNA-protein interactions and has been a great tool for studying epigenetic mechanisms. However, conventional ChIP assays require millions of cells for tests and are not practical for examination of samples from lab animals and patients. Automated microfluidic chips offer the advantage to handle small sample sizes and facilitate rapid reaction. They also eliminate cumbersome manual handling. In this chapter, we review recent works on microfluidic ChIP development and discuss the design criteria for these microfluidic assays.

Keywords Microfluidics • Chromatin immunoprecipitation • Next-generation sequencing • Epigenetics • Epigenomics • PCR • Histone

The human genome has been sequenced and completed over a decade ago. The information provided by the genomic map inspired numerous studies on genetic variations and their roles in diseases. However, genomic information alone is not always sufficient to explain important biological processes. Gene activation and expression are not only associated with alteration in the DNA sequence, but also affected by other changes to DNA and supporting proteins (i.e., histones). Epigenetics refers to the molecular mechanisms that affect gene expression and phenotypes without involving changes in the DNA sequence. Epigenetic changes occur throughout all stages of cell development or in response to environmental cues. They affect gene accessibility of regulatory proteins such as methyl-CpG-binding proteins, transcription factors, RNA polymerase II, and other components of the transcriptional machinery, eventually changing transcription patterns in a tissue/

Y. Zhu • C. Lu, Ph.D. (✉)
Department of Chemical Engineering, Virginia Tech, 235 Goodwin Hall, 635 Prices Fork Road, Blacksburg, VA 24061, USA
e-mail: yanzhu@vt.edu; changlu@vt.edu

© Springer International Publishing Switzerland 2016

C. Lu, S.S. Verbridge (eds.), *Microfluidic Methods for Molecular Biology*,
DOI 10.1007/978-3-319-30019-1_16

cell-specific fashion [1–6]. Epigenetics is the field to examine the environmental factors and determine how they alter gene regulations and what roles they play in disease development. The study of epigenetics at the genome scale is referred to as epigenomics [7].

With a large epigenomic database, one can formulate hypotheses about the causalities between epigenomic changes and phenotypical changes, link chromatin modifications with aberrant gene expressions in diseases, and define core regulatory mechanisms that occur in different tissues and in different developmental stages [8, 9].

16.1 Epigenetic Changes

There are a number of processes that affect gene expressions without changing DNA sequence. First, transcription factor binding is an important mechanism that affects gene activity. It is widely accepted that the molecular requirements for human cancer include loss of growth inhibition by ligands, self-sufficiency in growth stimulation, limitless replicative potential, and avoidance of apoptosis [10]. All of these characteristics have close ties to dysregulated transcription. A significant fraction of transcription factors have functions related to cell proliferation and differentiation. It is also well established that the differential gene expression patterns in cancer cells correlate with the recruitment of specific and general transcription factors on relevant promoters. Intracellular cell signaling pathways typically rely on activated transcription factors or proteins that activate transcription factors for relaying the signal into the nucleus. These signaling proteins may experience mutation or overexpression and these changes alter spatiotemporal transcription patterns. As a result, a number of transcription factors (e.g., STATs, NF-κB, β-catenin, and NICD) are well known to have increased activity in most human cancers and prevent apoptosis of tumor cells in these cases [11–13]. In tumor cells, for example, transcription factor NF-κB is normally sequestered in the cytoplasm through the association with an I-κB protein. When the cell is exposed to activation signals such as tumor necrosis factor-α (TNF-α) binding to cell surface receptors, the I-κB protein is phosphorylated on serine 32 and 36, then ubiquitinated, and broken down in proteasomes. After being freed from its association with I-κB, the NF- κB complex moves to the nucleus where it binds to specific sequences in the promoter/enhancer regions of genes to regulate transcription [14, 15].

In mammals, DNA methylation is the best studied epigenetic modification. DNA methylation typically refers to the addition of a methyl group at the carbon-5 position of cytosine residues within CpG dinucleotides, forming 5-methylcytosine (5mC). This process is catalyzed by enzymes called DNA methyltransferases. Human genome contains 0.7 % clusters of CpG sites ("CpG islands" or CGIs) and those CpG sites contain 7 % of all CpG dinucleotides. CpG islands are mostly enriched at gene promoters. There about 60 % of all mammalian gene promoters are

CpG rich [16]. CpG islands typically stay unmethylated. This opens up space on nucleosome for DNA to be easily transcribed. On the other hand, methylated CGIs are associated with gene repression [17, 18]. DNA hypermethylation or hypomethylation compared to normal tissues have been associated with a large number of human diseases. For example, transcriptional silencing of tumor-suppressor genes by CGI-promoter hypermethylation plays an important role in cancer development [19–21]. Genes that were critically involved in cancer biology, including the cell-cycle inhibitor p16-INK4a and the DNA-repair genes MLH1 and BRCA1, have been shown to undergo methylation-associated silencing in tumor cells [22, 23]. Typically, hypermethylation of tumor-suppressor genes and hypomethylation of oncogenes occur during cancer development [24].

Histones are proteins found in eukaryotic cell nuclei that are wrapped around by 147 bp DNA. The histone-DNA complex is called chromatin. Chromatin exists in two forms. One form, called euchromatin, is less condensed and can be transcribed. The second form, called heterochromatin, is highly condensed and typically not transcribed. The core histones (H2A, H2B, H3, and H4) are in globular shape, consisting of two H2A-H2B dimmers and an H3-H4 tetramer [25]. The histone tails are subjected to various posttranslational modifications including acetylation, methylation, phosphorylation, and ubiquitination. Of these modifications, histone acetylation and methylation are the best understood. In general, acetylation on lysine residues provides relaxed chromatin structure that allows binding of transcription factor and therefore increases gene expression. However histone methylation affects nucleosomal interactions and can either repress or activate transcription [26]. For example, trimethylation of histone 3 lysine 4 (H3K4me3) is associated with transcriptionally active chromatin, whereas trimethylation of histone 3 lysine 27 (H3K27me3) leads to compact chromatin and gene repression.

16.2 Chromatin Immunoprecipitation

Epigenetic regulations of gene expressions play pivotal roles in normal and disease development. Understanding of epigenetic regulations will improve various aspects of biomedicine, including strategies for manipulating cell fate for regenerative medicine, novel epigenetics-based disease markers and biomarkers, and novel therapeutics.

To date, chromatin immunoprecipitation (ChIP) assay is the gold standard for investigating in vivo epigenetic modifications and transcription factor binding at the genome scale. ChIP assays involve a number of steps. First, DNA is cross-linked with protein by cross-linking agents (e.g., formaldehyde), in order to freeze protein–DNA and protein–protein interactions. Subsequently, cells are lysed in order to release chromatin. Then, chromatin is sonicated to yield fragments of protein-bound DNA that are typically 200–600 base pair long. These fragments are then immunoprecipitated onto the surface of antibody-coated beads. The antibody here specifically targets a transcription factor, a specific modified form of histone, or a cytosine methyl group. Finally, the immunoprecipitated fraction is isolated. The cross-linking is reversed and the released DNA is assayed to determine the sequences [27]. The identification of the DNA sequences can be done by qPCR if there are known candidates. Alternatively, these binding sites can be mapped at the genome scale by sequencing (ChIP-seq) using high-throughput sequencing technology (e.g., Illumina HiSeq 2500 System) [28, 29].

Although ChIP-qPCR/seq has been tremendously useful for epigenetic/epigenomic studies, the technology suffers from serious limitations. First, the key limitation is that it requires a large number of cells ($>10^6$ cells per IP) [30–32]. This is feasible when using cell lines. However, such requirement has become a major challenge when primary cells are used because very limited amounts of samples can be generated from lab animals or patients. For example, there are around 10,000 naturally occurring T regulatory cells per murine spleen, and ~5000 per ml peripheral blood. For metastatic cancer patients, there are only about 1–10 circulating tumor cells per ml of whole blood. Furthermore, primary samples are typically mixture of different cell types. Isolating single cell type from a mixture brings further loss in the sample amount. Second, the outcome of a ChIP assay can be affected when a large cell number is used. Population heterogeneity may contribute to the large standard deviations among trials. Finally, most ChIP assays involve lengthy manual handling and they normally take 3–4 days or longer to finish. These cumbersome procedures not only create loss of materials but also introduce human/technical errors that lead to large variations between replicates [33, 34]. In general, ChIP assays with ultrahigh sensitivity and high degree of automation and integration are highly desirable.

16.3 Microfluidic ChIP Assays

Microfluidics enables controlling and transferring tiny quantities of liquids to allow biological assays to be integrated on a small chip. Microfluidic chips can be made from polydimethylsiloxane (PDMS) by soft lithography at low cost [35, 36]. Microfluidic chips offer reduction in sample amounts, high level of integration and automation, and high throughput. Microfluidics provides an automated platform for performing ChIP assays based on a low number of cells with a short assay time. Automation not only helps to standardize and simplify the ChIP process, but also eliminates errors from pipetting. Moreover, a tiny volume inside a microfluidic device tends to create a high concentration that benefits processes such as immunoprecipitation.

Although a significant amount of effort has been directed toward DNA/RNA analysis using microfluidics [37–42], the reports on microfluidics-based ChIP assays (or epigenetic/epigenomic studies) have been scarce until only recently.

Wu et al. developed an automated microfluidic ChIP ("AutoChIP") that used 2000 cells per reaction [43]. Figure 16.1 shows a labeled schematic of the device where different stages of ChIP are performed. In Ring 1, fixed cells were trapped and lysed and the DNA was fragmented with micrococcal nuclease (MNase) digestion. Antibody-bead preparation was loaded into Rings A to D and mixed with fragmented DNA using valve-actuated mixing in order to facilitate IP. After IP, the samples were washed in the columns stacked behind sieve valves SV6, SV7, and SV8 by passing washing buffer through beads, and collected from the device. Purification of DNA was conducted off-chip. The major improvement of AutoChIP when compared to conventional protocols was its short assay time. AutoChIP only took 2 h for IP instead of overnight in conventional assays. Fast mixing of the reagents and beads in ring mixers from A to D accelerated antibody-antigen binding. 400 nl microfluidic reaction chamber (compared with 1 ml in the conventional assays) created high bead density to effectively speed up ChIP. The device shown in Fig. 16.1 was also designed with four rings to accommodate four ChIP reactions at one time. As Wu et al. demonstrated in their later work, scaling up was easily accomplished by fabricating a chip with more mixing rings, resulting in even more parallel reactions. Figure 16.2 shows an upgraded device that has 16 rings to accommodate 16 ChIP reactions at one time [44]. Shen et al. also used very similar design for their ChIP-seq study on H3K4me3 epigenomic landscape using 1000 mouse early embryonic cells [45].

Our group (Geng et al.) designed a microfluidic device for ChIP-qPCR assay with the ultrahigh sensitivity [46]. Optimized microfluidic and ChIP conditions permitted the analysis of histone modifications from as few as ~50 cells without losing specificity. The whole ChIP-qPCR could be finished within ~8.5 h. As shown in Fig. 16.3, cells were first stacked against magnetic beads that were functionalized

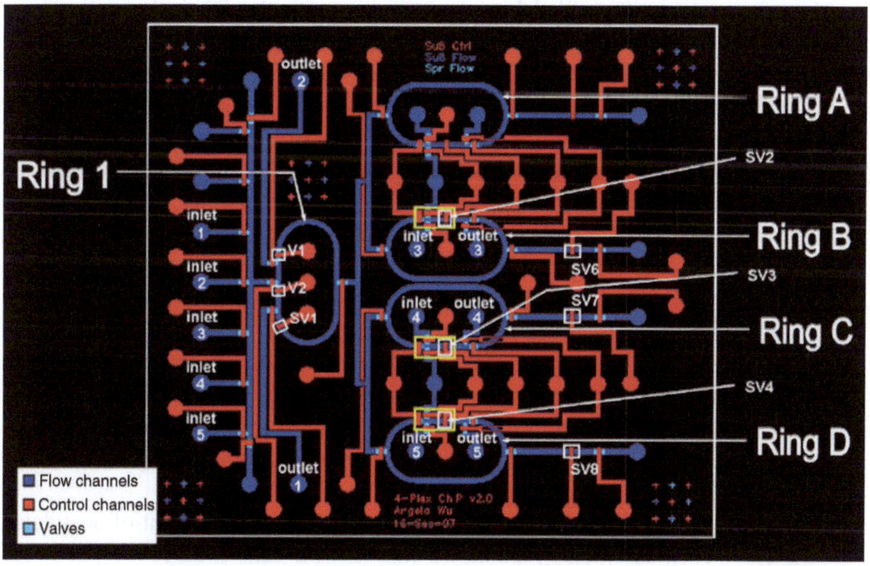

Fig. 16.1 The design of AutoChIP for four reactions. Reagent-containing flow channels are shown in *blue shades*, and valve-actuating control channels in *red*. Reproduced with permission from Ref. [43]

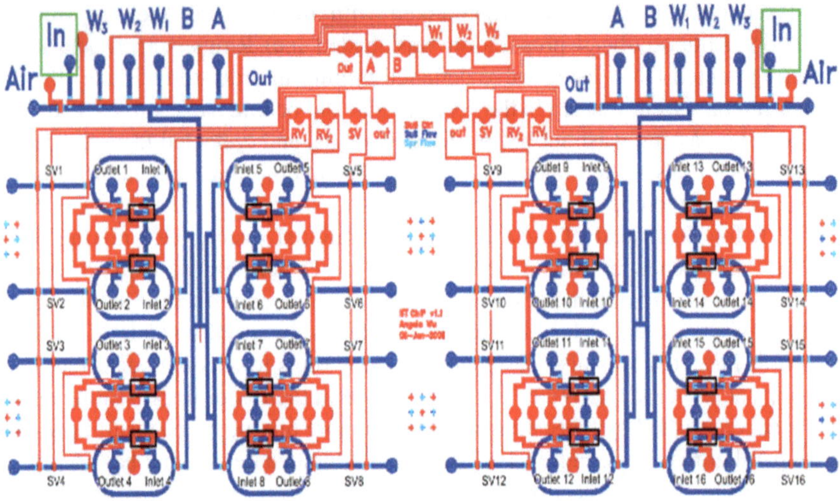

Fig. 16.2 Schematic of the microfluidic AutoChIP for 16 reactions. Reagent-containing flow channels are shown in *blue shades*, and valve-actuating control channels in *red*. Reproduced with permission from Ref. [44]

with anti-histone antibody. They were lysed to release chromatin and chromatin was then sheared by MNase digestion. The fragmented chromatin was then incubated with the antibody-coated magnetic beads with mixing of the beads by a moving magnet and a periodically actuated valve ("flapping valve"). Once the bead surface captured profiled histone (together with the DNA that it interacts with), the beads were washed to remove nonspecifically bound chromatin/DNA, while the magnet retained the beads and washing buffer passed through beads. The beads were then collected from an outlet of the device and taken off the chip for remaining purification and qPCR procedure.

Compared to AutoChIP, several measures were taken to improve the sensitivity. First, we reduced the reaction chamber for IP down to ~50 nl and ensured IP beads to occupy a large fraction of the reaction chamber. The close proximity among beads greatly increased the efficiency and rate for chromatin adsorption on the bead surface due to the short diffusion lengths involved. The adsorption of a chromatin molecule among beads was rapid given that travel time $\tau_D \sim w^2/D$, where w is diffusion distance between two beads, and D is diffusivity. Second, the magnetic beads were utilized to capture cells and perform immunoprecipitation and this may minimize nonspecific adsorption by the beads. Finally, the chamber for cell lysis and that for immunoprecipitation were adjacent and only separated by a valve with minimal material loss during transport.

We examined two loci (FR and bA-globin genes) for enrichment of acetylated histone H3. Previous studies had shown that in 6C2 cells the FR promoter was acetylated at histone H3 and the bA-globin promoter was unacetylated. Our result indicated that the percentage of DNA precipitation relative to input (i.e., percent input) increased at FR gene when the cell number decreased. We reasoned that

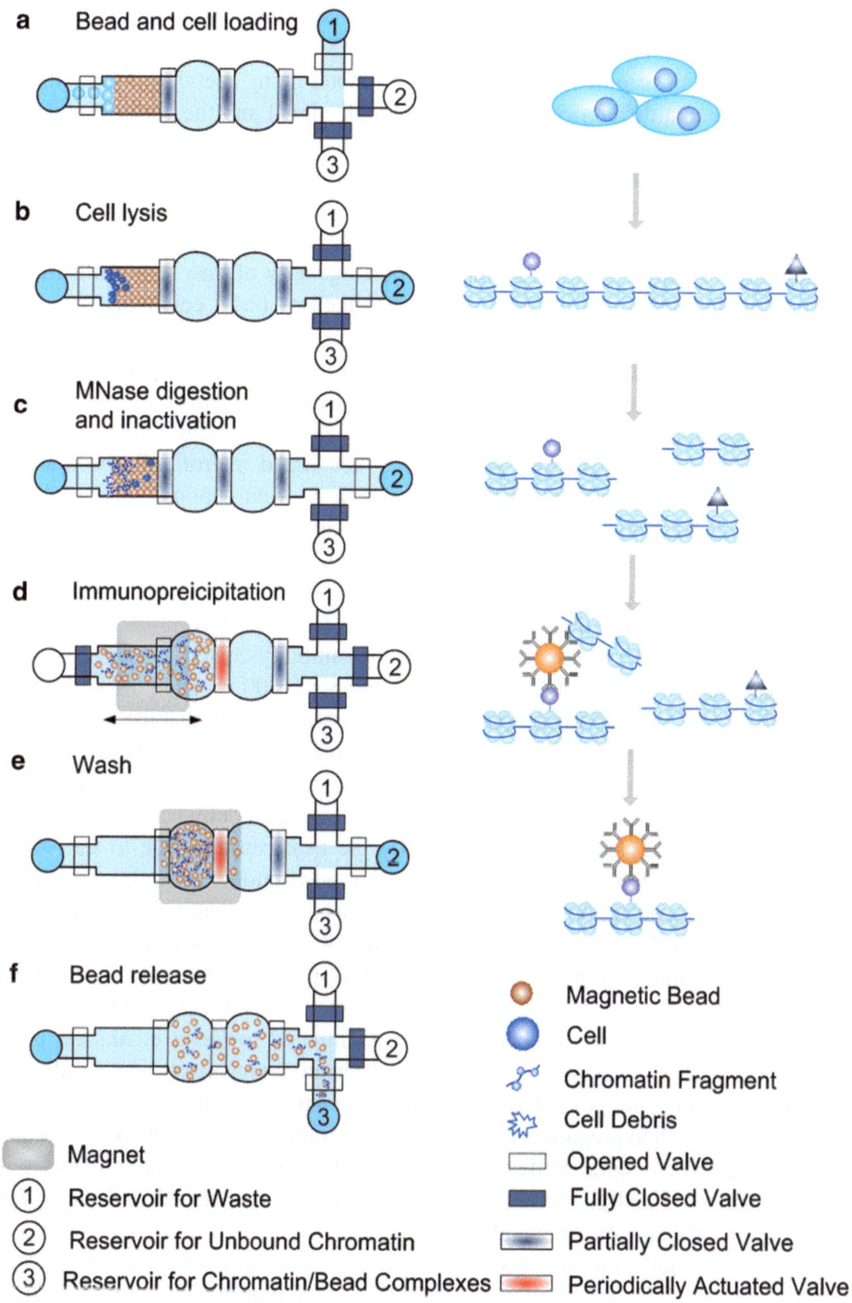

Fig. 16.3 The procedures involved in the microfluidic ChIP device from Geng et al. Reproduced with permission from Ref. [46]

antibody-antigen binding was more complete when there were less cells. The positive association of FR gene with acetylated H3 was confirmed with 100 cells. This result was consistent with the FR gene promoter being acetylated. In comparison, no significant difference between the sample signal and the control (with no antibody coated on the IP bead surface) at the bA-globin gene promoter was observed. This suggested that acetylated H3 was absent at the bA-globin promoter. These results demonstrated that the assay could effectively distinguish different acetylation profiles at different genes. This ultrahigh-sensitivity ChIP device has the potential to yield new insights into the heterogeneity of histone modifications and protein–DNA interactions within cell populations (<50 cells) that is often obscured in the results of bulk assays.

Genome-wide studies of histone modifications require ChIP-Seq. The state-of-the-art ChIP-seq protocols require at least 10,000 cells, and an additional whole-genome pre-amplification of immunoprecipitated DNA is typically needed before sequencing library construction. We recently developed microfluidic oscillatory washing-based ChIP-seq (MOWChIP-seq), which allowed genome-wide analysis of histone modifications using as few as 100 cells [47]. Two features of the technology permitted highly efficient ChIP DNA collection: (1) The use of a packed bed of beads for highly efficient adsorption and (2) microfluidic oscillatory washing (MOW) for effective removal of nonspecifically adsorbed DNA.

As shown in Fig. 16.4a, a suspension of antibody-coated magnetic beads (IP beads) was flowed into the microfluidic chamber (~800 nl) and packed against a partially closed valve to form a packed bed. It is worth noting that the packed bed took up to 30 % of the total chamber volume, whereas in conventional assays beads only take about 1 % of the total reaction volume. Sonicated chromatin was then flowed through bead-packed bed, taking advantage of drastically improved adsorption and capture of chromatin molecules by the bead surface, due to the small diffusion lengths. However, the high-efficiency adsorption likely also increased nonspecific DNA collection. Thus an effective washing was required for generating high-quality ChIP DNA. We applied alternating pressure pulses at two ends of the microfluidic chamber (regulated by a LabVIEW program) to create oscillatory washing. The oscillatory movement of the magnetic IP beads effectively removed nonspecific binding from the bead surface. Compared to Geng et al., the pure mechanical nature of MOW ensured that no IP bead aggregation occurred during washing. IP bead aggregation tends to occur when a magnetic field is used to move the beads and such aggregation leads to physical trapping. After MOW, the beads were retained by a magnet while neat buffer was flowed to wash removed DNA out of the chamber. Finally, the IP beads were collected into a vial and ready for downstream DNA purification and analysis.

The quality and amount of ChIP DNA were critically affected by several parameters including amount of beads, antibody concentration (used during coating of IP beads), and washing duration (Fig. 16.4b–d). We optimized these parameters by using qPCR to examine fold enrichment at known positive (UNKL and C9orf3) and negative (N1 and N2) loci. The relative fold enrichment was defined as the percent input at a locus divided by the percent input at the lowest negative locus.

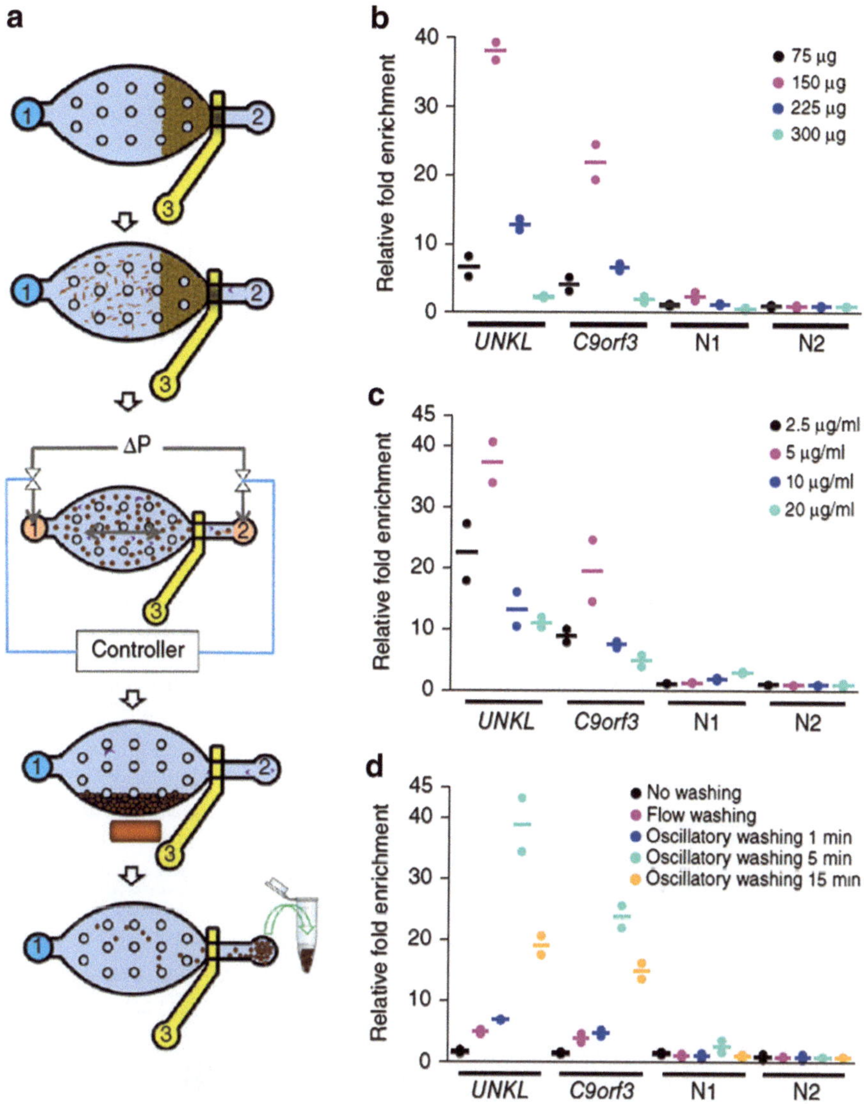

Fig. 16.4 MOWChIP and the optimization of conditions. (**a**) Illustration for the five major steps for MOWChIP. The relative fold enrichment (for H3K4me3) optimized on several parameters including the amount of beads in device chamber (**b**); concentration of antibody used for pre-coating IP beads (**c**); and washing duration with each of the two washing buffers (**d**). Reproduced with permission from Ref. [47]

Under optimized conditions, we obtained ~1.3 ng ChIP DNA from 10,000 cells and ~180 pg from 1000 cells for H3K4me3. This yield (5–6 % of the total chromatin) was within the range of the theoretical limit (2–8 % of the genome is marked by H3K4me3). We examined two histone marks (H3K4me3 and H3K27Ac) in a cell

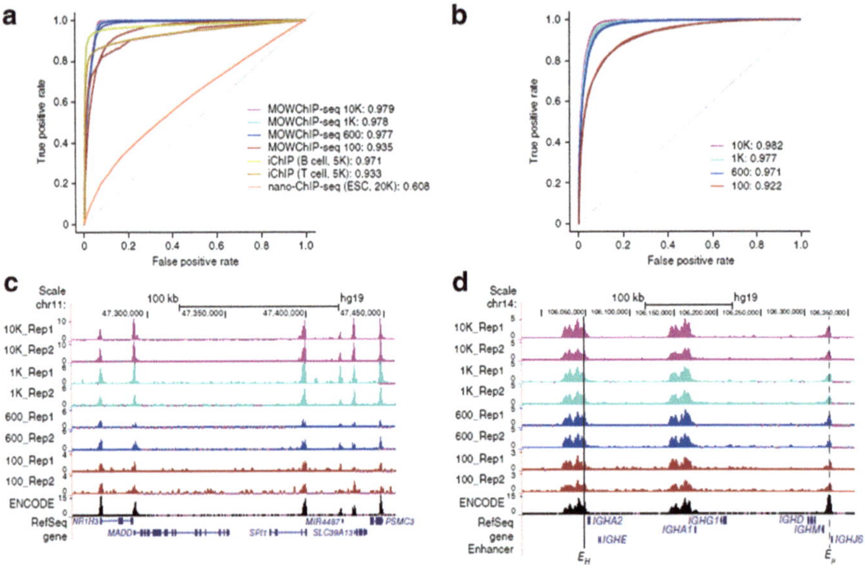

Fig. 16.5 Data from MOWChIP-seq and their comparison to nano-ChIP-seq and iChIP. (**a**) ROC curves for H3K4me3 data. (**b**) ROC curves for H3K27ac data generated by MOWChIP-seq. (**c**) Normalized H3K4me3 MOWChIP-seq signals at the SPI1 gene locus using data generated with various sample sizes. (**d**) Normalized H3K27ac MOWChIP-seq signals at the immunoglobulin heavy-chain locus. Reproduced with permission from Ref. [47]

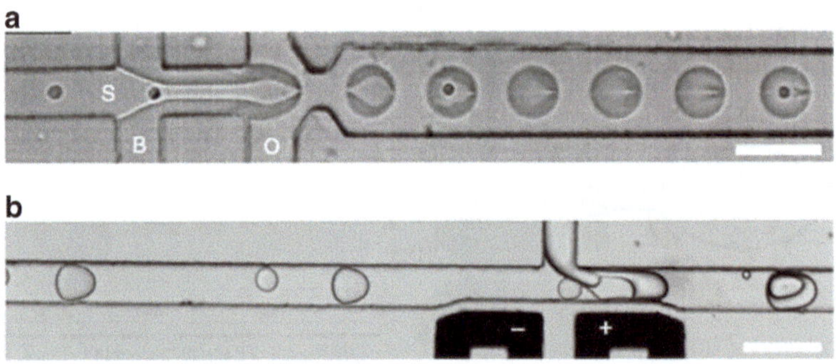

Fig. 16.6 Device and operation for Drop-ChIP. (**a**) Micrograph of the formation of cell-bearing drops. (**b**) Micrograph of three-point merging device with an induced electric field. Reproduced with permission from Ref. [48]

line and hematopoietic stem and progenitor cells (HSPCs) isolated from mouse fetal liver (FL). Our data reveal new enhancers and superenhancers that were critically involved in early hematopoiesis (Fig. 16.5).

Recently, Rotem et al. combined droplet microfluidics, DNA barcoding, and sequencing to process thousands of individual cells and use data to generate chromatin state maps for embryonic stem cells, fibroblasts, and hematopoietic progenitors with single-cell resolution [48]. As shown in Fig. 16.6a, they co-flowed cell suspension together with lysis buffer and MNase to form droplets in oil. In parallel, another droplet-maker encapsulated designed oligonucleotides (more than 1000 barcodes to ensure that more than 95 % barcodes are statistically unique to single cell) into droplets with each droplet containing multiple copies of the same barcode. They then merged a stream of nucleosome-containing droplets and a stream of barcode-containing droplets with an enzymatic buffer stream containing DNA ligase through a three-point merging device. Dielectrophoresis was used to trigger fusion of all three types of droplets/solution (Fig. 16.6b). Indexed chromatins from 100 cells were immunoprecipitated with carrier chromatin from a different organism. Enriched ChIP DNA was used to prepare sequencing library.

It is worth noting that their typical yield from a pool of 100 cells was seven million aligned reads. To avoid situations that one barcode ligated with two cells, or two barcodes ligated with two cells, they filtered the sequencing data to include only reads that represent one barcode with one cell and two barcodes with one cell. As the result, they obtained 500–10,000 reads per single cell.

As a validation, the aggregated chromatin profiles from 200 single-cell assays were comparable to bulk ChIP-seq assays (Fig. 16.7a). However the detail of the single-cell profiles was compromised by the low sequencing coverage (only ~800 peaks were detected per cell). The data still allowed them to distinguish single-cell

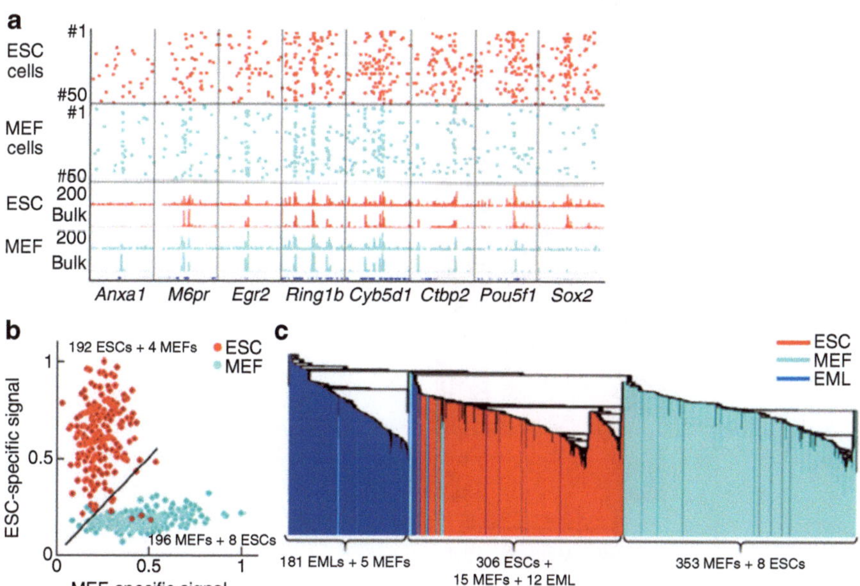

Fig. 16.7 Drop-ChIP data differentiate cell types (ESCs and MEFs). (**a**) Drop-ChIP data is shown for 50 ES cells and 50 MEFs across representative gene loci. (**b**) Scatterplot depicts normalized detection of ES cell-specific intervals versus MEF-specific intervals for 400 single-cell H3K4me3 profiles. (**c**) DIANA is used to discriminate between different cell types. Reproduced with permission from Ref. [48]

MEF profiles from single-cell ES cell profiles with nearly 100 % accuracy (Fig. 16.7b). Moreover, they separately barcoded ES cells, MEFs, and EML cells but mixed cells to acquire their profiles. They were able to discriminate between cell types with >95 % accuracy using an unsupervised divisive hierarchical clustering algorithm (Fig. 16.7c). In order to detect subsets within the same cell type (ES cells) with distinct regulatory patterns, they developed a signature-based clustering method to examine signal-to-noise ratio of these subsets or their "signatures." They identified 91 representative signatures from 314 clustered public ChIP-seq profiles based on the overlaps between enriched regions for histone modifications, transcription factors, and chromatin regulators. Figure 16.8a shows that ES cell subpopulations with correlated chromatin landscapes were distinguished by clustering of the signature matrix. They used multidimensional scaling (MDS) plot

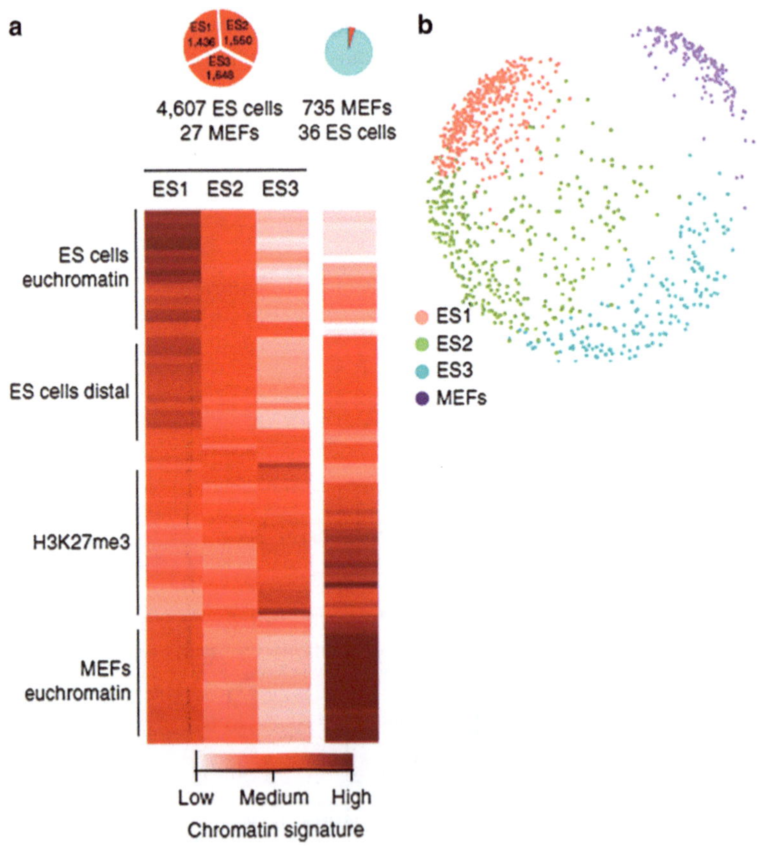

Fig. 16.8 Drop-ChIP shows a spectrum of ES cell subpopulations with variable chromatin signatures for pluripotency and priming. (**a**) Several prominent groups of cells with correlated chromatin landscapes were distinguished by clustering of the signature matrix. (**b**) MDS plot compares the chromatin landscapes of single ES cells and MEFs. Reproduced with permission from Ref. [48]

to compare the chromatin landscapes of single ES cells and MEFs. Figure 16.8b shows that the cell states were fairly dense for MEF cluster and ES1 cluster and this suggested homogeneity in the chromatin landscapes. In contrast, the scattered dots in ES2 and ES3 clusters reflected a strong gradient in the single-cell state. Their single-cell epigenomic profiles (although sparse in the number of reads obtained for each cell) revealed some aspects of epigenetic heterogeneity in a cell population.

16.4 Summary

Epigenetic regulations including histone modification, DNA methylation, and transcription factor binding affect how genes are turned on or off within a cell and serve as an integral part of regulatory systems that determine the phenotypes. In contrast to relatively constant genomic information, epigenetic/epigenomic regulations are highly specific to cell types and states. Thus studies of epigenomics using samples from patients and lab animals often require isolation of a specific population of cells. Such isolation process typically generates only cell samples with a low number of cells. Increased sensitivity for epigenetic/epigenomic assays is in high demand for these reasons. Microfluidics provides a powerful platform and potential solution for highly sensitive epigenetic analysis. Microfluidic epigenetic/epigenomic assays will open door to studies on scarce cell samples with direct biomedical relevance. Furthermore, these devices/assays will generate unprecedented insights into population heterogeneity that were not accessible previously with bench-scale devices.

Acknowledgements This work was supported by National Institutes of Health grants [CA174577, EB017235, EB017855, HG008623, and EB019123].

References

1. Ruthenburg AJ, Li H, Patel DJ, Allis CD (2007) Multivalent engagement of chromatin modifications by linked binding modules. Nat Rev Mol Cell Biol 8:983–994
2. Suzuki MM, Bird A (2008) DNA methylation landscapes: provocative insights from epigenomics. Nat Rev Genet 9:465–476
3. Egger G, Liang GN, Aparicio A, Jones PA (2004) Epigenetics in human disease and prospects for epigenetic therapy. Nature 429:457–463
4. Kashyap V et al (2009) Regulation of stem cell pluripotency and differentiation involves a mutual regulatory circuit of the NANOG, OCT4, and SOX2 pluripotency transcription factors with polycomb repressive complexes and stem cell microRNAs. Stem Cells Dev 18:1093–1108
5. Boland MJ, Nazor KL, Loring JF (2014) Epigenetic regulation of pluripotency and differentiation. Circ Res 115:311–324
6. Zhu J et al (2013) Genome-wide chromatin state transitions associated with developmental and environmental cues. Cell 152:642–654

7. Capell BC, Berger SL (2013) Genome-wide epigenetics. J Invest Dermatol 133:e9
8. Romanoski CE, Glass CK, Stunnenberg HG, Wilson L, Almouzni G (2015) Epigenomics: roadmap for regulation. Nature 518:314–316
9. Bernstein BE et al (2010) The NIH roadmap epigenomics mapping consortium. Nat Biotechnol 28:1045–1048
10. Hanahan D, Weinberg RA (2000) The hallmarks of cancer. Cell 100:57–70
11. Jimenez-Sanchez G, Childs B, Valle D (2001) Human disease genes. Nature 409:853–855
12. Darnell JE (2002) Transcription factors as targets for cancer therapy. Nat Rev Cancer 2:740–749
13. Baldwin AS Jr (2001) Series introduction: the transcription factor NF-kappaB and human disease. J Clin Invest 107:3–6
14. Beg AA, Finco TS, Nantermet PV, Baldwin AS Jr (1993) Tumor necrosis factor and interleukin-1 lead to phosphorylation and loss of I kappa B alpha: a mechanism for NF-kappa B activation. Mol Cell Biol 13:3301–3310
15. Nowak DE, Tian B, Brasier AR (2005) Two-step cross-linking method for identification of NF-kappaB gene network by chromatin immunoprecipitation. Biotechniques 39:715–725
16. Ehrlich M et al (1982) Amount and distribution of 5-methylcytosine in human DNA from different types of tissues of cells. Nucleic Acids Res 10:2709–2721
17. Blackledge NP, Klose R (2011) CpG island chromatin: a platform for gene regulation. Epigenetics 6:147–152
18. Deaton AM, Bird A (2011) CpG islands and the regulation of transcription. Genes Dev 25:1010–1022
19. Esteller M (2007) Cancer epigenomics: DNA methylomes and histone-modification maps. Nat Rev Genet 8:286–298
20. Jones PA, Baylin SB (2007) The epigenomics of cancer. Cell 128:683–692
21. Jaenisch R, Bird A (2003) Epigenetic regulation of gene expression: how the genome integrates intrinsic and environmental signals. Nat Genet 33(Suppl):245–254
22. Herman JG, Baylin SB (2003) Gene silencing in cancer in association with promoter hypermethylation. N Engl J Med 349:2042–2054
23. Feinberg AP, Tycko B (2004) The history of cancer epigenetics. Nat Rev Cancer 4:143–153
24. Gonzalo S (2010) Epigenetic alterations in aging. J Appl Physiol (1985) 109:586–597
25. Sarma K, Reinberg D (2005) Histone variants meet their match. Nat Rev Mol Cell Biol 6:139–149
26. Bernstein BE, Meissner A, Lander ES (2007) The mammalian epigenome. Cell 128:669–681
27. Dahl JA, Collas P (2008) μChIP—a rapid micro chromatin immunoprecipitation assay for small cell samples and biopsies. Nucleic Acids Res 36:e15
28. Park PJ (2009) ChIP-seq: advantages and challenges of a maturing technology. Nat Rev Genet 10:669–680
29. Wang J, Tan AC, Tian T (2012) Next generation microarray bioinformatics: methods and protocols. Methods Mol Biol 802:1–401
30. Collas P (2010) The current state of chromatin immunoprecipitation. Mol Biotechnol 45:87–100
31. O'Neill LP, VerMilyea MD, Turner BM (2006) Epigenetic characterization of the early embryo with a chromatin immunoprecipitation protocol applicable to small cell populations. Nat Genet 38:835–841
32. Nelson JD, Denisenko O, Bomsztyk K (2006) Protocol for the fast chromatin immunoprecipitation (ChIP) method. Nat Protoc 1:179–185
33. O'Neill LP, Turner BM (1996) Immunoprecipitation of chromatin. Methods Enzymol 274:189–197
34. Spencer VA, Sun JM, Li L, Davie JR (2003) Chromatin immunoprecipitation: a tool for studying histone acetylation and transcription factor binding. Methods 31:67–75
35. Thorsen T, Maerkl SJ, Quake SR (2002) Microfluidic large scale integration. Science 298:580–584

36. Unger MA, Chou HP, Thorsen T, Scherer A, Quake SR (2000) Monolithic microfabricated valves and pumps by multilayer soft lithography. Science 288:113–116
37. Burns MA et al (1998) An integrated nanoliter DNA analysis device. Science 282:484–487
38. Hong JW, Studer V, Hang G, Anderson WF, Quake SR (2004) A nanoliter-scale nucleic acid processor with parallel architecture. Nat Biotechnol 22:435–439
39. Pal R et al (2005) An integrated microfluidic device for influenza and other genetic analyses. Lab Chip 5:1024–1032
40. Witek MA, Llopis SD, Wheatley A, McCarley RL, Soper SA (2006) Purification and preconcentration of genomic DNA from whole cell lysates using photoactivated polycarbonate (PPC) microfluidic chips. Nucleic Acids Res 34:e74
41. Yeung SW, Lee TMH, Cai H, Hsing IM (2006) A DNA biochip for on-the-spot multiplexed pathogen identification. Nucleic Acids Res 34:e118
42. Cipriany BR et al (2010) Single molecule epigenetic analysis in a nanofluidic channel. Anal Chem 82:2480–2487
43. Wu AR et al (2009) Automated microfluidic chromatin immunoprecipitation from 2,000 cells. Lab Chip 9:1365–1370
44. Wu AR et al (2012) High throughput automated chromatin immunoprecipitation as a platform for drug screening and antibody validation. Lab Chip 12:2190–2198
45. Shen J et al (2015) H3K4me3 epigenomic landscape derived from ChIP-Seq of 1,000 mouse early embryonic cells. Cell Res 25:143–147
46. Geng T et al (2011) Histone modification analysis by chromatin immunoprecipitation from a low number of cells on a microfluidic platform. Lab Chip 11:2842–2848
47. Cao ZN, Chen CY, He B, Tan K, Lu C (2015) A microfluidic device for epigenomic profiling using 100 cells. Nat Methods 12:959–962
48. Rotem A et al (2015) Single-cell ChIP-seq reveals cell subpopulations defined by chromatin state. Nat Biotechnol 33:1165–1172

Index

© Springer International Publishing Switzerland 2016
C. Lu, S.S. Verbridge (eds.), *Microfluidic Methods for Molecular Biology*,
DOI 10.1007/978-3-319-30019-1

G

H

I

Q

R

MIX
Papier aus verantwortungsvollen Quellen
Paper from responsible sources
FSC® C105338

If you have any concerns about our products,
you can contact us on
ProductSafety@springernature.com

In case Publisher is established outside the EU,
the EU authorized representative is:
Springer Nature Customer Service Center GmbH
Europaplatz 3, 69115 Heidelberg, Germany

Printed by Libri Plureos GmbH
in Hamburg, Germany